Quantum Computing and Instrumentation for Smart Agriculture

Micro and Nano Technologies

Quantum Computing and Instrumentation for Smart Agriculture

Edited by

Shubham Mahajan

Assistant Professor, Amity School of Engineering and Technology,
Amity University, Gurugram, Haryana, India

Anand Nayyar

Professor, Scientist, Vice-Chairman (Research) and Director,
IoT and Intelligent Systems Lab,
Duy Tan University, Danang, Vietnam

Mohd Naved

Associate Professor, Jaipuria Institute of Management,
Noida, Uttar Pradesh, India

ELSEVIER

Elsevier
Radarweg 29, PO Box 211, 1000 AE Amsterdam, Netherlands
50 Hampshire Street, 5th Floor, Cambridge, MA 02139, United States

ISBN: 978-0-443-44247-6

For Information on all Elsevier publications visit our website at
https://www.elsevier.com/books-and-journals

Publisher: Matthew Deans
Acquisitions Editor: Ana Claudia Garcia
Editorial Project Manager: Thangminlun Kipgen
Production Project Manager: Maria Shalini A
Cover Designer: Arumugam Kothandan

Typeset by Aptara, New Delhi, India

Contents

PART VI Stategic vision

Contributors

Andi Adawiah
Department of Management, Faculty of Economics and Business, Universitas Lamappapoleonro, Soppeng, Indonesia

Bhimo Widyo Andoko
Faculty of Fine Arts and Design, Indonesia Institute of the Arts Yogyakarta, Yogyakarta, Indonesia

Ankit
Department of Horticulture, School of Agriculture, Lovely Professional University, Phagwara, Punjab, India

Sandhya Bansal
Department of Computer Science and Engineering, Maharishi Markandeshwar Engineering College, Maharishi Markandeshwar (Deemed to be University), Mullana, Ambala, Haryana, India

Deepa Bhadan
Research Scholar, Genetics and Plant Breeding, Chaudhary Charan Singh University, Meerut, Uttar Pradesh, India

Vipin Kumar Chaudhary
School of Computer Science and Engineering, Lovely Professional University, Phagwara, Punjab, India

Vishal Dattana
Oman College of Management and Technology, Muscat, Oman

Dian Suluh Kusuma Dewi
Department of Government Science, Universitas Muhammadiyah Ponorogo, Jawa Timur, Kabupaten Ponorogo, Indonesia

Deepika Ghai
School of Electronics and Electrical Engineering, Lovely Professional University, Phagwara, Punjab, India

Utpal Ghosh
Department of Computer Science, Sarojini Naidu College for Women, Kolkata, West Bengal, India

Elavarasi Gunasekaran
Department of Civil Engineering, Alagappa Chettiar Government College of Engineering and Technology, Karaikudi, Tamil Nadu, India

Harshita Gupta
Independent Researcher, Gurugram, Haryana, India

Kavita Gupta
Department of Computer Science and Engineering, Maharishi Markandeshwar Engineering College, Maharishi Markandeshwar (Deemed to be University), Mullana, Ambala, Haryana, India

Akhmad Habibullah
Department of Government Science, Universitas Pancasakti Tegal, Jawa Tengah, Kota Tegal, Indonesia

Shahul Hameed
Department of Civil Engineering, Terra Firma Ground Engineering, Kingdom of Saudi Arabia

Himanshu
Department of Horticulture, School of Agriculture, Lovely Professional University, Phagwara, Punjab, India

Bambang Irawan
Department of Public Administration, Universitas Mulawarman, Kalimantan Timur, Samarinda, Indonesia

Ramya Karuppasamy
Department of Agricultural Engineering, Hindusthan College of Engineering and Technology, Tamil Nadu, India

Amandeep Kaur
Department of Management, Advanced Centre for Research and Innovation, Chandigarh School of Business, Chandigarh Group of Colleges, Mohali, Punjab, India

Gaganpreet Kaur
Department of Management, Advanced Centre for Research and Innovation, Chandigarh School of Business, Chandigarh Group of Colleges, Mohali, Punjab, India

Rahmatullah Kazimi
Department of Horticulture, Daikundi Institute of Higher Education, Daikondi University, Daikundi, Afghanistan

Ashish Kumar
Operations Management and Decision Sciences Area, Jaipuria Institute of Management, Noida, Uttar Pradesh, India

R Lakshmana Kumar
Department of Artificial Intelligence and Machine Learning, KIT - Kalaignar Karunanidhi Institute of Technology, Coimbatore, Tamil Nadu, India

Shrabanti Kundu
Department of Computer Science & Engineering, MCKV Institute of Engineering, Howrah, West Bengal, India

Thanjaivadivel M
Department of Computer Science, Nandha Engineering College, Erode, Tamil Nadu, India

Halimah Abdul Manaf
UUM College of Law, Government, and International Studies, Universiti Utara Malaysia, Kedah, Malaysia

Kumaresan Mayakannan
Department of Mechatronics Engineering, Hindusthan College of Engineering and Technology, Tamil Nadu, India

Eliza Meiyani
Department of Government Affairs and Administration, Universitas Muhammadiyah Yogyakarta, Yogyakarta, Indonesia

Arunkumar Munimathan
Department of Mechatronics Engineering, Hindusthan College of Engineering and Technology, Tamil Nadu, India

Sonali Narbariya
T A Pai Management Institute, Manipal Academy of Higher Education, Manipal, Karnataka, India

Darshan Pandya
Centre of Management Studies, New Age Makers' Institute of Technology (NAMTECH), Gandhinagar, Gujarat, India

Rajib Roy
Jaipuria Institute of Management, Noida, Uttar Pradesh, India

Ramandeep Sandhu
School of Computer Science and Engineering, Lovely Professional University, Phagwara, Punjab, India

Deepika Saxena
Department of Horticulture, School of Agriculture, Lovely Professional University, Phagwara, Punjab, India

Anshu Sharma
School of Computer Science and Engineering, Lovely Professional University, Phagwara, Punjab, India

Anurag Sharma
Department of Electronics and Communication Engineering, CT Institute of Technology and Research, Maqsudan, Jalandhar, Punjab, India

Charu Sharma
Department of Computer Science and Engineering, Maharishi Markandeshwar Engineering College, Maharishi Markandeshwar (Deemed to be University), Mullana, Ambala, Haryana, India

Chirag Sharma
School of Computer Science and Engineering, Lovely Professional University, Phagwara, Punjab, India

N Sumalatha
ICT, Alagappa Chettiar Government College of Engineering and Technology, Karaikudi, Tamil Nadu, India

Abhishek Tanwer
Lincoln Institute for Agri-Food Technology, School of Agri-Food Technology and Manufacturing, University of Lincoln, Lincolnshire, United Kingdom

Sowmiya Velmurugan
Department of Agricultural Engineering, Hindusthan College of Engineering and Technology, Tamil Nadu, India

Ni Putu Tirka Widanti
Department of Public Administration, Universitas Ngurah Rai, Bali, Kota Denpasar, Indonesia

Muhammad Younus
Department of Government Affairs and Administration, Universitas Muhammadiyah Yogyakarta, Yogyakarta, Indonesia; Department of Product Research and Software Development, Trans Pacific Logistics (TPL) Pvt Ltd, Karachi, Pakistan

Arif Zainudin
Department of Government Science, Universitas Pancasakti Tegal, Jawa Tengah, Kota Tegal, Indonesia

Preface

Agriculture is entering a phase where the dominant bottleneck is no longer mechanization alone, but decision latency under uncertainty: growers and agri-enterprises must continuously optimize resource allocation, anticipate weather and pest dynamics, and coordinate logistics across volatile markets—while operating within stringent sustainability constraints. This decision problem is inherently high-dimensional and spatiotemporal: heterogeneous soil states, multimodal remote sensing streams, genomic and phenotypic signals, and nonlinear climate–crop interactions generate complex objective landscapes and tightly coupled constraints. Smart agriculture therefore demands computational paradigms that can ingest data at scale, represent uncertainty explicitly, and produce robust actions within operational time windows. In parallel, instrumentation is evolving from periodic sampling toward continuous, in situ measurement, pushing the frontier from "more data" to "better data": higher sensitivity, improved selectivity, tighter calibration, and traceable measurements that can support auditability and reproducible agronomic analytics.

Against this backdrop, quantum technologies—encompassing quantum computing, quantum-enhanced algorithms, and quantum/advanced sensing—are increasingly viewed as complementary tools for the next generation of precision agriculture pipelines. Near-term progress is expected to be hybrid rather than purely quantum: classical–quantum workflows in which quantum subroutines (e.g., kernel methods, variational circuits, sampling, or structured linear-algebra primitives) are coupled with classical optimization, model selection, and domain constraints. The practical system-level questions are as important as algorithmic novelty: how to benchmark quantum advantage under realistic noise; how to engineer reliable IoT–edge–cloud architectures for farms; how to integrate sensing modalities into digital twins of fields and supply chains; and how to maintain data integrity as quantum computing also threatens legacy cryptography. Real deployments must contend with noisy devices, imperfect labels, drift in sensor calibration, and the economics of field operations—making validation, traceability, and security first-class design requirements rather than afterthoughts.

The chapters in this volume collectively respond to these realities by moving from computational and predictive tasks to measurement and infrastructure, and then outward to governance, inclusion, and ethical deployment. The narrative links quantum computing methods (for optimization, machine learning, and simulation) with the instrumentation that supplies the state-of-the agro-ecosystem, emphasizing that robust decision support requires both: credible measurements and trustworthy inference. Across the book, recurring themes include hybridization (quantum plus classical), the need for careful benchmarking and error analysis, and the importance of domain-specific problem formulation—because the best algorithmic performance is realized only when agronomic constraints, sensing limitations, and cost models are embedded into the pipeline.

The volume also underscores that "field readiness" is not only technical maturity but sociotechnical viability. Data governance and cybersecurity are treated as enabling infrastructure, particularly as smart agriculture becomes a distributed cyberphysical system spanning sensors, UAVs, mobile devices, and cloud platforms. At the same time, the book recognizes that quantum agriculture must be evaluated through the lenses of equity, accessibility, and responsible innovation: benefits must extend to diverse farm contexts, including low-resource settings, and must respect farmer agency and community trust. By interweaving methods, instrumentation, security, and societal impact, the chapters assemble a coherent perspective on how quantum technologies can be translated into practical, accountable, and scalable agricultural intelligence.

The rest of this book is organized as: Chapter 1 titled "**Climate and Weather Prediction Using Quantum Computing**" frames climate and weather forecasting as a computationally demanding, multiscale inference and simulation problem, central to agricultural risk management. It discusses how quantum computing may accelerate or enhance components such as optimization, partial differential equation solvers, and data assimilation workflows. This chapter emphasizes the realities of nonlinearity, chaos, and uncertainty quantification in atmospheric systems, relating them to time-critical agricultural decision cycles. It positions quantum approaches as potential complements to high-performance computing, particularly where dimensionality and latency are limiting factors. Chapter 2 titled "**AgriQML: A Novel Quantum Machine Learning Framework for Agriculture Crop Yield Prediction**" presents a hybrid quantum machine learning pipeline for yield prediction under high-dimensional, nonlinear soil–weather–crop interactions. It combines quantum-enabled dimensionality reduction (via quantum PCA) with learning models that leverage quantum feature maps and variational circuits, reflecting current near-term strategies. The discussion links model expressivity and kernel methods to agricultural datasets that are noisy, heterogeneous and often limited in labeled samples. It also highlights practical training considerations for noisy quantum operations and the role of hybrid optimization in stabilizing convergence and performance. Chapter 3 titled "**Quantum Sensors in Soil Health Monitoring**" surveys quantum sensing principles and maps them onto the measurement needs of soil health diagnostics. It explains how quantum phenomena can support high-sensitivity detection of soil properties such as moisture, pH, nutrients, salinity, and microbial activity, with improved temporal resolution. This chapter connects sensing advances to precision agriculture by showing how better measurements translate into better models and more reliable interventions. It also foregrounds integration challenges—field deployability, calibration, and robust data acquisition—necessary for real-time soil monitoring systems. Chapter 4 titled "**Quantum Cryptography for Agri-Data Security Quantum Sensors for Soil Health Monitoring**" links continuous soil monitoring to the need for secure, dependable data pipelines, treating sensing and data integrity as coupled requirements. It discusses limitations of conventional soil diagnostics and motivates quantum sensing as a path to continuous, noninvasive, high-resolution soil characterization. In parallel, it introduces security considerations for the resulting data streams, emphasizing integrity and trust in decision support. This chapter proposes a field-scale monitoring perspective where sensing, communications, and security co-evolve to support sustainable, data-driven farming. Chapter 5 titled "**Quantum Cryptography for Securing Smart Agriculture: A Future-Ready Framework for Data Integrity and Resilience**" addresses cybersecurity as a structural necessity for smart agriculture, particularly as IoT and analytics platforms expand the attack surface. It explains why quantum computing pressures classical public-key cryptography and motivates quantum cryptography—especially quantum key distribution—as a physics-based mechanism for tamper-evident secure communications. This chapter relates these ideas to agricultural workflows such as yield analytics, soil diagnostics, and supply chain coordination, where compromised data can cause direct economic and safety impacts. It also considers resilience-oriented security design, aligning cryptographic choices with operational constraints in distributed farm environments. Chapter 6 titled "**Quantum Cryptography for Secure Agricultural Data: Applications and Impact**" further elaborates on quantum cryptography in the agricultural context, emphasizing confidentiality, integrity, and availability for data-intensive precision agriculture systems. It frames the transition toward quantum-resilient security as a forward-looking requirement given the prospective capability of quantum adversaries. This chapter connects QKD and related security mechanisms to real deployment settings involving remote sensing, big data analytics, and multi-stakeholder data sharing. It highlights how security architecture must be aligned with the realities of heterogeneous devices, bandwidth constraints, and the operational need for trustworthy decision support. Chapter 7 titled "**Analyzing Fresh Water Shortage Crises and the Need for Government Intervention to Manage Resources**" situates smart agriculture within a broader resource-governance problem, using freshwater scarcity as a critical driver of agricultural vulnerability. It examines policy and institutional intervention as essential levers for managing water resources and mitigating socioeconomic impacts. This chapter connects directly to precision agriculture by emphasizing how data, modeling, and monitoring can support allocation, conservation, and accountability. It provides context for why advanced analytics and instrumentation—potentially including quantum-enabled approaches—must ultimately interface with governance structures to produce system-level outcomes. Chapter 8 titled "**Frugal and Inclusive Quantum Agri-Tech Ecosystems for the Global South**" reframes quantum agriculture through the lens of frugality, inclusion, and context-aware innovation in low- and middle-income settings. It proposes that the pathway to impact depends on modular, affordable architectures and governance models rather than technology maximalism, emphasizing "quantum sufficiency" over spectacle. This chapter integrates themes of quantum optimization, quantum sensing, decentralized deployment, and responsible innovation to outline a whole-system approach. It connects technical feasibility to adoption realities—limited infrastructure, constrained data, and local adaptability—highlighting pathways for equitable scaling. Chapter 9 titled "**Quantum Agriculture and Its Ethical Futures: A Multidisciplinary Management Inquiry**" examines ethical questions that emerge when computational power and predictive capability outpace institutional and social preparedness. It discusses how quantum-enhanced decision systems can affect farmer autonomy, trust, and agency, and how governance must anticipate rather than retroactively respond to harms. This chapter positions ethics as integral to design—shaping data practices, accountability, and stakeholder inclusion—rather than as a compliance add-on. It connects multidisciplinary management perspectives to the practical deployment of quantum technologies in agri-business ecosystems and decision support. Chapter 10 titled "**Ethical Considerations and Social Impact in Quantum Agriculture**" expands the ethical and societal analysis by focusing on privacy, unequal access, digital divides, and the displacement or marginalization of traditional knowledge systems. It emphasizes that advanced instrumentation and quantum computing can amplify disparities if affordability, literacy, and infrastructure constraints are not addressed explicitly. This chapter discusses the role of governments, technologists, and regulators in constructing legal and policy frameworks for responsible adoption. It connects social impact considerations to sustainability goals, arguing that technological benefits must be distributed fairly to strengthen resilience and food-system stability. Chapter 11 titled "**Roadmap for Quantum Agriculture: Challenges and Future Directions**" provides a structured roadmap of the research and engineering challenges that must be resolved for quantum technologies to integrate meaningfully into smart agriculture. It highlights constraints such as hardware cost and complexity, the need for algorithms tailored to agricultural data, and the engineering work required to interface quantum methods with IoT instrumentation. This chapter emphasizes hybrid strategies, scalability, and benchmarking as near-term priorities, while outlining longer-term possibilities for quantum-augmented precision farming. It synthesizes opportunities across prediction, optimization, sensing, and supply chains, offering an agenda for field-relevant progress.

Taken together, this book is designed for readers who want more than a speculative tour: it offers a technically grounded view of how quantum computing and advanced instrumentation can be integrated into the measurement-model-decision loop of smart agriculture, while remaining attentive to security, governance, and ethics. For researchers, it helps clarify problem formulations, validation expectations, and realistic hybrid pathways; for practitioners and policymakers, it illuminates what must be engineered—and what must be governed—for trustworthy deployment. These chapters invite the reader to treat quantum agriculture as an accountable, interdisciplinary discipline where rigor, practicality, and responsibility advance together.

Shubham Mahajan
Anand Nayyar
Mohd Naved

Acknowledgment

With the divine blessings of Shri Guru Maharaj Ji, this book, *"Quantum Computing and Instrumentation for Smart Agriculture,"* has come to completion. His guidance and grace have given me strength, clarity, and faith throughout this journey.

I sincerely express my deepest gratitude to my respected parents, **Sh. Manoj Kumar Gupta** and **Smt. Madhu Gupta**, for their unconditional love, constant encouragement, and countless sacrifices. Whatever I have achieved is rooted in the values and support they have given me.

I am also thankful to my sister, **Sunidhi,** and my jiju, **Er. Shiv Saroop** for their continuous support, positivity, and motivation. Their presence has always inspired me to move forward with confidence.

My heartfelt thanks to all my friends, mentors, colleagues, and well-wishers who encouraged me, shared their insights, and stood by me in different ways. Every contribution, big or small, has played a meaningful role in completing this work.

To everyone who has been part of this journey—directly or indirectly—I offer my sincere thanks and gratitude.

With respect and appreciation,

Dr. Shubham Mahajan

PART

I

Foundations of quantum agriculture

1

Climate and weather prediction using quantum computing

Kavita Gupta[a], Sandhya Bansal[a], Charu Sharma[a] and Vishal Dattana[b]

[a]DEPARTMENT OF COMPUTER SCIENCE AND ENGINEERING, MAHARISHI MARKANDESHWAR ENGINEERING COLLEGE, MAHARISHI MARKANDESHWAR (DEEMED TO BE UNIVERSITY), MULLANA, AMBALA, HARYANA, INDIA [b]OMAN COLLEGE OF MANAGEMENT AND TECHNOLOGY, MUSCAT, OMAN

1.1 Introduction

Climate change and extreme weather conditions can be regarded as one of the most significant scientific and social problems of the twenty-first century. The increased global temperatures, increased atmospheric instability, increased intensity of the tropical cyclones, the long term droughts, and the unpredictable weather patterns have all increased the need to have proper and timely climate and weather predictive systems. The effects of such are extremely hazardous to ecosystems, infrastructure, food security, human health, and economic stability on the global scale. Thus, it is necessary that good modeling and forecasting should be in place to enable the creation of early warning mechanisms, climate adaptation, and good policy-making (Tennie & Palmer, 2023).

Weather and climate prediction in the contemporary world is a computation issue. Global climate models (GCMs) and numerical weather prediction (NWP) are tools used to simulate the dynamics of the Earth system based on large-scale, coupled, nonlinear partial differential equations (PDEs), which are used to describe the atmospheric circulation, ocean dynamics, radiation transfer, land–atmosphere interactions, and biogeochemical processes. These models are based on huge amounts of non-homogeneous observational data received by satellites, radar, ocean buoys, weather stations, and on novel sensor networks. In spite of the unceasing improvement in high-performance computing (HPC), classical computational techniques have intrinsic limitations in scales, resolution, uncertainty and real-time requirements of such simulations (Ho et al., 2024).

Technically, climate and weather models have limitations in the form of computational cost, energy usage and algorithmic complexity. As spatial resolution or ensemble size increases, the computational cost grows exponentially and typically making either physical realism or

Quantum Computing and Instrumentation for Smart Agriculture. DOI: https://doi.org/10.1016/B978-0-443-44247-6.00009-9

operational feasibility tradeoffs. Furthermore, the nature of climate systems is also chaotic i.e. minor changes in initial conditions or numerical models can lead to big deviations in the predictions. The aggravation of these issues is due to data assimilation, the quantification of the uncertainty, and parameter optimization, leading to long-term predictions and short-term forecasting of anything that is computationally expensive and subject to error (Frolov, 2017).

In this case, quantum computing is an attractive computational paradigm, and it can address some of the significant bottlenecks of classical approaches. It is based on the physics of quantum mechanics, and aims to utilize qubits which can be in a superposition state and which can be entangled. Qubits can be used to search large search spaces and complex correlations among variables, unlike classical bits, which are used to express information in binary form. The above properties make quantum computers particularly handy in the solution of issues with high-dimensional linear algebra, probabilistic modeling, optimization, and sampling, that underlie climate and weather forecast systems (Zeema et al., 2025).

Lately, the progress in quantum hardware and software has given a boost to the desire to implement quantum computing to solve scientific problems including chemistry, materials science, finance and, most recently, climatic science. Theoretical speedups Quantum algorithms Harrow Hassidim Lloyd (HHL) algorithm to solve linear systems, quantum annealing to optimize, quantum Monte Carlo, and variational quantum algorithms (VQAs) in practice are used to solve problems that are in the heavy workload of atmospheric models. Specifically, they have been demonstrated to be promising in rapid solution of discretized PDEs, efficient ensemble forecasting, and noisy and incomplete environmental data assimilation (Rahman et al., 2024).

This chapter is directly aimed to investigate the possible application of quantum computing within the field of climate and weather prediction by trying to solve the computational bottlenecks that constrain the existing modeling systems. The main discussion to note is that classical computing systems are inefficient to effectively scale to the increasingly complex, data-intensive, and uncertain demands of Earth system models (ESM). These constraints become even more eminent as climate simulations become kilometer-scale resolved and fully coupled models of the Earth system, where other or supplementary computational methods must be developed (Saiyed, 2025).

Another promising field that is studied in this chapter is quantum machine learning (QML). Using quantum-enhanced learning frameworks, scientists can enhance their mode of pattern recognition in high-dimensional climate data, discover predictors of extreme weather phenomena, optimize model parameters, and further estimate forecast uncertainty. The techniques are especially useful in cases of the sparse/noisy/under-sampled observational data presence, a typical feature of the climate monitoring tasks (Fan & Han, 2022).

Simultaneously, this chapter is aware of the operational limitations of the existing NISQ systems, such as the number of qubits, decoherence, and gates errors. Quantum computing is the long-term goal of full fault tolerance. Consequently, the focus is on the hybrid quantum-classical computing models as a viable near-term direction, whereby quantum processors are deployed to speed up certain computational subtasks in the existing classical climate modeling pipelines (Eswaran et al., 2024). Fig. 1.1 gives an overview of quantum computing in climate science.

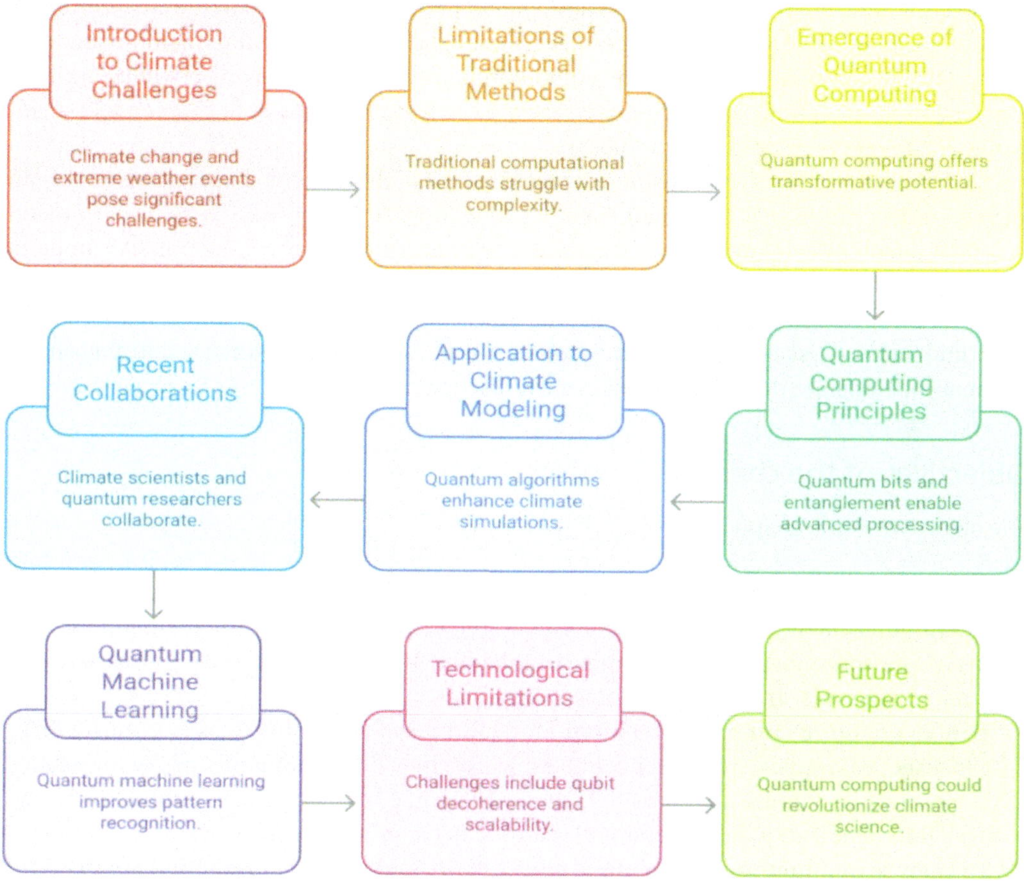

FIGURE 1.1 Quantum computing in climate science.

1.1.1 Purpose and Scope

This chapter is aimed at introducing a wide range of the intersection of quantum computing and climate science. It tackles the computational challenges of climate and weather modeling, provides the underlying principles of quantum computing as applied to the computational challenges, and surveys the newer quantum algorithms and applications in this field. Having combined the insights and visions offered by atmospheric science and quantum technologies, this chapter offers a perspective of the future in the realm of quantum computing improving predictive accuracy, climate resilience strategies, and more global response to climate change mitigation and adaptation (Singh et al., 2024).

Interdisciplinary collaborations between quantum computing and climate science have started to develop in recent years, with some researchers considering the future of the field of quantum-accelerated environmental science. QML algorithms are under development to enhance the process of satellite data assimilation, early indicators of extreme weather events, and derive meaningful patterns of large-scale climate data. Also, VQAs and quantum annealing methods have potential to improve the computational efficiency of solving complex optimization problems, including cloud micro-physics and earth atmospheric energy exchange calculations, among others (Rietsche et al., 2022).

This chapter starts with establishing the weaknesses of the current computational tools and analysis of the core issues of climate/weather modeling. It then gives a general introduction to fundamental ideas in quantum computing. The available research development, simulation studies, and theoretical developments are also examined in order to identify the possible role of quantum computing in climate modeling and forecasting. The current technological limitations including hardware scaling and qubit decoherence are also critically analyzed in this chapter and realistic timelines on practical deployment discussed.

Objectives of the chapter

The objectives of the chapter are:

1. **To assess the computational issues of climate and weather modeling:** This goal is aimed at determining the shortcomings of the classical HPC systems to solve the scale, complexity, and uncertainty of atmospheric and climate simulations. There is a focus on the data assimilation, numerical solvers, and ensemble forecasting issues;
2. **To take quantum computing techniques to climate science:** This chapter is meant to be a focused, easy-going, and all-encompassing overview of the fundamentals of quantum computing, i.e., qubits, superposition, entanglement, and quantum algorithms, and how each finds application in addressing climate-related computational issues;
3. **To analyze new quantum algorithms and climatic and weather prediction applications:** This chapter is meant to be a focused, easy-going, and all-encompassing overview of the fundamentals of quantum computing and how each finds application in addressing climate-related computational issues;
4. **And, to evaluate existing constraints and the research opportunities of the future:** This chapter assesses the limitations of Noisy Intermediate-Scale Quantum (NISQ) systems, such as the ability to scale hardware or decoherence, and identifies future research possibilities, hybrid quantum-classical models, and paths to realistic implementation.

Organization of the chapter

The rest of the chapter is organized as: Section 1.2 enlightens introduction to climate modeling and computational challenges. Section 1.3 discusses fundamentals of Quantum Computing. Section 1.4 illustrates the importance of Quantum Computing for Climate and Weather modeling. Section 1.5 highlights Hybrid Quantum-Classical Modeling Frameworks. Section 1.6 focuses on key areas of research in quantum climate forecasting, followed by challenges, opportunities and the future of climate forecasting. And, finally, Section 1.8 concludes the chapter.

Scientists working with quantum computing and those working with climate issues have recently begun collaborating to examine how environmental modeling could change under quantum-accelerated methods. Quantum learning algorithms are under development to enhance the assimilation of satellite data, early warnings about the onset of extreme events are being developed, and pattern recognition using climate data is also under investigation. Variational algorithms and quantum annealing potentially could help resolve optimization problems such as modeling cloud formation or calculating energy exchange in the Earth's atmosphere faster and with greater efficiency (Rietsche et al., 2022).

It begins with defining the limitations of the computational tools available and examines the fundamental problems of modeling the climate and weather. Then, it provides an accessible overview of the basics of quantum computing relevant to this subject, including qubits, superposition, entanglement, and of note, quantum algorithms. Offering present progressions, simulation research, and theory, the leading discussion of the chapter points to the prospective use of quantum computing in climate and forecasting. The limits that are presently imposed by technology and become obvious as well are hardware scalability and qubit decoherence, and the time they will become practical is a function that we will estimate.

1.2 Introduction to climate modeling and computational challenges

Climate modeling is one of the major scientific methods of learning the climate system of the Earth and predicting the manner in which it can evolve in future as a result of natural and anthropogenic aspects. With climate change persisting to affect all the ecosystems, economies, infrastructure, and human society everywhere in the world, accurate climate modeling and forecasting have emerged as an effective risk assessment and policy formulation model as well as disaster preparedness. The forecasts of global warming, extreme weather patterns, elevation and long-term sustainability of the sea level, and climate models have their concerns. However, the system is hard to simulate because of scientific and computational challenges (Frolov, 2017). At the most basic level, climate modeling is founded on mathematical models of physical processes that govern the earth system, including atmospheric circulation, ocean processes, interactions between the cryosphere and land-surface processes, biogeochemical cycles. These models are developed upon the complex of connected, nonlinear PDEs that are developed on the background of the fundamental principles of fluid mechanics, thermodynamic, radiative diffusion, and chemical reaction kinetics. The basic equations that are employed to describe it include the NavierStokes equations of atmospheric and oceanic flow, the mass and energy conservations equations and sub-grid-scale process parameterizations such as cloud formation, convection, and turbulence. Climate system is nonlinear, chaotic, and of such a nature that small modifications in the initial conditions of a model can lead to enormous diversification of the long-term model outcomes. This sensitivity involves simulations and probabilistic forecasting methods in an ensemble to quantify uncertainty. In addition, additional model resolution, physical realism, and heterogeneous observational data assimilation also require additional calculation. This has resulted to the fact that the traditional high-performance models of computing are inclined to be scaling, energy, and real-time performance limited and as such new models of computations have to be considered in light of the next generation of climate modeling (Upama et al., 2022).

1.2.1 Understanding weather and climate systems

Weather and climate are interchangeably used, although they are not the same and yet related to each other as the aspects of how the atmosphere system of the earth looks like. Weather refers to the immediate state of atmosphere at a given place and at a given time that is represented by rapidly varying parameters such as temperature, pressure, humidity, the velocity of the wind, cloudy coverage, and precipitation. These phenomena occur on time scales ranging between minutes and days and are governed by mesoscale and synoptic scale atmospheric processes, which include frontal systems, convective activities, and cyclonic circulations. Climatic in contrast is applied to describe the statistical behavior of weather over long-term scales, usually on the basis of multidecadal averages, variability, and extremes of atmospheric variables on regional and global scales. Climate analysis takes trends, seasonal cycles, and distributions of frequency of weather conditions in the 30 years or longer intervals of time that meet the standards of the World Meteorological Organization (Gill et al., 2022). The interaction of the many components of the earth system including hydrosphere, cryosphere, lithosphere, atmosphere, and biosphere controls weather and climate. These subsystems are firmly constrained by exchange of energy, momentum, mass, and chemical constituents. Radiative forcing on solar energy leads to atmospheric circulation that is governed by surface albedo, concentration of greenhouse gases, interactions between aerosols and clouds, and cloud–radiation interactions. Thermohaline circulation and oceanic decadal variations of sea surface temperature, among other oceanic processes, have a significant role in the decadal regulation of atmospheric dynamics. These coupled systems are chaotic and nonlinear and, because of multiscale interactions and feedback, these systems are extremely challenging to numerically model. Initial state or perturbation of the boundary condition may be scaled up to create massive forecast uncertainty. Therefore, proper modeling and forecasting of weather and climate require dynamic structures of numerical models, high-resolution computer models, ensemble modeling, and sophisticated data assimilation techniques, which underscores the computational intensity and the scientific complexity of atmospheric prediction systems (Ho et al., 2024).

1.2.2 The role of prediction in climate science

Correct forecasting is the center of contemporary climate science. In weather forecasting, immediate and accurate forecasts are essential to agriculture, transport, and safety of the people. Long-term projections are used in laying down futures policy, economy, and sustainable development in regard to climate change (Fan & Han, 2022).

The most common climate predictions include future projections of the temperature increase, sea-level changes, extreme climate patterns, or effects to ecology on timelines decades to many hundreds of years. Such projections are done through complex models that mimic the physical processes of the earth following basic laws of physics, e.g., thermodynamics, fluid dynamics, and radiation transfer (Sodhi, 2018).

Prediction also involves data assimilation where actual observational information on satellites and buoys as well as sensors in real time are included in the models in order to enhance accuracy. The amount and type of data are exponentially expanding, which also results in increasing computational load.

1.2.3 Classical computational approaches and their limitations

General circulation models (GCMs) and ESMs are simulations of the system of the earth by scientists. Such models divide the planet into 3D grids, and then, they model how the atmosphere, ocean, and land variables change with time. The activities depicted by the grid cells are the formation of clouds, the reflection of solar radiation, the evaporation, or the exchange of energies (Serrano et al., 2022).

These are models that are carried to supercomputers and which require immense computing capacity. Regardless of the advances of the computing power they possess, classical systems share significant limitations:

- **Resolution vs. Computation trade-off**: Finer local weather phenomena such as hurricanes or tornadoes can be represented with high-resolution models, but at a very high cost in terms of computation.
- **Uncertainty propagation**: The dynamical complexities of the atmosphere can cause significant differences between both short-term and long-term forecasts with even the slightest uncertainty in initial conditions.
- **Owen reducing the physics**: The physical processes at the sub-grid scale (e.g. cloud microphysics, vegetation feedback) are also required to be parameterized, with some error or approximation.
- **Scalability issues**: Classical systems lack scalability due to inefficiencies in the process of processing and simulating the immense amounts of data generated by global monitoring systems and satellites.

As the amount of data needed in climate modeling continues to increase and more precision is sought by the models, classical computing techniques are finding their performance limited.

1.2.4 The need for next-generation computing solutions

The enormous complexity, scale, and uncertainty of climate and weather models are forcing climate science to turn to next-generation computational paradigms to mitigate the mounting constraints of classical HPC to such models. The alternative designs can be helpful as new models such as quantum computing, neuromorphic computing, and artificial intelligence-based models of simulation are introduced. The paradigms are destined to deal with energy efficiency bottlenecks, scalability, and computational throughput that do not scale up to traditional numerical models.

Quantum computing is one of those, and it is a particularly disruptive technology. Taking advantage of the quantum systems and the quantum mechanical principles of entanglement and superposition, quantum systems can execute computations in high-dimensional state spaces in parallel and by means inaccessible to classical computers. Applications quantum algorithms have the capability to accelerate significantly important computations in climate, including high-dimensional optimization, quantification of uncertainties, and large-scale linear and nonlinear systems. Moreover, quantum-enhanced data assimilation technologies can allow combining a wide range of different and large volumes of observational data to predictive models in a short time and, therefore, improve the initiation of predictions and accuracy.

The application of the next-generation computing platforms results into the enhanced efficiency and accuracy of the climatic modeling, such as the availability of real-time forecast and advanced early warning in the event of extreme weather in case of the latter. More broadly speaking, introducing quantum computing to artificial intelligence and the Earth system science is a necessary revolutionization of climate research and the idea of changing the predictive abilities and improving the readiness of the world to climate change during the fast-tracking period (Chander, 2024).

1.3 Fundamentals of Quantum Computing

Quantum computing is a young paradigm, which involves the use of quantum mechanical properties to process information differently in a fundamentally new way. In contrast to classical computing that is based on binary bits, quantum computing utilizes qubits that are able to exist in a superposition of states, something only possible by obeying quantum mechanics through superposition and entanglement. Quantum computers can solve some computational problems exponentially quicker than classical mechanisms, partly because these properties open up interesting avenues in areas such as cryptography, machine learning, and scientific modeling (including climate modeling) (Moguel et al., 2022).

1.3.1 Basic principles: Qubits, Superposition, Entanglement

1.3.1.1 Qubits

The unit of quantum information is known as a qubit. As opposed to a classical bit which can be either 0 or 1, the qubit can be put into superposition between 0 and 1. The state of a qubit can be written mathematically as it is in eq. (1.1).

$$\psi\rangle = \alpha|0\rangle + \beta|1\rangle \tag{1.1}$$

Where:

- $|\psi.\rangle$ is the quantum state of the qubit.
- $|0.\rangle$ and $|1.\rangle$ are the basis states.
- α and β are complex numbers representing the probability amplitudes.

- The probabilities must satisfy eq. (1.2):

$$|\alpha|^2 + |\beta|^2 = 1 \tag{1.2}$$

This allows qubits to store and process a large amount of information with just a few units.

1.3.1.2 Superposition

Superposition allows a quantum computer to evaluate many possibilities simultaneously, making it particularly powerful for optimization problems (e.g., route planning, resource allocation), search problems (e.g., Grover's algorithm finds targets faster), and simulation physical systems (e.g., molecules, quantum materials) (Long, 2001).

1.3.1.2.1 Creating Superposition

Superposition is typically introduced using the Hadamard gate as shown in eq. (1.3):

$$H|0\rangle = \frac{1}{\sqrt{2}}(|0\rangle + |1\rangle) \tag{1.3}$$

This creates an equal superposition, where the qubit has a 50% chance of being measured as 0 or 1.

1.3.1.2.2 Limitations and considerations

- While a qubit can be in many states at once, we can't directly read all the information at once—measurement gives only one outcome.
- To take advantage of superposition, quantum algorithms need to be produced with a keen eye to amplify the correct ones and suppress the incorrect ones, through quantum interference.

Superposition helps quantum systems to go through various pathways in parallel, which gives them the computational parallelism that gives quantum computers their advantage. It forms the basis of much quantum algorithms and is the point that makes quantum computing so radically transformative in terms of science, industry, and technology (Serrano et al., 2022).

1.3.1.3 Entanglement

It describes the situation in which the state of one qubit is defined with dependence on the state of another even though they may be separated by arbitrary distances. This nonclassical correlation allows strong multi-qubit operations and make quantum computers significantly faster according to the complexity of the problem at hand than classical computers through exponential time.

1.3.1.3.1 Properties of entanglement

1. **Nonlocal correlation**
   a) The state of one qubit instantly affects the state of the other.
   b) The two qubits act like one combined system, not two independent ones.
2. **No cloning or communication**
   a) Entanglement doesn't allow you to transmit information faster than light.

   b) Measuring one qubit changes the system but does not reveal useful information without comparing both qubits.
3. **Entangled vs. Separable States**
   a) A separable state (not entangled) as shown in eq. (1.4):

$$|\psi\rangle = |0\rangle \otimes |1\rangle \tag{1.4}$$

   b) An entangled state cannot be written as a simple product as shown in eq. (1.5):

$$|\psi\rangle = \frac{1}{\sqrt{2}}(|01\rangle + |10\rangle) \tag{1.5}$$

1.3.1.3.2 Applications of Entanglement

1. **Quantum** Computing
   a) **Speeds up computation**: It speeds up computation by allowing qubits to store and process more complex information through correlations.
   b) **Quantum Teleportation**: Transmits quantum states over distance using entanglement.
   c) **Error Correction**: Entanglement helps detect and correct quantum errors in fragile qubit states.
2. **Quantum Communication**
   a) Quantum **key distribution (QKD)**: Secure communication protocols like **BB84** and **E91** use entangled photons to detect eavesdropping.
3. **Quantum Teleportation**
   a) **You can t**ransmit the state of a qubit using entangled particles, a classical message, and quantum entanglement—without physically moving the particle.

1.3.2 Quantum Gates, Circuits, and Measurement

1.3.2.1 Quantum Gates

Quantum gates are mathematical operations applied to qubits to manipulate their state. They are the quantum analog of classical logic gates (AND, OR, and NOT) but with some major differences:

- Quantum gates are unitary (they preserve probabilities).
- They operate on qubits that can be in superposition or entangled states.
- Gates can act on single or multiple qubits at a time.

Each quantum gate corresponds to a **unitary matrix,** and the action of the gate is applied via matrix multiplication to the qubit's state vector (Singh et al., 2024).

1.3.2.1.1 Quantum gates are of following different types:

1. **Single-Qubit Gates**

   These gates operate on a single qubit and can rotate or flip its state on the Bloch sphere.

a) Pauli Gates (X, Y, Z)
b) Hadamard Gate (H)
   i. Creates superposition as shown in eq. 1.6:

$$H|0\rangle = \frac{1}{\sqrt{2}}(|0\rangle + |1\rangle) \tag{1.6}$$

   ii. Matrix: $\text{H}=\frac{1}{\sqrt{2}}\begin{bmatrix} 1 & 1 \\ 1 & -1 \end{bmatrix}$

c) Phase Gates (S and T)
   i. **S Gate** ($\sqrt{z}$ gate): adds a phase of $\frac{\pi}{2}$

$$S = \begin{bmatrix} 1 & 0 \\ 0 & i \end{bmatrix}$$

   ii. **T Gate ($\pi/4$ phase gate)**: adds a phase of $\frac{\pi}{4}$

$$\text{S} = \begin{bmatrix} 1 & 0 \\ 0 & e^{i\pi/4} \end{bmatrix}$$

2. Multi-Qubit Gates

These gates operate on multiple qubits and are necessary for **entanglement** and conditional operations.

a) **CNOT Gate (Controlled-NOT)**
   i. Flips the second qubit (target) only if the first qubit (control) is $|1\rangle$.
   ii. Truth table (Refers to Table 1.1):

Table 1.1 Truth table.

Control	Target (before)	Target (after)
0	0	0
0	1	1
1	0	1
1	1	0

Matrix :

$$\text{CNOT} = \begin{bmatrix} 1 & 0 & 0 & 0 \\ 0 & 1 & 0 & 0 \\ 0 & 0 & 1 & 0 \\ 0 & 0 & 0 & 1 \end{bmatrix}$$

b) **Controlled-Z (CZ) and Controlled Gates**
   i. Applies a **Z** operation to the target if the control is $|1\rangle$.

   ii. More generally, controlled gates perform an operation only if certain conditions are met.

c) **Toffoli Gate (CCNOT)**
   i. 3-qubit gate: flips the third qubit if the first two are 1.
   ii. It's universal for classical computing inside a quantum computer.

1.3.2.2 Quantum Circuits

Quantum circuits are quantum gates on a collection of qubits. These circuits quantum algorithms are being built to execute. The qubits are set in a starting state, can undergo transformations under the gates, and are finally measured. Such circuits are based on the principle of superposition, entanglement,; and interference to achieve computation traditionally intractable on classical computers (Fingerhuth et al., 2018).

1.3.2.3 Measurement

Measurement destroys this superposition of a qubit in a probabilistic collapse and leaves it in a basis state (0 or 1). As an example, a qubit in the state $\alpha|0. + \beta|1.$ will be observed to be in the state $|0.$ probability$|\alpha|^2$ and in $|1.$ with probability $|\beta|^2$.

1.3.3 Quantum algorithms

These are the computational processes that take advantage of the basic quantum mechanical laws of superposition, entanglement, and quantum interference to address the problem more effectively than the classical algorithms. These algorithms are capable of searching large solution spaces in parallel and enhancing the good solutions as a result of constructive interference by working on quantum states representing many possibilities at the same time. Consequently, a number of quantum algorithms have provable exponential or even polynomial speedups as compared to their classical counterparts. Early examples are Shors algorithm of integer factorization, Grovers algorithm of unstructured search, and the HHL algorithm of solving linear systems. The algorithms below are fundamental to most of the contemporary quantum applications in science and engineering fields (Long, 2001).

1.3.3.1 HHL algorithm (Harrow–Hassidim–Lloyd)

Solves linear systems of the form Ax = b Ax = b

Input:

Sparse, Hermitian matrix AA, vector bb

Steps:

1. **State Preparation:**
   i. Encode the vector b into a quantum state $|b\rangle$.
2. **Quantum Phase Estimation (QPE):**
   i. Apply QPE on matrix A to estimate its eigen values λ_i and decompose $|b\rangle$. into eigen states $|u_i\rangle$.

3. **Controlled Rotation**
   i. Perform a controlled rotation on an ancilla qubit using $\lambda_i^{-1}$to encode $1/\ \lambda_i$.
4. **Uncompute QPE:**
   i. Apply the inverse QPE to disentangle the eigenvalue register.
5. **Measurement:**
   i. Measure the ancilla. If successful, the remaining system collapses to the solution state $|b \ltimes A^{-1}|b$.

1.3.3.2 Grover's algorithm

It performs a search over an unsorted database of size N with time complexity $O\left(\sqrt{N}\right)$ time.

Input:

Oracle function f(x) marking the target state

Steps:

1. **Initialization:**
   i. Prepare an equal superposition of all NN states using Hadamard gates.
2. **Oracle Application:**
   i. Apply the oracle to invert the phase of the target state.
3. **Amplitude Amplification:**
   i. Reflect amplitudes about the mean to amplify the target state.
4. **Iteration:**
   i. Repeat steps 2–3 approximately $\pi/4N\pi/4N$ times.
5. **Measurement:**
   i. Measure the register to get the target state with high probability.

1.3.3.3 Quantum Fourier Transform

Transforms a quantum state into its frequency representation

Input:

Quantum state $|x.$of n qubits

Steps:

1. **Apply Hadamard Gate:**
   i. Apply a Hadamard gate to the first qubit.
2. **Controlled Phase Rotations:**
   i. Apply controlled phase shifts between qubits.
3. **Repeat for All Qubits:**
   i. Continue steps 1–2 for all remaining qubits.
4. **Swap Qubits:**
   i. Reverse the qubit order to complete the transformation.
5. **Output State:**
   i. Obtain the Fourier-transformed quantum state.

QFT is the quantum analog of the discrete Fourier transform (DFT) and is crucial for problems involving periodicity. It transforms a quantum state from the computational basis into the Fourier basis.

1.3.3.3.1 Mathematical definition

For an n-qubit state |*x*., the QFT maps it to:

$$QFT|x\rangle = \frac{1}{\sqrt{2^n}} \sum_{y=0}^{2^n-1} e^{2\pi ixy/2^n} |y\rangle \tag{1.7}$$

Eq. (1.7) is a unitary transformation that can be implemented with $O(n^2)$ gates.

1.3.3.3.2 Algorithm

To implement QFT on an n-qubit system:

- Apply Hadamard to the first qubit.
- Apply a series of controlled phase shift gates to entangle it with all other qubits.
- Repeat for remaining qubits.
- Use SWAP gates to reverse the qubit order.

1.3.4 Overview of Quantum Hardware and Software Platforms

Quantum computing is rapidly evolving from theoretical constructs to physical machines and accessible programming environments. This development is driven by advances in both quantum hardware—the physical systems that embody qubits—and software platforms—the tools that allow users to build, simulate, and run quantum algorithms.

1.3.4.1 Quantum Hardware Platforms

Quantum computing is fast becoming a theoretical framework and a physical system and a programming platform. It has become possible using the advances in quantum hardware, or the physical systems that actually make qubits, and software platforms, or the tools that enable users to implement, simulate, and execute quantum algorithms.

1.3.4.1.1 Superconducting Qubits

1. **Technology:** Use Josephson junctions at cryogenic temperatures.
2. **Principle:** Quantum states are stored in microwave resonators with near-zero resistance.
3. **Leaders:**
   a) **IBM Quantum** (IBM Q)
   b) **Google Sycamore**
   c) **Rigetti Computing**
4. **Advantages:** Fast gate speeds, integration with classical control systems.
5. **Challenges:** Requires dilution refrigerators (~15 mK), decoherence over microseconds.

1.3.4.1.2 Trapped ion Qubits

1. **Technology:** Use electromagnetic fields to trap individual ions.
2. **Principle:** Quantum gates are implemented using laser pulses.
3. **Leaders:**
   a) **IonQ**
   b) **Honeywell Quantum (now Quantinuum)**
4. **Advantages:** Long coherence times (up to seconds), high gate fidelity.
5. **Challenges:** Complex laser systems.

1.3.4.1.3 Photonic quantum computers

1. **Technology:** Use light particles as qubits.
2. **Principle:** Information is encoded in properties of photons such as polarization or path.
3. **Leaders:**
   a) **PsiQuantum**
   b) **Xanadu (Borealis processor)**
4. **Advantages:** Operate at room temperature, easy transmission via fiber optics.
5. **Challenges:** Photon loss, difficulty in creating deterministic entanglement.

1.3.4.1.4 Spin qubits and quantum dots

1. **Technology:** Use the spin of electrons in semiconductor quantum dots.
2. **Leaders:**
   a) **Intel (Horse Ridge)**
   b) **UNSW Sydney (silicon-based qubits)**
3. **Advantages:** Compatibility with existing semiconductor fabrication.
4. **Challenges:** Complex to scale, sensitive to material imperfections.

1.3.4.1.5 Topological qubits

1. **Technology:** Based on exotic particles called anyons.
2. **Leader:**

a. **Microsoft Azure Quantum (StationQ)**

4. **Advantages:** Theoretically immune to certain types of errors.
5. **Challenges:** Still largely experimental, not yet demonstrated at scale.

1.3.4.2 Quantum Software Platforms

Quantum software platforms allow developers, researchers, and students to write, simulate, and deploy quantum algorithms on simulators.

1.3.4.2.1 Qiskit (IBM)

1. **Language:** Python based
2. **Features:**
   a) Circuit creation and visualization
   b) QASM simulator
3. **Strengths:** Open-source, strong community support, robust documentation.

1.3.4.2.2 Cirq (Google)

1. **Language:** Python
2. **Features:**
   a) Optimized for NISQ devices
   b) Emphasizes circuit compilation and control
3. **Strengths:** Integration with Google's Sycamore processor, useful for hardware research

1.3.4.2.3 PennyLane (Xanadu)

1. **Focus:** Quantum machine learning (QML)
2. **Features:**
   a) Hybrid quantum-classical optimization
   b) Interfaces with TensorFlow and PyTorch
3. **Unique Point:** Supports photonic circuits and VQA.

1.3.4.2.4 Q# and azure quantum (Microsoft)

1. **Language:** Q# (custom language for quantum programming)
2. **Platform:** Azure quantum cloud services
3. **Features:**
   a) Quantum Development Kit
   b) Resource estimation tools
   c) Integration with Visual Studio and Python

1.3.4.2.5 Braket (Amazon)

1. **Interface:** Python SDK
2. **Cloud Access:** Amazon-managed simulators and real hardware (IonQ, Rigetti, and OQC)
3. **Features:**
   a) Integration with AWS services
   b) Pay-per-use quantum computing access
   c) Hybrid algorithm support

1.3.4.2.6 Other notable tools

1. **Strawberry Fields** (Xanadu)—for photonic quantum computing
2. **Ocean SDK** (D-Wave)—for quantum annealing and optimization
3. **ProjectQ**—Python framework for compiling to different backends
4. **QuTiP**—for quantum simulation and open quantum systems

1.3.4.3 Hardware-Software Integration

Cloud quantum platforms enable access to real quantum processors in modern quantum platforms, in the form of API and SDK. This allows it to be more democratic to acquire and even quick in developing and the experimentation can be done without special laboratories. Hardware is ever approaching the limits of supporting the qubits and fidelity, and software is being opened to quantum programming. The convergence of these aspects is rendering the principle of viable quantum computing within the application in fields such as cryptography, optimization, and machine learning (Frolov, 2017).

1.4 Quantum Computing for Climate and Weather Modeling

This section focuses on the application of quantum computing to climate and weather science exclusively on how quantum computing can be applied to climate and weather science, emphasizing practical operations, domain-specific formulations, experimental efforts, and algorithmic workflows relevant to atmospheric and Earth system modeling.

1.4.1 Quantum Acceleration of Climate System Simulations

Climate models are governed by large coupled systems of PDEs that describe atmospheric circulation, ocean dynamics, radiative transfer, land–surface interactions, and cryospheric processes. These equations are discretized into massive sparse linear or nonlinear systems whose dimensionality increases rapidly with spatial resolution and ensemble size.

Quantum computing offers acceleration of climate-relevant numerical kernels, including:

- Solving discretized elliptic PDEs arising from pressure correction and Poisson equations.
- Performing spectral transforms for wave dynamics and climate oscillations.
- Accelerating ensemble simulations for climate uncertainty quantification.
- Improving data assimilation workflows that integrate satellite observations into dynamical models.

1.4.2 Quantum solution of Climate Differential Equations

Climate and weather evolution is governed by PDE systems refers to eq. (1.8):

$$\frac{\partial x}{\partial t} = f(x, \nabla \mathrm{x}, t) \tag{1.8}$$

where ∇x represents state variables (temperature, humidity, and velocity), and F is nonlinear.

1.4.2.1 Quantum workflow for climate PDEs

Step 1: Discretization

Finite difference, finite volume, or spectral discretization transforms PDEs into:

$$\mathrm{Ax} = \mathrm{b} \tag{1.9}$$

Step 2: Quantum Encoding

Encode b into amplitude states (refers to eq. 1.10):

$$|b\rangle = \sum_i b_i \left| i \right. \tag{1.10}$$

Map operator AA to a Hermitian Hamiltonian
Step 3: Quantum Processing
Use HHL or variational solvers to compute:

$$|x\rangle \approx A^{-1}|b\rangle \tag{1.11}$$

Apply quantum Fourier transform (QFT) for spectral PDE solvers
Step 4: Measurement
Extract observables such as mean temperature, vorticity, or pressure gradients

1.4.3 Quantum Algorithms for Climate and Weather PDEs

1.4.3.1 Linear solvers for Discretized Atmospheric Models

After discretization:

$$\mathrm{Ax} = \mathrm{b} \tag{1.12}$$

Quantum solvers produce the quantum state (refers to eq. 1.13):

$$|x\rangle = \frac{A^{-1}|b|}{\|A^{-1}\ b\|} \tag{1.13}$$

This is used for:

- Pressure correction in primitive equations
- Heat diffusion models
- Groundwater-climate interaction models

Experimental Perspective:

IBM quantum and IonQ have demonstrated small-scale linear solvers for PDE systems using 4–10 qubits to approximate diffusion equations.

1.4.3.2 Spectral Climate Solvers using QFT

Climate wave dynamics are represented as:

$$\mathrm{u}(x,t) = \sum_{k=0}^{n} u_k(t)e^{ikx} \tag{1.14}$$

The QFT efficiently computes uik enabling:

- Fast Laplacian diagonalization
- Efficient simulation of Ross by waves
- Spectral solution of the heat equation:

$$\frac{\partial T}{\partial t} = \alpha \nabla^2 T \tag{1.15}$$

Algorithm Sketch:

1. Encode temperature field into quantum amplitudes
2. Apply QFT to transform into frequency space
3. Multiply by diagonal evolution operator
4. Apply inverse QFT to recover spatial solution

1.4.3.3 Nonlinear climate PDE solvers

Nonlinear systems (e.g., Navier-Stokes, moist convection) require iterative schemes.

(a) Linearization + Quantum solvers

Newton iteration:

$$x_{k+1} = x_k - J^{-1}(x_k)F(x_k) \tag{1.16}$$

Each iteration step uses a quantum linear solver.

(b) Variational quantum algorithms

Define a residual loss:

$$\ell(\theta) =) =$$
$$\mathrm{L}(\theta)\|\mathrm{D}\psi\theta - \mathrm{f}\|2\mathrm{L}(\theta) = \|\mathrm{D}\psi\theta - \mathrm{f}\|2$$

Optimize a parameterized quantum circuit $\psi\theta$ to approximate the PDE solution.

Practical Example:

Variational solvers have been tested for Burgers' equation and diffusion PDEs using 6–12 qubits on NISQ hardware.

1.4.4 Quantum Modeling of Atmospheric and Fluid Dynamics

1.4.4.1 Navier–Stokes Equations

$$\frac{\partial u}{\partial t} = (u.\nabla)\mathrm{u} = -\nabla\mathrm{p} = \nu\nabla^2 u = f \tag{1.17}$$

- Quantum computing accelerates
- Poisson solves for pressure
- Spectral turbulence modeling
- Vorticity-stream function transformations

1.4.4.2 Primitive Atmospheric Equations

$$\frac{Du}{Dt} = -\nabla - \mathrm{fk} \times \mathrm{u} \tag{1.18}$$

Quantum algorithms target:

- Coupled momentum-thermodynamic solvers
- Energy flux modeling
- Fast ensemble propagation

1.4.5 Practical applications in climate and weather modeling

Table 1.2 represents the different quantum operations performed based on di climate components.

Table 1.2 Climatic Component vs Quantum Operations.

Climate component	Quantum operation
Pressure correction	Quantum linear solvers
Wave dynamics	QFT-based spectral solvers
Turbulence models	Variational solvers
Ensemble forecasting	Quantum parallel state prep
Data assimilation	Quantum Bayesian inference
Scenario optimization	QAOA / quantum annealing

1.5 Hybrid Quantum-Classical Modeling Frameworks

With the dawn of quantum computing, an increasing number of people are beginning to see in hybrid quantum-classical computing the practical application of quantum computing in high-stakes tasks such as weather forecasting and climate modeling. Current quantum hardware capabilities such as the number of qubits are incompressible, as well as, noise of quantum gates, decoherence, circuit depth, and so on. Hybrid forms are taking its place as the form of the best solution to follow in the case of the immediate impact. These architectures use quantum processors as special purpose accelerators to extend HPC pipelines, or a particular computational bottleneck that is the most common in climate and weather simulations, e.g., large linear system solvers, high-dimensional optimization, uncertainty quantification, and data assimilation.

Weather prediction and operations climatology make use of numerical models of the discrete approximations of the primitive equations and the NavierStokes equations, examples being ECMWF: the Integrated Forecasting System (IFS) and National Oceanic and Atmospheric Administration (NOAA): the Global Forecast System (GFS) and Weather Research and Forecasting (WRF) model. These solvers consist of repetitive studies of expensive solvers, e.g., pressure Poisson solvers, spectral transforms, ensemble propagation, and variational data assimilation. Quantum modules can be easily added to the existing workflow, including hybrid quantum-classical, without changing the entire model structure.

1.5.1 Integrating quantum modules into existing weather models

Most operational atmospheric models today, such as IFS of ECMWF, GFS of the US National Oceanic Deposition, and the WRF model, have their foundations in large-scale numerical solvers that solve combinations of the primitive equations of atmospheric dynamics, thermodynamics, and mass conservation on a global or regional grid. The models are as well grounded on physical parameterization schemes, which are the model of the unresolved subgrid processes,

Table 1.3 Hybrid Architectures of Quantum Computing.

Module	Quantum equivalent/enhancement
Linear solvers (e.g., Poisson eq.)	HHL algorithm or variational solvers
Spectral transforms	Quantum Fourier transform (QFT)
Optimization in data assimilation	Quantum annealing or QAOA
Covariance matrix updates	Quantum matrix inversion

which consist of cloud microphysics, convection, transport of radiation, and turbulence. These frameworks have proven to be utterly successful but are suffering because they are becoming computationally bottlenecked as the spatial resolution is refined, the size of the ensemble is also increased, and the data assimilation loops are becoming shorter.

With this existing design, quantum algorithms can be added as specialized CPU accelerators to compute-intensive sub-elements, without re-architecting the entire modeling system. Indicatively, the pressure–velocity coupler and the elliptic solvers in semi-implicit time stepping reiterate the answer of a notable sparse linear framework. Quantum linear solvers or VQAs can more cost-effectively estimate these solutions to save time to solution in high-resolution forecasts. Similarly, any spectral aspect of climate models, including the one applied to solve planetary waves and large-scale circulation, can have the luxury of QFT-related methods to accelerate spectral decomposition and operator diagonalization.

The data assimilation is one of the fundamental processes in weather prediction where the state vectors of high dimensions are optimized to place observational data and model predictions into correspondence. Variational data assimilation (4D-Var) and ensemble Kalman filtering would be run more quickly using quantum optimization algorithms and quantum-enhanced Bayesian inference and improve forecast initializations and measure uncertainty. QML could be applied in climate modeling to assist in downscaling a coarse-resolution simulation, identifying emergent climate behavior, and parameterization calibration.

Such hybrid architectures (refers to Table 1.3) can be directly utilized to enable faster forecast cycles, superior ensemble forecasting, and more precise long-term climate predictions in that they integrate quantum subroutines within classical HPC workflows. It is an effective method to apply quantum computing to maximize the quality of weather and the fidelity of climate modeling (Fan & Han, 2022).

1.5.2 Use of quantum simulators for forecast prototyping

Before running algorithms on quantum hardware, they are often tested on quantum simulators that mimic qubit behavior using classical resources. These are critical for:

- Developing quantum circuits for PDE solvers
- Performing error analysis and noise modeling
- Running small-scale forecast models for proof-of-concept

Available quantum simulators are discussed below in Table 1.4.

Table 1.4 Quantum Simulators.

Simulator	Platform	Features
Qiskit Aer	IBM	High-performance simulation, noise models
Cirq Simulator	Google	NISQ-optimized simulation
PennyLane DefaultQubit	Xanadu	Supports variational/hybrid algorithms
Amazon Braket Simulators	AWS	Integrated with cloud pipeline

Table 1.5 Evaluation Metrics.

Metric	Description
Quantum speedup	Compare runtime vs. best classical algorithm
Fidelity	Compare quantum forecast to ground truth
Noise resilience	Test accuracy under varying noise models
Hybrid overhead	Evaluate classical–quantum communication cost

1.5.3 Ensemble prediction and uncertainty quantification

In the past, ensemble is a well-known method of weather forecasting which attempts to model uncertainty due to chaotic processes and inaccurate measurements. These ensembles can be enhanced, or accelerated, with quantum techniques.

1.5.3.1 Quantum enhancements

Quantum amplitude estimation (QAE): With an accuracy level of desired, a Monte Carlo sampling method can be optimized to achieve a quantum sampling efficiency of or higher, with QAE. It is also applicable in computing the ensemble mean, measure of risk, or variance.

Quantum sampling of initial conditions: Quantum circuits can sample initial condition probability distributions or set of parameter distributions and calculate many weather scenarios simultaneously using quantum superposition.

Quantum Bayesian inference: The estimation of the posterior probability can be supported by quantum algorithms based on VQAs that can be more predictive and capable of more effectively incorporating satellite or radar data.

1.5.4 Benchmarking and performance evaluation

To justify the inclusion of quantum modules in mission-critical forecasting systems, it's essential to benchmark there. Various benchmarking startegies are discussed in Table 1.5.

- Accuracy
- Scalability
- Execution time
- Hardware cost-efficiency

1.5.4.1 Tools and frameworks

- **Qiskit Ignis** (for error characterization)
- **Pennylane metrics** (for variational circuit convergence)
- **Cross-validation with real meteorological datasets**

Hybrid quantum-classical modeling frameworks are a crucial step toward recognizing the benefits of quantum computing in operational weather forecasting. By identifying modules with computational bottlenecks and replacing them with quantum-enhanced alternatives, researchers can push the boundaries of resolution, uncertainty estimation, and forecast speed. While quantum advantage is not yet universal, these hybrid approaches provide a practical path forward in integrating quantum computing into atmospheric sciences, especially in the NISQ era.

1.6 Key areas of active research in quantum climate forecasting

Coming advances in quantum computing will offer a platform to remove ongoing bottlenecks in computational processes in climate modeling and weather prediction. The multiscale, climate studies are also the studies that are concerned with the dynamics of the short-term, real-time, and high-resolution atmospheric processes in weather prediction compared to climate studies where the focus is on the dynamics of the Earth system at long-terms. Quantum computing offers complementary services to the two areas because it accelerates numerical solvers, improves the rate of data assimilation, quantifies uncertainty more rapidly, and provides more efficiency to a greater number of ensemble simulations. This section describes how quantum algorithm development, data encoding schemes, and development of hardware will help in supporting climate and weather studies.

1.6.1 Algorithm development

The progress of effective quantum algorithms to address the problem of complexity of climate systems is among the most dynamic and frequently studied topics. Climate and weather models are large-scale, nonlinear, and multivariate processes that are difficult to compute classically. Linear system solvers, 3D parameter space optimization, better quantification of uncertainty, and better data assimilation are being addressed on quantum computers. Such advances will help minimize the cost of computations and augment the accuracy of models and their scalability to support the next-generation systems of climate and weather prediction (Eswaran et al., 2024).

1.6.1.1 Quantum linear solvers

Large sparse linear systems are the outcomes of the discretization of PDEs of motion, which are important to the atmosphere and in the ocean, both in climate models and in NWP systems. To speed up such computations, quantum linear solvers, including HHL algorithm, are being optimized. Such solvers can optimally complement spectral and finite-difference models in GCM in climate modeling and rapid-update cycle models in weather forecasting. Some of the important research problems are the minimization of the depth of the circuit, enhancement of numerical stability, and noise resilience.

1.6.1.2 Variational quantum algorithms of PDEs

VQAs, such as the variational quantum eigensolver (VQE) and the quantum approximate optimization algorithm (QAOA), are being extended to nonlinear and time-dependent differential equations by minimizing physics-based cost functions. These techniques are especially appealing as a means of weather forecasting because they can be run on low gate depth, and they can be run on NISQ devices. VQAs have a potential in climate applications as they can offer slow-evolving processes such as ocean circulation and radiative balance.

1.6.1.3 Quantum machine learning of climate and weather intelligence

Methods of QML are being researched to identify patterns and exceptions of large-scale atmospheric data. QML in weather prediction, including the early warning of extreme weather, storm tracking, and nowcasting. QML is applicable in climatic studies to downscale the output of coarse resolution model, identify climate regimes, and to more quantitatively represent uncertainty. The algorithms, which are the focus of such works, are quantum support vector machines and quantum-enhanced neural networks.

1.6.1.4 Quantum sampling and amplitude estimation

Both the climate projections and the weather projections need to quantify the uncertainty using the measure of probabilistic simulations and ensemble projections. QAE is a quantum algorithm that has a quadratic speed advantage to classical Monte Carlo simulation and can potentially have faster ensemble-based simulation. Its application is particularly in the climate sensitivity analysis and probabilistic weather forecasting.

1.6.2 Climate and weather data encoding for quantum processing

One of the challenges that are needed is the effective encoding of high-dimensional climate and weather data into quantum states. The financial requirements of climatic applications include that they demand small-sized descriptions of the long-term and multivariate data, in contrast to weather forecasting, which involves the encoding of observational data over time. The hybrid preprocessing and quantum state preparation need to be developed to enable scalable quantum-based environmental modeling (Singh et al., 2024).

1.6.2.1 State preparation techniques

Quantum state study protocols are currently under development to encode atmospheric parameters such as temperature, wind speed, pressure, and humidity. Techniques such as amplitude encoding, basis encoding, and angle encoding are discussed, and applications in climate, such as an average field over long periods of time, and weather, such as high-resolution and localized data, are discussed.

1.6.2.2 Compression of data and exploitation of sparseness

The datasets of climate are usually high in density and dimension, whereas the weather data may demand quick processing of local information. Encoding algorithms are becoming more and more sparse, based on reduced-order encoding and compressed sensing, to be encoded with a small number of qubits.

1.6.2.3 Time-series encoding and spatio-temporal encoding

Weather forecasting captures fast changing temporal dynamics appropriately, whereas longer-term trend and variability are prioritized in climate modeling. Quantum recurrent model, quantum auto-regressive structures, and quantum hybrid temporal encoding schemes are under study to find application in both.

1.6.2.4 Hybrid preprocessing pipelines

Climate and weather data are run on classical algorithms such as grid coarsening, anomaly detection, and dimensionality reduction and processed using quantum processors of hybrid quantum-classical pipelines. This type of methodology guarantees near term hardware viability, but physical interpretability.

1.6.3 Quantum hardware advancements supporting climate and weather modeling

Progress in quantum hardware is critical to translating theoretical algorithms into practical tools for atmospheric science.

1.6.3.1 Qubit scaling and fidelity

The additional number of qubits with high fidelity is important to climate-scale simulations and high-resolution weather models. Qubits based on superconducting, trapped-ion, and photonic systems are the leading systems, with their advantages and disadvantages based on the workload.

1.6.3.2 Noise mitigation and error management

Noise remains a significant barrier on NISQ devices. For climate and weather applications, researchers are developing error mitigation techniques such as shallow circuit designs and early-stage quantum error correction to improve reliability.

1.6.3.3 Quantum–classical integration

The short-term deployment will be based on the hybrid architecture where quantum processors will be tightly coupled with classical HPC systems. These architectures enable the divide of the task, the rapid communication of information with low latency, and the reuse of quantum circuits in the functioning weather and climate services.

1.6.3.4 Benchmarking and calibration for atmospheric workloads

Benchmarking of quantum hardware can use climate- and weather-specific workloads (spectral solvers, ensemble generators, and data assimilation kernels), which are realistic workloads of hardware performance, and benchmarking the hardware as well as hardware calibration to Earth system workloads.

1.7 Challenges, opportunities, and the future of climate forecasting

The most important part of climate forecasting is in addressing the global problems of disaster preparedness, agricultural planning, and sustainability policy. However, the Earth systems, nonlinear systems as well as chaotic systems, large-scale system, and time scale system require much of the classical computing. The radical change of the paradigm is the potential of quantum computing with its radically different model of information processing. Quantum-enhanced climate forecasting is, however, not yet put into practice. This section outlines the challenges, existing opportunities, and a future roadmap for applying quantum technologies in climate science.

1.7.1 Challenges

Quantum computing holds great promise for accelerating and enhancing climate forecasting, but several major challenges need to be addressed before widespread adoption becomes practical as shown in Fig. 1.2:

1. ***Hardware limitations (NISQ Era)***
   a) NISQ devices have limited qubits (tens to hundreds), short coherence times, and high error rates.
   b) Lack of fault tolerance restricts the complexity and depth of quantum circuits currently executable.
2. ***Scalability and resolution***
   a) High-resolution climate models require encoding millions of spatial and temporal variables—far beyond the current capability of quantum hardware.
   b) Mapping real-world PDE-based models like the Navier–Stokes or primitive equations into quantum circuits remains a major technical barrier.
3. ***Data encoding bottlenecks***
   a) Loading large-scale atmospheric data (e.g., from satellites or reanalysis systems) into quantum systems is slow and resource intensive.
   b) Quantum input/output is still inefficient and limited in practical scenarios.
4. ***Lack of specialized algorithms***
   a) Existing quantum algorithms (like HHL) are primarily designed for linear systems and may not generalize well to the nonlinear, chaotic nature of climate models.

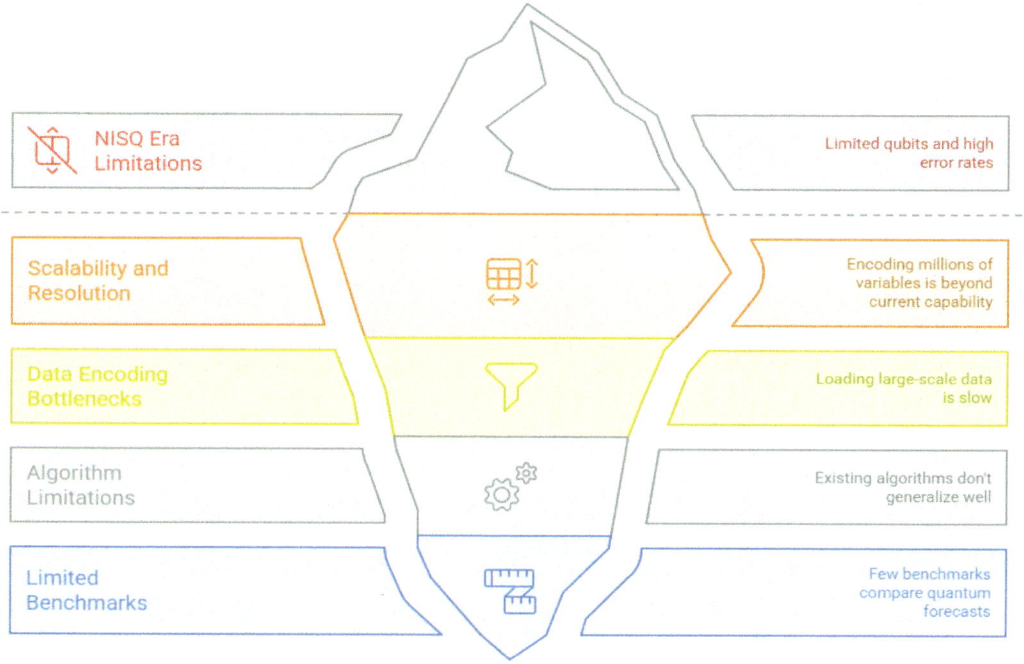

FIGURE 1.2 Quantum computing challenges in quantum computing.

  b) Quantum algorithms for uncertainty quantification, ensemble modeling, or long-range simulations are still in early development.

5. ***Limited benchmarks and standards***
  a) Few real-world benchmarks exist comparing quantum and classical climate forecasts.
  b) There's a lack of consensus on performance metrics for hybrid or quantum-enhanced models.

1.7.2 Research gaps and multidisciplinary collaboration

1.7.2.1 Key research gaps

1. **Quantum PDE solvers for geophysical models**
  a) Lack of scalable algorithms for nonlinear PDEs in quantum form
  b) Few real-world tests of quantum solvers on fluid or radiative transfer models
2. **Hybrid algorithm design**
  a) Need for optimized partitioning between classical and quantum components
  b) Limited benchmarking data comparing classical vs. hybrid workflows
3. **Uncertainty quantification in quantum forecasting**
  a) Few methods exist to integrate ensemble techniques, stochastic modeling, or error propagation on quantum hardware

Table 1.6 Enhanced Quantum Approaches.

Domain	Role
Quantum computing	Algorithm and circuit design, hardware optimization
Climate science	Physical modeling, validation, data interpretation
Numerical methods	Discretization, solver optimization, model scaling
Computer science	Software engineering, cloud integration, benchmarking
Ethics and policy	Responsible deployment, bias auditing, data governance

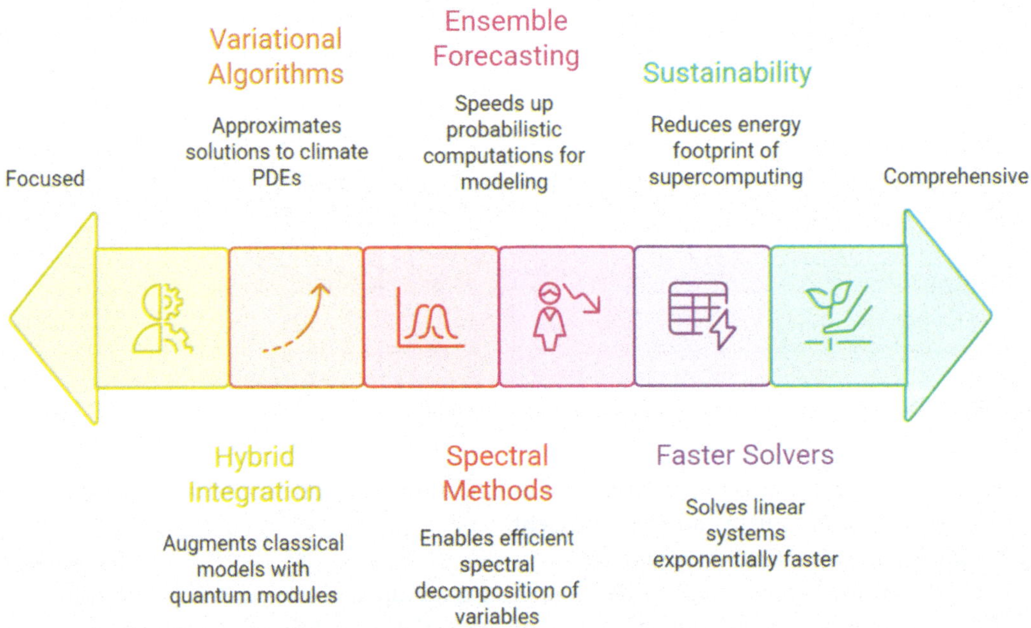

FIGURE 1.3 Quantum computing integration ranges from focused to comprehensive.

4. **Data assimilation with quantum methods**
   a) Quantum Kalman filtering and Bayesian inference are still experimental.

1.7.2.2 Need for multidisciplinary teams

Quantum-enhanced climate prediction requires integration of expertise as discussed in Table 1.6.

1.7.3 Opportunities

Despite current challenges, quantum computing offers several unique and transformative opportunities for climate science. Fig. 1.3 shows quantum computing integration ranges from focused to comprehensive.

1. *Faster linear solvers*
   a) To some linear systems, quantum algorithms including the HHL algorithm are capable of providing the solutions with exponential speedups than classical methods.
   b) These could benefit parts of climate models that involve solving large sparse matrices (e.g., pressure–velocity coupling).
2. *Enhanced ensemble forecasting*
   a) Quantum algorithms like quantum amplitude estimation (QAE) can speed up probabilistic computations, improving efficiency in ensemble modeling and uncertainty estimation.
3. *Spectral methods with QFT*
   a) The QFT enables efficient spectral decomposition, which is commonly used in global atmospheric models for transforming spatial variables.
4. *Hybrid quantum-classical Integration*
   a) Using quantum computing modules for subtasks (e.g., optimization, data assimilation, and sensitivity analysis) while retaining classical simulation engines.
   b) Supports progressive integration without overhauling existing climate modeling frameworks.
5. *Variational algorithms for PDEs*
   a) VQA (e.g., VQE, QAOA) can be used to approximate solutions to differential equations, including simplified climate PDEs, with reduced circuit depth.
6. *Sustainability and efficiency*
   a) Quantum computing can reduce the energy footprint of supercomputing, aligning with environmental sustainability goals.

1.7.4 Ethical, societal, and environmental considerations

While quantum computing holds technical promise, its application in climate forecasting must consider broader impacts as follows:

1. *Ethical and societal questions*:
   a) **Access equity**: Will quantum-enhanced forecasting benefit only wealthy nations or private corporations?
   b) **Algorithmic bias**: QML models must be audited for data bias or misrepresentation of vulnerable regions.
   c) **Transparency**: Complex hybrid systems should maintain interpretability to build public trust.
2. *Environmental impact:*
   a) **Energy efficiency**: Quantum computers consume far less energy than exascale classical systems for certain tasks.
   b) **Sustainable forecasting**: Efficient simulations can reduce energy use in research, transportation, agriculture, and disaster mitigation.

1.7.5 Roadmap toward quantum-enhanced climate prediction

Achieving quantum advantage in climate modeling requires a staged and realistic roadmap:

1. *Short Term (1–3 years):*
    a) Develop small-scale prototype models (e.g., 1D diffusion, low-resolution convection)
    b) Use quantum simulators for benchmarking
    c) Apply variational quantum solvers to simplified PDEs
    d) Build interdisciplinary teams and training programs
2. *Medium Term (3–7 years):*
    a) Integrate quantum modules (e.g., solvers, optimizers) into existing forecasting frameworks
    b) Employ hybrid cloud platforms (e.g., Amazon Braket, Azure Quantum) for simulation
    c) Standardize benchmarking protocols for quantum climate algorithms
    d) Collaborate with climate policy stakeholders for targeted use cases (e.g., flood risk, heat waves)
3. *Long Term (7–15+ years):*
    a) Deploy fault-tolerant quantum computers for full-resolution climate modules
    b) Enable real-time quantum-assisted forecasting
    c) Support global, equitable access to quantum infrastructure
    d) Integrate quantum climate models with digital twin Earth systems

Quantum computing opportunities can radically change the future of climate forecasting through the creation of fundamentally new opportunities that can be applied to process, simulate, and interpret the dynamics of the complex Earth system. Its capability of searching high-dimensional solution space and speeding up the various important numerical computations such as optimization, sampling, and data assimilation can be useful in enhancing both short-term weather forecasting and long-term climatic forecasting. Nevertheless, the way ahead would require realistic ambitions since the present quantum hardware is limited to few qubit counts, noise, and decoherence. Such constraints require specific studies on hybrid quantum classical models, noise-resistant algorithms, and usage-inspired benchmarking as opposed to supplanting classical models wholesale.

Ethical implementation is also important. The quantum-enhanced climate models should be prepared transparently, reproducibly, and scientifically verified so that the outputs of such models can be relied upon by the climate science community as well as the policymakers. To convert the theoretical progress to the practical means, the multidisciplinary teamwork of atmospheric scientists, quantum, computer scientists, and engineers will be crucial to achieve the progress. Simultaneously, ethical issues, such as fair access to high-tech computing tools and the sustainable utilization of predictive information, should condition the use of technologies. Through a methodical solution to hardware challenges, climate-aware quantum algorithms, and integration of innovation with societal requirements, quantum computing will become a change agent of climate resilience, knowledgeable policymaking, and sustainable development in a highly uncertain climate future.

1.8 Conclusion

Quantum computing has tremendous potential in changing climate and weather forecasting by making more and more complex processes in the atmosphere simulated. As the model resolution, not to mention the volume of data, is steadily expanded, the traditional supercomputers are increasingly being challenged by the large scale, coupled, nonlinear, PDEs that form the atmosphere processes, oceanic circulation, and land-atmosphere interactions. An interesting escape of such limitations is quantum algorithms, which are able to execute high-dimensional state spaces, and carry out tasks such as solving linear systems, optimization, and probabilistic sampling.

Although quantum hardware is still in the NISQ stage and is not yet capable of operating on large-scale scale forecasting, there is much underway in both the design of algorithms and error reduction and quantum-classical hybrid workflows. Such strategies can be used to scale quantum devices in the near term to provide complement to classical HPC by running specific subroutines rapidly, such as data assimilation and uncertainty quantification and pattern detection of climate data.

The future of quantum computing in climatic/weather prediction is very extensive. As fault-tolerant quantum computers, scalable qubit systems, climate-conscious quantum computations, and many other new approaches are developed, it might become feasible to model Earth systems ultrahigh resolution, to predict climate on real-time ensembles, and to predict climate more accurately in the long term. Lastly, quantum-enhanced forecast systems will potentially play significant roles in maximizing disaster early warning mechanisms, climate adaptation, and evidence-based policy to environmental policy adjustment to the rapid climatic change.

References

Chander, S. (2024). The current landscape of quantum hardware development-an overview. *Intersect: The Stanford Journal of Science, Technology, and Society (pp. 5-8), 17*(3).

Eswaran, U., Eswaran, V., Murali, K., & Eswaran, V. (2024). Quantum-based predictive modeling for extreme weather events. In C. K. K. Reddy, A. Nag, M. Ouaissa, B. Bhushan, & M. M. Hanafiah (Eds.), *The rise of quantum computing in Industry 6.0 towards sustainability: Revolutionizing smart disaster management* (pp. 123–140). Switzerland: Springer Nature.

Fan, L., & Han, Z. (2022). Hybrid quantum-classical computing for future network optimization. *IEEE Network, 36*(5), 72–76.

Fingerhuth, M., Babej, T., & Wittek, P. (2018). Open source software in quantum computing. *PloS one, 13*(12), e0208561.

Frolov, A. V. (2017). Can a quantum computer be applied for numerical weather prediction? *Russian meteorology and hydrology, 42*(9), 545–553.

Gill, S. S., Kumar, A., Singh, H., Singh, M., Kaur, K., Usman, M., & Buyya, R. (2022). Quantum computing: A taxonomy, systematic review and future directions. *Software: Practice and Experience, 52*(1), 66–114.

Ho, K. T. M., Chen, K. C., Lee, L., Burt, F., Yu, S., & Lee, P. H. (2024). Quantum computing for climate resilience and sustainability challenges. In *2024 IEEE International Conference on Quantum Computing and Engineering (QCE): 2* (pp. 262–267). IEEE.

Long, G.-L. (2001). Grover algorithm with zero theoretical failure rate. *Physical Review A, 64*(2), 022307.

Moguel, E., Rojo, J., Valencia, D., Berrocal, J., Garcia-Alonso, J., & Murillo, J. M. (2022). Quantum service-oriented computing: Current landscape and challenges. *Software Quality Journal, 30*(4), 983–1002.

Rahman, S. M., Alkhalaf, O. H., Alam, M. S., Tiwari, S. P., Shafiullah, M., Al-Judaibi, S. M., & Al-Ismail, F. S. (2024). Climate change through quantum lens: Computing and machine learning. *Earth Systems and Environment, 8*(3), 705–722.

Rietsche, R., Dremel, C., Bosch, S., Steinacker, L., Meckel, M., & Leimeister, J. M. (2022). Quantum computing. *Electronic Markets, 32*(4), 2525–2536.

Saiyed, A. I. (2025). Hybrid quantum-classical cryptographic protocols: Enhancing security in the era of quantum supremacy (pp. 1-7). *Spectrum of Research, 5*(1).

Serrano, M. A., Cruz-Lemus, J. A., Perez-Castillo, R., & Piattini, M. (2022). Quantum software components and platforms: Overview and quality assessment. *ACM Computing Surveys, 55*(8), 1–31.

Singh, P., Dasgupta, R., Singh, A., Pandey, H., Hassija, V., Chamola, V., & Sikdar, B. (2024). A survey on available tools and technologies enabling quantum computing. *IEEE access, 12*, 57974–57991.

Singh, T. M., Reddy, C. K. K., & Lippert, K. (2024). An empirical study into quantum machine learning for precise. *The Rise of Quantum Computing in Industry 6.0 Towards Sustainability: Revolutionizing Smart Disaster Management*, 177–191).

Sodhi, B. (2018). Quality attributes on quantum computing platforms (1st ed.). arXiv preprint arXiv:1803.07407. https://arxiv.org/abs/1803.07407

Tennie, F., & Palmer, T. N. (2023). Quantum computers for weather and climate prediction: The good, the bad, and the noisy. *Bulletin of the American Meteorological Society, 104*(2), E488–E500.

Upama, P. B., Faruk, M. J. H., Nazim, M., Masum, M., Shahriar, H., Uddin, G., Ahamed, S. I., & Rahman, A. (2022). Evolution of quantum computing: A systematic survey on the use of quantum computing tools. In *2022 IEEE 46th Annual Computers, Software, and Applications Conference (COMPSAC)* (pp. 520–529). IEEE.

Zeema, J. L., Deepa, S., Kirubanand, V. B., Jose, T., & Kalpana, P. (2025). Exploring quantum computing in weather forecasting: Leveraging optimization algorithms for long-term accuracy. *Quantum Computing and Artificial Intelligence: The Industry Use Cases*, 139–162.

2

AgriQML: A novel quantum machine learning framework for agriculture crop yield prediction

Deepa Bhadan[a], Thanjaivadivel M[b] and R Lakshmana Kumar[c]

[a]RESEARCH SCHOLAR, GENETICS AND PLANT BREEDING, CHAUDHARY CHARAN SINGH UNIVERSITY, MEERUT, UTTAR PRADESH, INDIA [b]DEPARTMENT OF COMPUTER SCIENCE, NANDHA ENGINEERING COLLEGE, ERODE, TAMIL NADU, INDIA [c]DEPARTMENT OF ARTIFICIAL INTELLIGENCE AND MACHINE LEARNING, KIT - KALAIGNAR KARUNANIDHI INSTITUTE OF TECHNOLOGY, COIMBATORE, TAMIL NADU, INDIA

2.1 Introduction

Karthikeyan and Murugan (2024) and Jabed and Murad (2024) forecasted that farm crop yields is a crucial component of food security, economic planning, and sustainable agricultural development. Accurate yield estimation supports informed decision-making for farmers, supply-chain stakeholders, and policy planners while mitigating risks associated with climate variability, resource scarcity, and market instability. In recent years, the advent of intelligent agriculture has significantly transformed traditional farming practices by integrating Internet of Things (IoT) sensors, satellite-based remote sensing platforms, unmanned aerial vehicles, and automated weather monitoring systems. These technologies continuously generate massive volumes of heterogeneous data describing soil composition, moisture levels, meteorological conditions, temperature fluctuations, vegetation indices, and crop phenology (Burdett & Wellen, 2022; Rashid et al., 2021).

Sadenova et al. (2023) and Joshi et al. (2023) stated that although such data availability enables data-driven farming strategies, effectively extracting meaningful knowledge from these high-dimensional, nonlinear, noisy, and strongly correlated sources remains an open challenge for conventional computational approaches. Agricultural data often exhibit complex temporal dependencies, spatial variations, missing values, and nonlinear interactions across multiple environmental variables. These characteristics significantly complicate predictive modeling tasks, particularly in scenarios with limited labeled data, heterogeneous geographic regions, and dynamically changing climatic patterns.

Quantum Computing and Instrumentation for Smart Agriculture. DOI: https://doi.org/10.1016/B978-0-443-44247-6.00006-3

Nosratabadi et al. (2020) and Logeshwaran et al. (2024) showed that recent advances in machine learning (ML) had led to notable improvements in agricultural prediction systems, enabling automated learning from historical yield records, soil properties, and climate observations. Nevertheless, most classical ML models rely on Euclidean feature representations and linear or kernel-based transformations, which limit their ability to capture intricate correlations and higher-order interactions among agricultural variables. As a consequence, predictive performance often degrades when data dimensionality increases or when relationships between variables become highly nonlinear and interdependent.

Setiadi et al. (2024) forecasted that quantum machine learning (QML) had recently emerged as a promising paradigm that integrates principles of quantum computing with learning algorithms to address such computational bottlenecks. By exploiting quantum superposition, entanglement, and high-dimensional Hilbert space representations, QML offers new opportunities for solving complex optimization and pattern recognition problems more efficiently than classical approaches. In the agricultural domain, QML can facilitate richer feature representations, improved generalization in sparse datasets, and enhanced learning of nonlinear dependencies between environmental and biological factors. Consequently, applying QML to crop yield forecasting opens new avenues for precision agriculture, early risk assessment, and adaptive agricultural policy planning (Filippi et al., 2022; Elbasi et al., 2023).

Sivanantham et al. (2022) stated that quantum-based modeling enables multidimensional data transformations that better represent uncertainties, hidden interactions, and stochastic environmental behavior. These capabilities are particularly valuable for agricultural systems, where crop productivity depends on a delicate balance between soil chemistry, water availability, temperature, solar radiation, and management practices. As a result, quantum-aided learning frameworks can potentially surpass traditional ML techniques, especially in noisy, incomplete, and weakly structured datasets frequently observed in real-world farming environments. The growing synergy between agricultural technology and quantum computing, therefore, highlights the need for intelligent, scalable, and adaptive predictive systems capable of handling complex agrometeorological dynamics (Elavarasan & Vincent, 2020).

Bian et al. (2022) demonstrated that a wide range of classical approaches has been employed for crop yield prediction using historical observations, soil parameters, and climatic indicators. Linear regression, ridge regression, and Lasso regression are widely adopted due to their simplicity and interpretability; however, they exhibit limited performance when modeling nonlinear relationships and complex variable interactions (Date & Potok, 2021). Decision tree-based algorithms and ensemble methods, such as Random Forest and XGBoost, generally achieve higher predictive accuracy but struggle to effectively represent deeply correlated and high-dimensional feature spaces (Song et al., 2023). Moreover, these models often lack transparency in capturing temporal dependencies within crop growth cycles and are not easily adaptable to rapidly changing environmental conditions.

Jin and Xu (2024) stated that deep learning architectures, including long short-term memory (LSTM), bidirectional LSTM, and gated recurrent units (GRU), had been explored to model sequential crop and climate data. While such networks can learn temporal patterns, they require substantial labeled datasets and computational resources, and their performance often deteriorates when applied to heterogeneous agricultural regions or sparse data conditions.

Their reliance on classical computational infrastructures further restricts their ability to encode complex superposition-like correlations inherent in multisource agricultural datasets.

Li et al. (2025) found that kernel-based techniques such as support vector regression (SVR) and Gaussian process regression (GPR) offer flexibility in nonlinear learning but are highly sensitive to kernel selection, demand extensive feature engineering and exhibit poor scalability for large datasets. Convolutional neural networks and attention-based transformers have demonstrated success in satellite-based crop monitoring tasks (Lu et al., 2023); however, these models are computationally expensive and lack quantum-enhanced state representations (Setiadi et al., 2024). Existing frameworks generally do not incorporate quantum feature encoding, variational quantum circuits (VQC), or hybrid classical-quantum learning mechanisms capable of extracting structured information from entangled, high-dimensional agricultural variables (Dang et al., 2021).

Jin et al. (2023) stated that consequently, current predictive systems frequently suffer from underfitting in sparse datasets and overfitting in heterogeneous large-scale scenarios, limiting their reliability and robustness in operational agricultural deployments. Linear and polynomial regression models are particularly vulnerable to multicollinearity and poor generalization across geographical zones (Ansarifar et al., 2021; Shah et al., 2020). Tree-based ensembles face difficulties in handling strongly correlated features and dynamic environmental changes (Keerthana et al., 2021; Noorunnahar et al., 2023; Nti et al., 2023), while recurrent neural networks are computationally intensive and prone to instability under limited training samples (Bandar & Coşkunçay, 2024; Kumar et al., 2024; Magallanes-Quintanar et al., 2024). Similarly, SVR and GPR models encounter scalability issues and reduced robustness in the presence of noise and missing sensor readings (Ghosh et al., 2022; Servia et al., 2022).

This study aims to address these methodological limitations by developing a scalable and robust QML-based crop yield prediction framework capable of learning from heterogeneous agricultural data sources under uncertain and dynamically evolving conditions. The scope of this work encompasses quantum-assisted dimensionality reduction, hybrid feature representation, and predictive modeling across diverse environmental settings. To this end, a hybrid classical-quantum architecture is proposed that integrates quantum feature encoding, VQC, and classical neural layers for accurate crop yield estimation. Angle-based quantum embedding is employed to represent complex data distributions, while the hybrid learning pipeline combines the interpretability of classical ML with the expressive power of quantum entanglement. Furthermore, quantum-based dimensionality reduction techniques are utilized to mitigate the curse of dimensionality, and noise-aware optimization strategies are incorporated to enhance model stability under weather uncertainty and sensor variability. These contributions collectively enable improved generalization, reduced data dependency, and enhanced predictive accuracy, thereby establishing a practical foundation for quantum-enabled precision agriculture.

Objectives of the chapter

The objectives of this chapter are to analyze the limitations of traditional ML techniques in encoding entangled and nonlinear patterns in high-dimensional agricultural data and to design an efficient quantum-enhanced prediction framework. Quantum principal component analysis (qPCA) is applied to effectively reduce dimensionality and extract significant features from complex crop yield datasets. Dual-mode quantum feature encoding techniques, including angle and amplitude encoding, are employed to transform classical agricultural variables into quantum-representable states. Furthermore, a hybrid quantum–classical architecture integrating qPCA, quantum feature encoding, VQC, and QSVM is developed for accurate crop yield prediction. The predictive performance of the proposed quantum models is evaluated using MAE, RMSE, R^2, and accuracy and is systematically compared with classical approaches such as random forests, XGBoost, and support vector machines (SVMs).

Organization of the chapter

The rest of the chapter is organized as: Section 2.2 enlightens literature review. Section 2.3 focuses on Materials and Methods. Section 2.4 highlights proposed methodology. Section 2.5 discusses Experimentation, Results and Analysis. And, finally, section 2.6 concludes the chapter with future scope.

2.2 Literature survey

Incudini et al. (2023) proposed a quantum variational generator based on VQC to model continuous probability distributions using a hybrid quantum–classical adversarial framework. The novelty of the study lay in integrating quantum encoding with classical discriminators to learn complex data distributions. Experimental results demonstrated improved representation capability compared to classical generators. However, the model suffered from scalability limitations due to high measurement overhead and increased shot complexity as circuit depth increased.

Rajagopa (2021) developed a deep belief network combined with VGG-Net and chick swarm optimization for crop yield prediction using local agricultural datasets. The key contribution was the introduction of swarm-based hyperparameter optimization for improving classification accuracy. The experimental evaluation reported 97% accuracy on small datasets. Nevertheless, the approach exhibited severe overfitting and high computational cost due to bio-inspired optimization.

Maheshwari et al. (2022) presented a variational quantum classifier (VQC) employing angle-based quantum feature encoding for binary agricultural data classification. The novelty lay in transforming classical crop features into a quantum Hilbert space to enhance separability. Experiments indicated moderate accuracy gains over classical SVM models. However, barren plateau effects, noise sensitivity, and the overhead of exponential gradient estimation limited scalability to high-dimensional datasets.

Metawei et al. (2022) implemented a QSVM to classify crop stress using hyperspectral imagery. The study introduced quantum kernel mapping to increase class separability in complex spectral data. Results demonstrated improved classification margins compared to classical kernels. However, the method required $O(n^2)$ quantum circuits, leading to impractical runtimes under limited qubit availability.

Lou et al. (2024) proposed a transformer-based attention framework for multimodal crop yield prediction integrating weather, soil, and satellite data. The novelty was the use of self-attention to capture long-range spatial dependencies. Experimental results showed improved prediction accuracy over CNN and RNN baselines. Despite this, the model demanded large datasets and extensive GPU resources, restricting scalability.

Gong et al. (2021) introduced an attention-based recurrent neural network for sequential climate feature modeling in agricultural forecasting. The method improved temporal pattern learning in rainfall and soil moisture data. Experiments validated higher forecasting accuracy than conventional RNNs. However, vanishing gradients and poor cross-regional generalization reduced model robustness.

De Oliveira et al. (2021) proposed YieldNet, a convolutional neural network for multi-crop yield estimation from remote sensing images. The contribution involved multi-target learning and transfer learning across crop types. Results demonstrated consistent improvements over single-crop CNN models. Nonetheless, the system required large, labeled datasets and failed to explicitly capture seasonal dynamics.

Song (2024) developed a variational Monte Carlo framework using deep neural network ansatzes for quantum state estimation. The study contributed efficient wavefunction approximation techniques for quantum learning tasks. Simulation results validated high representational accuracy for low-dimensional systems. However, convergence difficulty and sampling complexity restricted applicability to large-scale agricultural datasets.

Bhagwakar et al. (2024) proposed a quantum-accelerated k-means clustering algorithm using Grover's search for unsupervised learning. The novelty was the quadratic speedup in centroid assignment operations. Experimental simulations showed reduced computational complexity. Nevertheless, the approach was sensitive to near-term quantum devices (NISQ) hardware noise and lacked robustness to missing sensor values, which are common in agricultural environments.

Existing studies had demonstrated that classical deep learning models require massive labeled datasets and substantial computational resources and exhibit limited generalization across regions and seasons. QML approaches showed promise for handling complex feature spaces but faced qubit scarcity, measurement overhead, noise sensitivity, and scalability constraints. Most works applied either QSVM or VQC independently without integrating dimensionality reduction techniques. High-dimensional crop datasets led to barren plateaus and unstable training in variational circuits. Current quantum kernels were computationally expensive and impractical for real-world deployment. Transformer and CNN models lacked quantum acceleration and remained constrained by classical hardware limitations. No existing framework jointly optimized feature extraction, quantum encoding, and classification in a unified pipeline. Furthermore, hybrid architectures are rarely evaluated for robustness against noisy agricultural sensor data. The absence of integrated qPCA with dual-mode quantum encoding limited representational efficiency. These technical gaps motivated the design of a hybrid quantum–classical framework combining qPCA, angle-amplitude encoding, and quantum classifiers to enable scalable, accurate, and resource-efficient crop yield prediction.

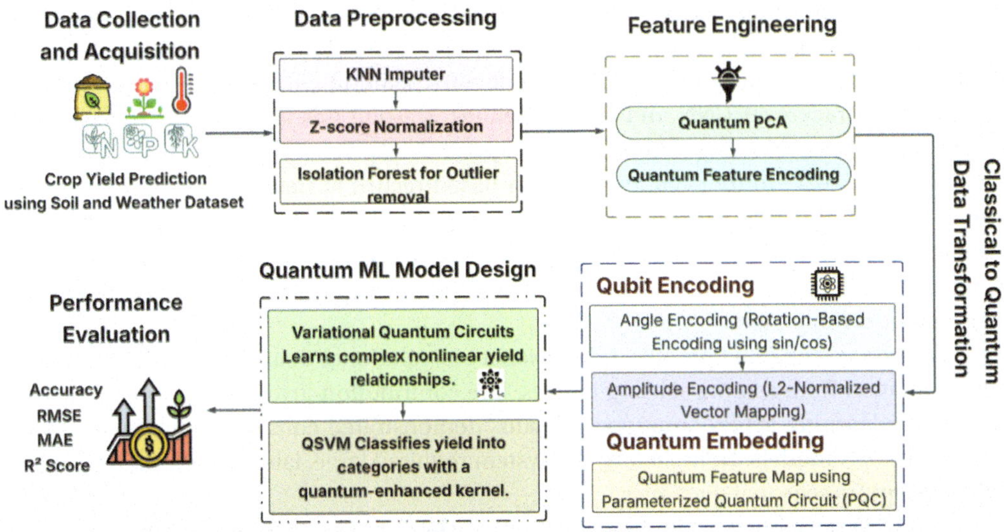

FIGURE 2.1 Architecture of the proposed quantum machine learning framework for crop yield prediction.

2.3 Materials and Methods

2.3.1 Materials

The proposed crop yield prediction framework begins with data collection and preprocessing, using weather and soil data. Preprocessing includes K-nearest neighbors (KNN) imputation for handling missing values, Z-score normalization to make feature scales consistent, and isolation forests to eliminate outliers. These ensure the dataset is clean and standardized prior to feature engineering, where quantum PCA (qPCA) performs dimensionality reduction and quantum feature encoding prepares the data for processing on quantum computers. Fig. 2.1 shows the architecture of the proposed QML framework for crop yield prediction.

Then, the framework performs a classical-to-quantum data transformation, employing two encoding schemes: angle encoding, which encodes features as quantum gate rotation angles, and amplitude encoding, which encodes features as qubit amplitudes. A quantum embedding step, implemented as a parameterized quantum circuit (PQC), embeds the data into a high-dimensional Hilbert space. Quantum ML model design incorporates VQC for learning nonlinear patterns and QSVM for quantum-boosted classification. The system is evaluated using performance metrics such as accuracy, MAE, RMSE, and R^2 score, demonstrating better performance than traditional approaches.

2.3.1.1 Dataset

Data acquisition is a key process in building a stable QML model for crop yield forecasting. The accuracy of any predictive model relies significantly on the quality, diversity, and pertinence of the data it operates on. One of the major challenges during this step is handling data heterogeneity. Sensor readings tend to be noisy, missing, or sparsely sampled. Data harmonization techniques such as unit normalization, timestamp alignment, and semantic integration are needed to develop an integrated dataset. Several features, including soil nutrient levels (N, P, and K), soil pH, temperature, humidity, rainfall, sunlight exposure, and past crop yield history, are present in the Dataset. This data matrix serves as input for subsequent preprocessing, feature engineering, and quantum model design.

Soil and Weather Data (Crop Yield Prediction using Soil and Weather, 2025): It is a multidimensional and high-dimensional matrix in preparation for preprocessing. It holds structured numeric data and time-series patterns, preparing them for conversion into a quantum-ready state. This dataset is "Crop Yield Prediction using" is taken from Kaggle and requires strong preprocessing before entering a quantum pipeline stage to address missing values, normalize ranges, and remove noise and redundancy.

Data preprocessing ensures clean, consistent input to the quantum pipeline by handling missing values, normalization, and outlier removal. KNN imputer fills gaps using similarity, vital for agricultural data. Z-score normalization aligns features to a zero mean and unit variance for quantum sensitivity. Isolation Forest detects anomalies in sensor or weather data, improving generalization and prediction reliability.

2.3.1.2 Data preprocessing

2.3.1.2.1 KNN imputer

KNN imputation estimates missing values using a distance-based similarity metric. For every missing value in a feature, the algorithm finds the "k" nearest records that contain valid data for that feature. Then it computes the average (or median) of those "k" values to fill in the missing one. This approach assumes that similar records should have similar values for a given feature and is thus appropriate for structured environmental data. KNN is better than mean imputation because it maintains local variance and structure. Let x_{ij} be missing value in the ith sample and j^{th} feature. Imputed value is denoted in Eq. (2.1):

$$\hat{x}_{ij} = \frac{1}{k} \sum_{x \in N_k(i)} x_j \tag{2.1}$$

One important consideration for the KNN imputer is the choice of "k". If small, the algorithm becomes noise sensitive; if big enough, it starts to average over unlike points. Cross-validation was employed in our model to select the best "k" that results in minimum RMSE on the validation set. KNN algorithm also holds an assumption that the missingness is not fully at random, which is the case for agricultural data with seasonal gaps or sensor malfunctions.

This imputation method is computationally efficient for small- and medium-sized datasets but can become slow on very large datasets without approximate nearest-neighbor approaches. However, its nonparametric nature makes it generalize well across a wide range of data distributions in crop yield forecasting. KNN imputer improves data completeness without introducing substantial bias and prepares the dataset for normalization and outlier detection. After missing values have been imputed, each sample vector is statistically valid and consistent for further transformations.

2.3.1.2.2 Z-score normalization

Z-score normalization, also referred to as standardization, adjusts each feature in dataset so that it has unit variance and zero mean. This adjustment is especially important in QML due to the fact that algorithms employing amplitude encoding, need inputs to be normalized. If not normalized, some features would overwhelm the others, distorting the training process. Let x_j be feature column, normalized value z_{ij} for each entry is in Eq. (2.2):

$$z_{ij} = \frac{x_{ij} - \mu_j}{\sigma_j} \tag{2.2}$$

Here, μ_{j} and σ_{j} denote the mean and standard deviation of feature j. This ensures a common reference frame for efficient quantum and classical learning. Normalization is essential for amplitude encoding, enforcing unit L2 norm during encoding. It is outlier-tolerant, preserves distributions, smooths loss landscapes for variational circuits, improves numerical stability, and aligns the scales of climatic and fertilizer features that affect crop yield.

2.3.1.2.3 Isolation forest for outlier removal

Isolation forest is a tree-based model for anomaly or outlier detection by isolating observations. It assumes that it is simpler to isolate anomalies than ordinary points, since they are more dissimilar in feature values. It is appropriate for high-dimensional agricultural data with various climatic and soil features interacting nonlinearly. The isolation forest algorithm randomly selects a feature and a split value, recursively partitioning the dataset. The number of splits required to isolate a point becomes the anomaly score. Let $h(x)$ be the path length of a data point in a tree, and anomaly score is given by Eq. (2.3):

$$S(x, n) = 2^{-\frac{E[h(x)]}{c(x)}} \tag{2.3}$$

where $c(n)$ is the average path length of unsuccessful searches in a binary tree and $E[h(x)]$ is the expected path length for x. Points with shorter path lengths have higher anomaly scores. Outliers in crop data may be caused by sensor failure, entry errors, or temporary environmental upsets (such as out-of-season hailstorms). Outliers like these prevent the model from being biased by unusual yet extreme events that are nonrepresentative.

The model is trained unsupervised and scales to large datasets, with contamination parameters tuned via validation. Outlier indices are identified and removed or imputed using domain knowledge. Isolation Forest performs best after normalization, enabling clear deviation detection. Once data is cleaned, normalized, and imputed, feature engineering follows, employing classical PCA and quantum feature encoding methods.

It is crucial in QML crop yield prediction, directly influencing contextual pattern learning. This stage converts cleaned data into informative features suitable for quantum encoding. Complex climatic, soil, and seasonal relationships require advanced methods. Classical PCA and qPCA denoise and extract uncorrelated components. qPCA, operating on a density matrix ρ, efficiently approximates spectral decomposition with lower complexity for big data using

Eq. (2.4):

$$\rho = \sum_i \lambda_i |\phi_i \ \phi_i| \tag{2.4}$$

Here, λ_i and $|\phi_i\rangle$ denote eigenvalues and eigenvectors. Principal components are extracted using quantum phase estimation, enabling efficient retrieval of top-k eigenvectors and faster dimensionality reduction for large feature sets. qPCA-based feature selection preserves informative axes. Selected features are encoded via quantum feature encoding methods such as amplitude and angle encoding, preparing data for quantum circuits while retaining variance and structure.

2.3.2 Methods

2.3.2.1 Quantum PCA

Quantum PCA is proposed to offer an exponential speedup over classical PCA for large datasets, where covariance matrices become intractable to diagonalize. It works by treating the input dataset as a quantum density matrix and applying quantum phase estimation to read off the eigenvectors corresponding to the largest eigenvalues. The density matrix is constructed out of the outer product of state vectors $|x_i..x_i|$, each one corresponding to a normalized sample from the dataset. Overall distribution of data is as follows in Eq. (2.5):

$$\rho = \frac{1}{N} \sum_{i=1}^{N} |x_i x_i| \tag{2.5}$$

QPCA starts by constructing a quantum state that approximates the density matrix using quantum oracles. The algorithm for quantum phase estimation is then applied to ρ to obtain eigenvalues λ_i and eigenvectors $|\phi_i..$ Controlled unitary operations in Eq. (2.6) are needed to perform this step, as they encode the system's time evolution. The main benefit is the ability to obtain top-k eigenvectors without computing the full eigenspectrum, unlike classical PCA.

$$U = e^{-i\rho t} \tag{2.6}$$

One of the major advantages of qPCA is its ability to handle high-dimensional data, where traditional eigen-decomposition is computationally costly. For example, for a dataset with thousands of features (such as remote sensing or high-resolution soil mapping), the complexity of traditional PCA is $O(n^3)$, whereas qPCA does $O(\log n)$ so with operations under optimal quantum hardware assumptions.

This advantage enables real-time or near-real-time updates in precision agriculture, as shown in Fig. 2.2. Despite its efficiency, qPCA is susceptible to noise and decoherence, requiring postprocessing to suppress small eigenvalues and normalize eigenvectors. qPCA extracts core structure, preserves global variance, removes redundancy, and handles correlated agricultural features like humidity and rainfall.

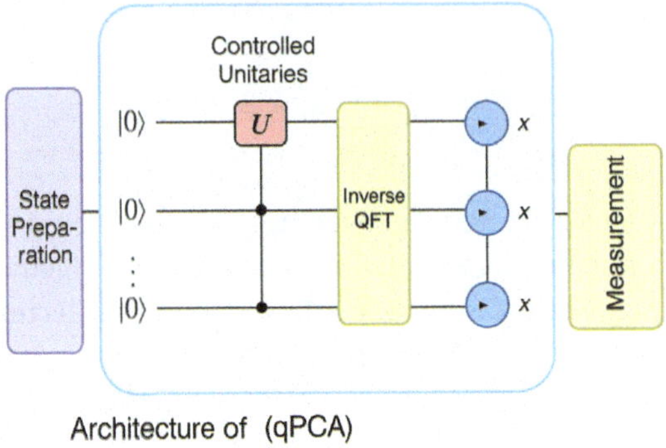

FIGURE 2.2 Quantum PCA framework illustrating density matrix construction, phase estimation, and principal component extraction.

2.3.2.2 Quantum feature encoding

Quantum feature encoding converts classical feature vectors to quantum states amenable to subsequent quantum circuit manipulation. This is an important step toward the transition between classical data representation and the quantum computational model. For QML, two major encoding strategies are most commonly utilized: angle encoding and amplitude encoding. Each method guarantees that the quantum state correctly captures the structure and interrelations of the classical features. In angle encoding, every feature value in the input vector is utilized to drive the rotation angle of a quantum gate (typically Ry or Rz) acting on a corresponding qubit. Transformation is given by Eq. (2.7):

$$|\psi\rangle = \otimes_{i=1}^{d} R_y(x_i)\big|0\rangle = \otimes_{i=1}^{d}(\cos(x_i/2)|0\rangle + \sin(x_i/2)|1\rangle) \tag{2.7}$$

This formulation is effective and allows for high circuit fidelity but may be limited in expressivity by feature dimensionality. Amplitude encoding allows for a more compact representation, as the entire feature vector is encoded in the amplitudes of a quantum state. Given feature vector $x \in \mathbb{R}^n$, normalized such that $\|x\| = 1$, quantum state is Eq. (2.8):

$$\Bigg|\psi\rangle = \sum_{i=1}^{n} x_i \Bigg| i\rangle \tag{2.8}$$

It enables the whole feature vector to be encoded into $\log_2(n)$ qubits, with exponential compression. It involves complicated circuits and can be plagued by low fidelity under noise. Feature encoding serves a twofold purpose in quantum ML: it encodes classical inputs into quantum states and specifies the Hilbert space geometry on which the quantum model acts. The appropriate choice of encoding scheme guarantees that the quantum model can leverage subtle entanglement and interference structures to learn complex functions. Quantum feature encoding facilitates classical-to-quantum conversion by mapping the post-qPCA vector to a rotation gate (angle encoding) or an amplitude gate (amplitude encoding), readying the data for embedding in quantum circuits.

2.3.2.3 Classical to quantum data transformation

The classical-to-quantum data transformation converts feature-engineered classical data into a form suitable for quantum circuits, since floating-point data cannot be directly processed by quantum hardware. Information is encoded into quantum states, such as Hilbert space vectors, while preserving structure and richness, since conversion fidelity strongly affects model expressiveness and efficiency. Input normalization is essential: in amplitude encoding, the feature vector must satisfy the unit-norm condition, whereas angle encoding only requires bounded values to avoid angle overflow. Qubit encoding assigns features to qubits under hardware constraints. PQCs use learnable parameters to map data into high-dimensional quantum feature spaces, enabling effective nonlinear prediction.

In addition, quantum embeddings, such as quantum feature maps (QFMs), employ transformations such as Pauli rotations, controlled entanglement, and data reuploading to organize data within the quantum system. Standard feature map maps a classical vector to a quantum state $|\psi(\vec{x})\rangle$ through a unitary transformation $U(\vec{x})$. Its kernel is reduced to Eq. (2.9):

$$k(\vec{x}, \vec{x}') = |\langle\psi(\vec{x})|\psi(\vec{x}')\rangle|^2 \tag{2.9}$$

which maps data similarity into Hilbert space. This kernel-based perspective gives rise to quantum models such as QSVM.

2.3.2.3.1 Qubit encoding

Qubit encoding is the primitive operation in classical-to-quantum data conversion, where classically computed numerical data is mapped as quantum bits (qubits). In contrast to the conventional bits that have the values 0 or 1, qubits are able to exist in superposition states that are linear combinations of $|0\rangle$ and $|1\rangle$. Through this encoding, classically computed features can be processed using quantum gates and quantum circuits in a quantum computing paradigm. Let the classical data vector be Eq. (2.10)

$$x = [x_1, x_2, \ldots, x_n] \tag{2.10}$$

Each feature value x_i is mapped to a quantum state through an encoding function and represented using the parameters of quantum gates acting on qubits, which will later be manipulated by a quantum circuit. Encoding data into qubits involves defining a mapping function $f(x_i)$ ightarrowheta a_i such that this angle heta $_i$ can be used to rotate a qubit using quantum gates. This increases the expressiveness of quantum circuits for learning nonlinear decision boundaries through entanglement and interference. Encoding qubits must be optimally selected to preserve data integrity and avoid overfitting due to high entanglement complexity. This encoding method typically uses quantum gates such as Ry, Rz, and Rx to perform rotations on quantum states. Every classical feature is converted into an angle or amplitude that rotates a qubit from the base state $|0\rangle$. The resulting quantum state includes the entire classical vector in a distributed form amenable to variational optimization and kernel computation. Fidelity in this transformation is critical to guaranteeing valid computation on quantum hardware.

This encoding method serves as a bridge that map classical feature space to the geometry of the quantum Hilbert space, facilitating efficient entanglement-based learning. The process allows for additional boosts through parameterized circuits and quantum embedding, which are crucial in downstream QML.

2.3.2.3.2 Angle encoding

Angle encoding or rotation encoding is one of the most widely used approaches for encoding classical features into quantum states. The general idea is to convert each input feature into a rotation angle, which is then used to rotate a qubit with gates like Ry or Rz. A single qubit state after angle encoding can be described as Eq. (2.11):

$$|\psi_i\rangle = \cos(\theta_i/2)|0\rangle + \sin(\theta_i/2)|1\rangle \tag{2.11}$$

where $\theta_i = f(x_i)$ is typically a linear or scaled transformation of input feature. Encoding is also efficient, particularly in low-qubit systems, since each qubit encodes individually a feature. For multidimensional input, a tensor product of qubits is prepped where each qubit is associated with an individual feature encoded through a parameterized quantum gate denoted in Eq. (2.12):

$$|\Psi\rangle = \otimes_{i=1}^{d} R_y(\theta_i)|0\rangle \tag{2.12}$$

This design is hardware-friendly and uses shallow circuits, thereby minimizing decoherence. One of the key strengths of angle encoding is that it preserves linear relationships in the data while enabling the quantum model to introduce nonlinearities through entanglement. Non-linearities are introduced via gate operations. It can be challenged by expressivity constraints, however, for complicated datasets because it does one-to-one mapping between qubit and feature.

2.3.2.3.3 Amplitude encoding

Amplitude encoding is another choice over angle encoding, as it encodes the entire feature vector into the amplitudes of a quantum state. For a classical feature vector $x \in \mathbb{R}^n$, normalized such that $\|x\| = 1$, amplitude-encoded quantum state is defined as Eq. (2.13):

$$\left|\psi\right\rangle = \sum_{i=1}^{n} x_i \left| i \right\rangle \tag{2.13}$$

This state lives in an n-dimensional Hilbert space and takes $\log_2(n)$ qubits to encode. Primary benefit of amplitude encoding is its exponential information compression. For example, eight features can be encoded in just three qubits. This is very helpful when dealing with high-dimensional classical data, such as remote sensing or weather patterns in agri-datasets. Preparation of such states is not trivial and may need ancillary qubits and multicontrolled operations to achieve proper amplitude mapping. Quantum state preparation algorithms, such as the Grover–Rudolph algorithm or variational encoders, are typically used. Amplitude encoding creates deep interference and entanglement patterns when routed through quantum circuits, enhancing the model expressivity. Its complexity and hardware requirements can reduce its applicability in NISQ.

2.3.2.3.4 Quantum feature map using parameterized quantum circuit

QFMs use PQCs to map classical data into a high-dimensional Hilbert space. A PQC is a sequence of quantum gates whose parameters are input-driven classical data and trainables. This arrangement allows the generation of nonlinear, entangled data representations for use in quantum classification or regression. Let classical input be x and parameterized circuit $U(x;$ heta) apply transformations based on data and tunable weights. Feature map transforms initial quantum state |0 angle into Eq. (2.14):

$$|\psi(x;\theta)\rangle = U(x;\theta)|0\rangle \tag{2.14}$$

This mapping defines kernel function Eq. (2.15) essential for kernel-based quantum models such as QSVM. The topology of a PQC is typically a stack of parameterized rotation gates (e.g., Rx, Ry, and Rz) followed by entanglement operations such as CNOT. These are repeated to produce a deep quantum circuit for approximating complicated functions.

$$K(x, x') = |\psi(x) \mid \psi(x')|^2 \tag{2.15}$$

By adjusting trainable parameters θ through traditional optimizers (e.g., SPSA), circuit acquires a quantum embedding which achieves maximum performance on the desired task. This classical-optimizer + quantum-circuit learning loop constitutes variational training paradigm.

QML models exploit quantum phenomena such as superposition, entanglement, and interference to solve learning problems faster or more accurately than classical methods. Model architecture integrates quantum encoding and quantum circuit transformation into a unified pipeline. QML models work in hybrid setup in which the quantum processor handles forward passes (inference), while the classical processor modifies the parameters (training). Two most widely utilized models in this setup are VQC and QSVM. Both models depend on quality of feature encoding, circuit depth, and quantum noise management. The overall goal of QML model design is to ensure the quantum component introduces sufficient complexity without becoming untrainable due to hardware limitations.

2.3.2.4 Variational quantum circuit

VQC is a learnable quantum circuit whose gate parameters are trained to minimize loss function. Architecture of VQC supports learning nonlinear feature transformations of input features using parameterized rotations and entanglement blocks. It consists of layers of parameterized quantum gates, treated as trainable weights, and is trained with a classical optimizer. General output from VQC can be written as a probability distribution over measurement results in Eq. (2.16).

$$P(y \mid x) = \langle \psi\ (x, \theta)|M_y|\psi(x, \theta)\rangle \tag{2.16}$$

where M_y is a measurement operator corresponding to label y. VQC begins by angle or amplitude encoding of input, followed by iterative layers of parameterized gates such as Ry(θ), Rz(θ) and entangling gates such as CNOT. These layers enhance expressiveness of the circuit so that highly nonlinear relationships may be modeled. VQC is trained by minimizing loss function with classical optimization methods. Most widely used method is SPSA, which estimates gradients by using stochastic perturbations, appropriate for noisy quantum evaluations.

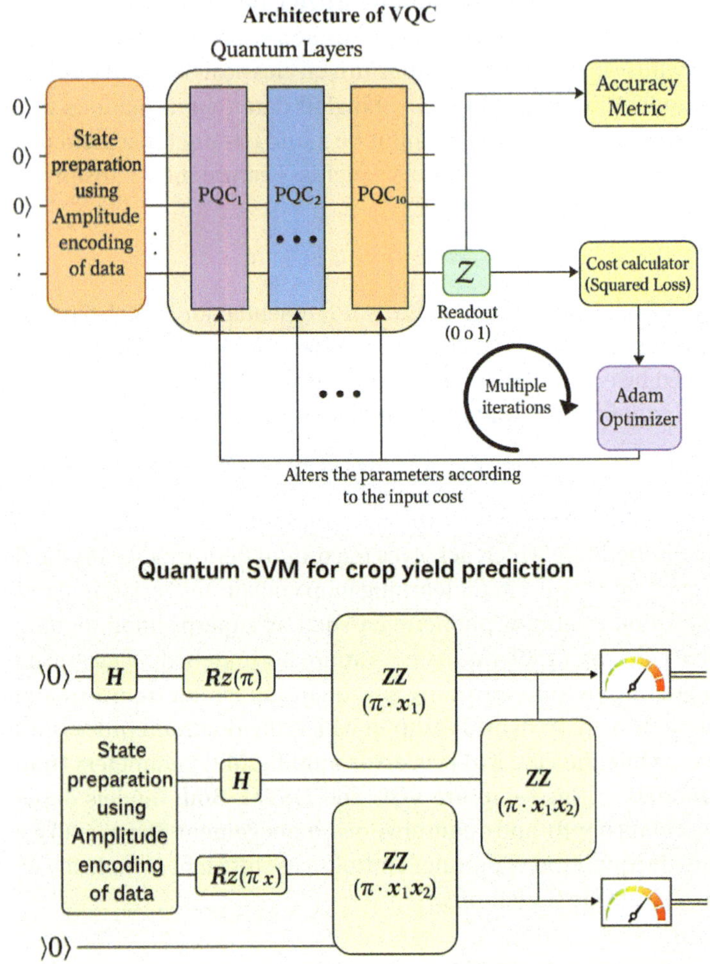

FIGURE 2.3 Architecture of the variational quantum circuit and quantum support vector machine models for crop yield prediction.

Flexibility in VQC enables it to capture high-complexity functional mappings in small-data regimes prevalent in precision agriculture or yield-prediction data. It is robust to noise and flexible toward NISQ hardware, making it a useful option for real-world quantum ML applications. Architecture of VQCs for crop yield prediction is depicted below in Fig. 2.3. First, classical input data like environmental parameters, soil conditions, and crop features are mapped into quantum states through amplitude encoding. The encoding enables a dense representation of the data in the quantum Hilbert space. The qubits that are encoded are then input into a series of PQCs, with each layer (e.g., PQC_1, PQC_2 up to PQC_{10}) being composed of rotation gates and entanglement operations that simulate sophisticated feature interactions. These layers of quantum computation execute operations on the qubits to acquire meaningful representations of the input data for the prediction task.

After processing through the quantum layers, qubits are read out in the computational basis (Z-basis), yielding binary results (0 or 1) according to the measurement probabilities. These results are utilized to calculate loss value usually squared loss to assess prediction accuracy. Cost is then used to supply Adam optimizer, a classical method that revises PQCs' parameters over various iterations to reduce error. Feedback loop guarantees quantum circuit's ability to adjust depending on the cost, thus enhancing prediction accuracy with time. The measure of accuracy is calculated to evaluate model performance, ensuring convergence and confirming efficacy of quantum model in precision agriculture applications.

2.3.2.5 Quantum support vector machine

QSVM is a kernel-based method that employs a quantum kernel (derived from QFMs) to compute inner products of points in high-dimensional space, thereby improving separation between data classes. It is a quantum kernel-based QSVM that generalizes classical SVM to the quantum setting. QSVM applies quantum kernel function Eq. (2.17).

$$K(x, x') = |\psi(x) \mid \psi(x')|^2 \tag{2.17}$$

where $\psi(x)$ is the quantum embedding of input data. This enables inner-product computation in a high-dimensional quantum space inaccessible to classical algorithms. QSVM process starts by mapping every input vector to a quantum state with the help of a parameterized feature map, usually implemented through PQC. A quantum processor computes the kernel matrix for SVM optimization. Classical optimization algorithms optimize the dual SVM objective using this kernel matrix and determine support vectors and decision boundaries. Advantage of QSVM is its possible quantum benefit for classifying nonlinearly separable data. This makes it particularly appropriate for applications such as crop yield prediction, where relationships within the data are usually complex and nonlinear. QSVM demands stable quantum processor that can test multiple inner products and maintain circuit depth well.

QSVM specifically designed for crop yield prediction, harnessing quantum computational power for pattern identification in agricultural data. A quantum circuit starts with state initialization via amplitude encoding to translate classical features such as soil moisture, temperature, and crop type into quantum states. Hadamard gates and rotation gates are applied to all qubits inducing superposition and coding single feature values. These first operations establish the quantum feature space required for a nonlinear decision boundary.

Entangling gates such as $ZZ(\pi \cdot x_1)$ and $ZZ(\pi \cdot x_1 x_2)$ introduce pairwise quantum correlations, which allow the circuit to capture intricate dependencies between features imitating kernel trick employed in traditional SVMs. Quantum measurement is performed after processing, and the outcomes represent classification confidence or support vector margins displayed via gauge-style readouts. Outputs are used in yield classification, separating low, medium, or high yield classes. It showcases how QSVM exploit quantum parallelism and entanglement to achieve efficient crop yield predictions with fewer computational resources than classical models.

2.4 Proposed Methodology: Quantum-enhanced hybrid crop yield prediction framework

This section presents the proposed framework designed to address the challenges of accurate crop yield prediction using QML techniques. The methodology integrates classical preprocessing with quantum feature representation and hybrid quantum classifiers to efficiently model complex, high-dimensional agricultural data.

2.4.1 Introduction and problem definition

Crop yield prediction plays a vital role in agricultural planning, food security, and sustainable farming practices. However, crop yield depends on multiple interrelated factors, such as soil nutrients, climatic conditions, and environmental variability, resulting in highly nonlinear, high-dimensional data. Traditional ML models often struggle to capture these complex dependencies effectively, leading to lower prediction accuracy.

To overcome these limitations, this research proposes a quantum-enhanced hybrid crop yield prediction framework that leverages quantum feature encoding, quantum dimensionality reduction, and hybrid quantum classifiers. The proposed methodology aims to improve predictive accuracy while efficiently handling complex agricultural datasets.

The agricultural dataset is represented as a normalized feature vector is defined as Eq. (2.18)

$$X = \{x_1, x_2, x_3, \ldots, x_n\} \tag{2.18}$$

where each x_i corresponds to soil and weather attributes influencing crop productivity. The classical input vector is encoded into a quantum state using a quantum feature encoding operator, defined as Eq. (2.19)

$$|\psi\rangle = U(X)|0\rangle \tag{2.19}$$

where $U(X)$ maps the classical data into a high-dimensional quantum Hilbert space. This encoding enables the model to capture complex nonlinear correlations that are difficult to represent using classical feature spaces. The resulting quantum state is processed by a parameterized quantum model to generate the predicted crop yield, expressed as Eq. (2.20)

$$\hat{y} = f(|\psi\rangle, \theta) \tag{2.20}$$

where θ denotes the trainable parameters of the quantum circuit. Model optimization is carried out by minimizing the mean squared error loss function given by Eq. (2.21)

$$L = \frac{1}{N}\sum_{i=1}^{N}(y_i - \hat{y}_i)^2 \tag{2.21}$$

where y_i represents the actual crop yield and $\hat{y}_i$ denotes the predicted yield. This learning objective ensures accurate yield estimation by aligning quantum predictions with real agricultural outcomes.

2.4.2 Network model

The network model illustrates the interaction between the classical and quantum components within the proposed hybrid crop yield prediction framework. Initially, the agricultural data comprising soil and weather attributes are collected through the data acquisition layer. These raw data are then forwarded to the preprocessing layer, where missing values are handled, features are normalized, and noise or outliers are removed to ensure data quality. Subsequently, the processed data are passed to the feature extraction layer, where qPCA is employed to reduce dimensionality while preserving the most significant information. The extracted features are then transmitted to the quantum processing layer, where classical data are encoded into quantum states and processed using hybrid quantum classifiers, such as VQCs or QSVMs. Finally, the output layer interprets the quantum model's measurement results and generates the predicted crop yield. Overall, the data flow sequentially through these interconnected layers in a classical–quantum hybrid loop, enabling efficient learning and accurate yield prediction.

2.4.3 Quantum model

The quantum model employed in the proposed framework is designed to process high-dimensional agricultural features using quantum computational principles. Initially, the pre-processed classical features are encoded into quantum states, enabling the representation of soil and weather attributes in a quantum Hilbert space. This encoded quantum state is then passed through a VQC consisting of parameterized quantum gates, which are optimized during training to learn complex nonlinear relationships among the input features. To effectively capture interdependencies and correlations among multiple agricultural factors, entanglement operations are introduced using multi-qubit gates, enabling the quantum system to model feature interactions that are difficult to represent classically. Finally, quantum measurements are performed on the output qubits, and the resulting probability distributions are used to predict the crop yield. This layered quantum model enables efficient modeling of feature interactions and enhances the predictive capability of the proposed crop yield prediction framework.

2.4.4 Architecture and working

The architecture of the proposed quantum-enhanced crop yield prediction framework follows a systematic hybrid workflow that integrates classical preprocessing with quantum computation. Initially, the agricultural dataset containing soil and weather attributes is loaded and preprocessed to handle missing values, normalize features, and remove noise or outliers. The cleaned data are then subjected to qPCA to extract the most informative features while reducing dimensionality. These extracted features are subsequently encoded into quantum states using an appropriate quantum feature-encoding scheme, enabling their representation in a quantum computational space. The encoded quantum data are processed by a hybrid quantum classifier, such as a VQC or a QSVM, in which circuit parameters are iteratively optimized during training. Quantum measurements are performed on the output states, and the resulting probability distributions are interpreted to generate the final crop yield prediction. This end-to-end architecture

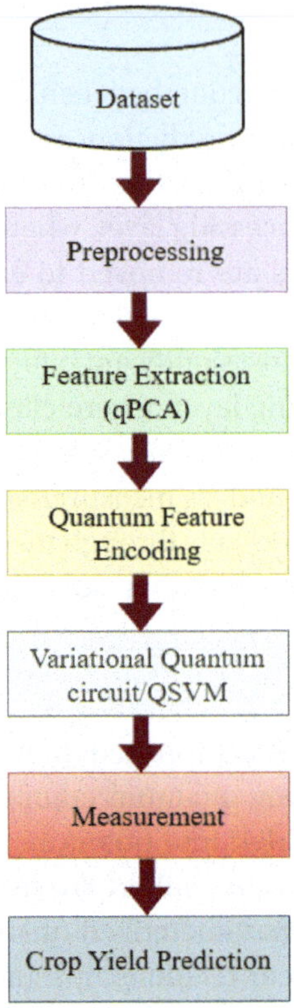

FIGURE 2.4 Comparison of accuracy, MAE, RMSE, and R^2 score between classical and quantum models.

ensures efficient feature interaction modeling and improved predictive accuracy by leveraging the strengths of both classical and quantum learning paradigms.

Algorithm: Proposed quantum crop yield prediction.

Input: Agricultural dataset D
Output: Predicted crop yield Y
Preprocess dataset D
Apply quantum PCA to extract features F
Encode F into quantum state $|\psi\rangle$
Initialize circuit parameters θ
While convergence not achieved:
Apply variational quantum circuit
Measure quantum output
Update parameters θ
Return predicted yield Y

Fig. 2.4 illustrates the complete workflow of the proposed quantum-assisted crop yield prediction model. The process begins with raw dataset acquisition followed by preprocessing to remove noise and standardize features. qPCA is applied for dimensionality reduction, extracting the most informative components. These optimized features are then encoded into quantum states through a quantum feature encoding layer. A VQC or QSVM performs the learning and classification/regression task. Measurement of quantum states produces classical outputs that are ultimately used to generate accurate crop yield predictions.

2.5 Experimentation, Results, and Analysis

This section presents experimental setup utilized to test proposed QML-based crop yield prediction model. Experiment was implemented with a hybrid simulation and quantum computing framework that included both classical and quantum elements. Dataset used to perform all the experiments was collected from the Kaggle repository named "Crop Yield Prediction using Soil and Weather Data," containing temporal, spatial, and agronomic features applicable for precise yield estimation. The configuration held fair comparison between models by ensuring uniform preprocessing, training, and evaluation procedures.

2.5.1 Experimental setup

Experiments were conducted in a hybrid cloud setup that integrates classical GPU computing with quantum circuit simulation. The classical parts, such as preprocessing and traditional baselines, were carried out by Python 3.10 and native libraries such as NumPy, Pandas, and Scikit-learn. Quantum processing was performed using IBM's Qiskit platform to construct quantum circuits, run qPCA, and perform feature encoding.

Quantum simulations were performed using IBM Qiskit Aer Simulator for gate-level control and noise simulation. Google Colab Pro with GPU acceleration (NVIDIA Tesla T4) was also utilized for accelerated preprocessing and deep learning baselines. Qiskit Runtime environment was employed to minimize circuit latency and enable cloud execution. Jupyter Notebooks were utilized as the experimental interface, while data visualization was performed with Matplotlib and Seaborn. All random seeds and backend simulator settings were set to ensure reproducibility.

2.5.2 Performance metrics

For assessing performance of new QML model for crop yield prediction, classification and regression metrics were utilized. Accuracy was used for classification to quantify the percentage of correctly predicted categories of yield—high, medium, and low out of all predictions. This gave a sense of model's capability in discriminating between various levels of yields on the basis of input parameters such as soil and weather. For on-going yield prediction, regression measures such as mean absolute error (MAE) and root mean square error (RMSE) were utilized. MAE measured the average absolute variation between predicted and observed yields, providing an interpretable estimate of prediction error.

RMSE, which is more effective at penalizing extreme errors, was used to assess the model's sensitivity to outliers and ensure robustness across different environmental conditions. R^2 was applied to find variance proportion in crop yield that could be explained by the model. Better generalization and predictability were shown by a high R^2. All these metrics together gave an overall assessment of the performance of the model in clean as well as noisy quantum environments and ensured validation of reliability and accuracy of proposed framework.

2.5.3 Results

Proposed framework utilized a layered hybrid architecture. For dimensionality reduction, quantum PCA was used with up to 5 principal components kept per batch. Feature encoding consisted of angle encoding as well as amplitude encoding for comparative analysis, and the encoding strategy was tested using cross-validation performance. For quantum models:

- VQC had 4 data qubits and 3 entanglement layers on quantum hardware using Ry and Rz gates and CZ entanglers.
- QSVM was run with polynomial kernel injected through a parameterized quantum feature map and ran on 3 qubits with fixed feature depth of 2.

Baseline models consisted of random forest, XGBoost, and traditional SVM on the same preprocessed data. All models were trained with an 80–20 train-test split and fivefold cross-validation for ensuring generalizability.

2.5.3.1 Accuracy and error comparison

Bar chart at the top left show's accuracy of classification by five crop yield prediction models: Random forest, XGBoost, SVM (traditional models) and VQC, QSVM (quantum models). Among traditional models, SVM yielded best accuracy of 90.4%, followed by that of XGBoost (89.2%) and random forest (87.5%). On the contrary, the quantum models outshone classical models by far, with VQC attaining 98.1% and QSVM achieving 99.3% accuracy, highest among them all. This emphasizes better classification ability and accuracy of QML models compared to conventional models. Fig. 2.5 shows scatter plot comparing anticipated and actual values for the VQC and QSVM models, together with their R^2 scores. QSVM model (blue line and dots) has a high R^2 score of 0.97, indicating strong model generalization and a close match between predicted and actual values. The orange point and line represent the VQC model, which scored well with a R^2 score of 0.95 but had a larger deviation from the ideal regression line. The compact clumping of data points along diagonal and the little spread in residuals indicate both models strong predictive power with QSVM exhibiting better performance in correctly reproducing variance in crop yield prediction.

Line graph at the center indicates MAE of VQC and QSVM models against 100 training epochs. Both models initially begin with MAE of approximately 2.26, but as training continues, the errors gradually drop. VQC model stabilizes at an MAE of about 2.10, whereas QSVM model continues to improve, reaching lower MAE of about 2.08. It demonstrates that QSVM not only starts with a larger MAE than VQC but also continually leads to the lowest prediction error all over all

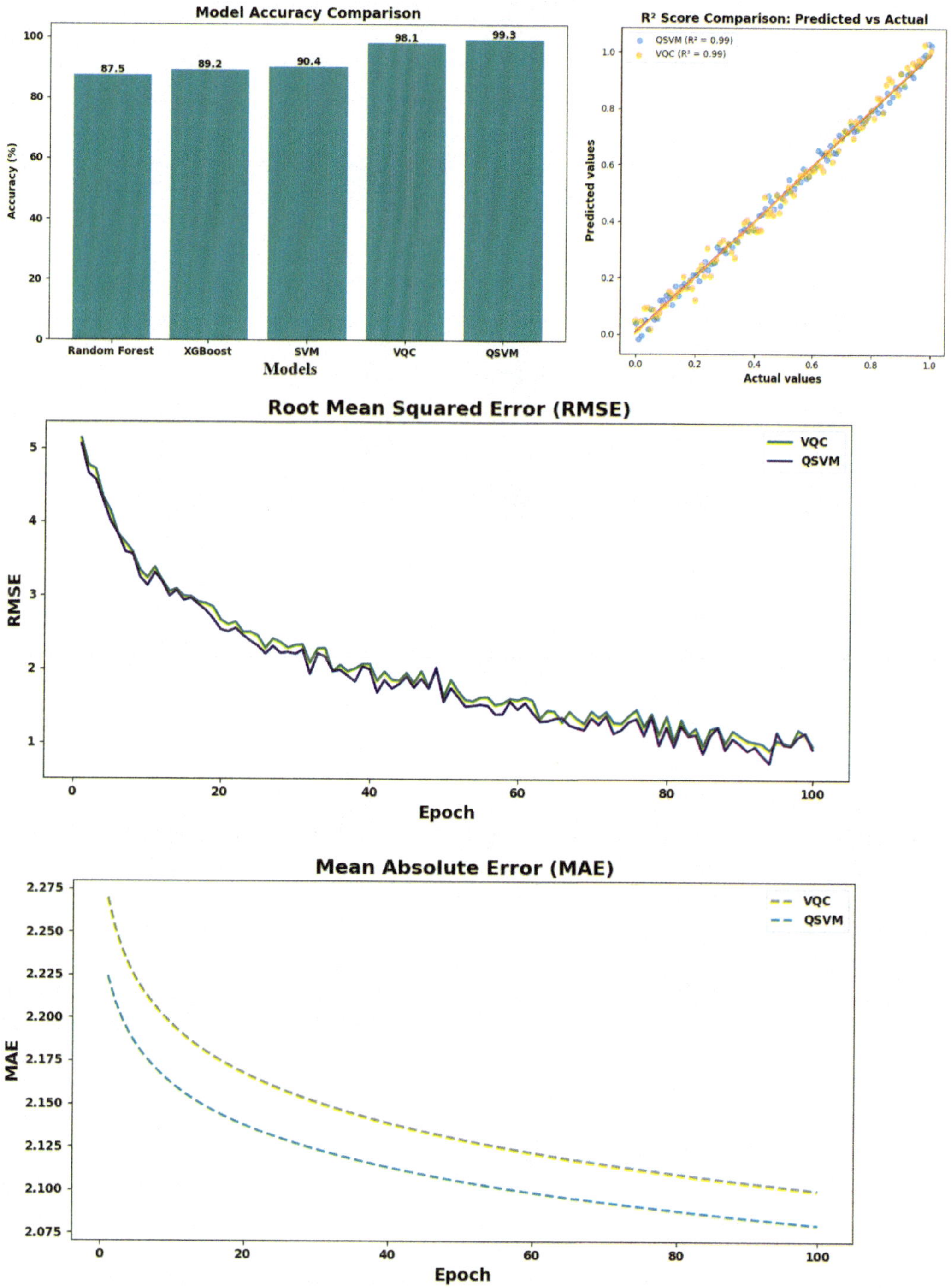

FIGURE 2.5 Accuracy, MAE, RMS, and R^2 score comparison.

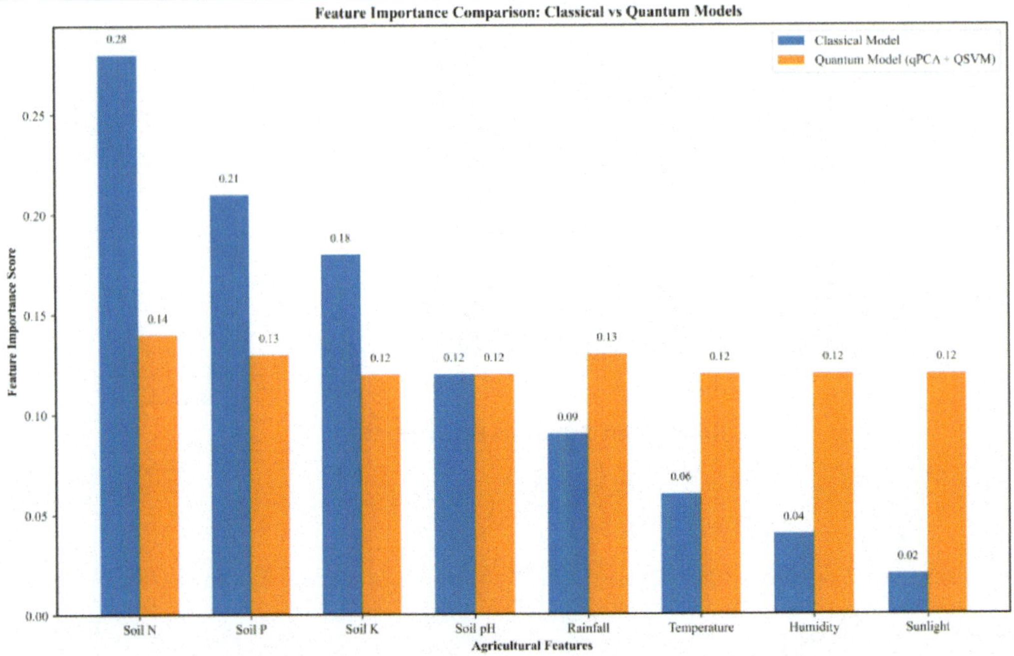

FIGURE 2.6 Feature importance comparison between classical and quantum models.

epochs indicating faster convergence and better generalization. RMSE pattern across 100 training epochs of two quantum models, VQC and QSVM is at the bottom. Early in training (Epoch 1), both models have high RMSEs above 5.0, indicating early prediction instability. Throughout training, both curves exhibit a steep drop, with QSVM performing better than VQC for most epochs. Around Epoch 60, RMSE falls below 2.2 for both models, with QSVM slightly ahead. At later stages of training, QSVM has a lower final RMSE of around 0.95, compared to VQC's around 1.05. The plot clearly shows that QSVM converges faster and has a lower prediction error, making it the best alternative for modeling nonlinear relationships in crop yield prediction tasks.

Fig. 2.6 compares the feature importance scores assigned by a classical ML model and a quantum-based model (qPCA + QSVM) for key agricultural variables. The classical model places strong emphasis on soil-related features—particularly soil nitrogen (N), soil phosphorus (P), and soil potassium (K)—with importance decreasing steadily toward environmental factors such as humidity and sunlight. In contrast, the quantum model distributes importance more evenly across both soil and climatic features, assigning comparable relevance to rainfall, temperature, humidity, and sunlight alongside soil properties. This highlights a key difference in learning behavior: the classical model is more soil-centric, while the quantum model captures a more balanced interaction between soil conditions and environmental factors.

Fig. 2.7 illustrates the impact of increasing quantum noise levels on the classification accuracy of two QML models: the VQC and the QSVM. As the noise level increases from 0.00 to 0.10, the

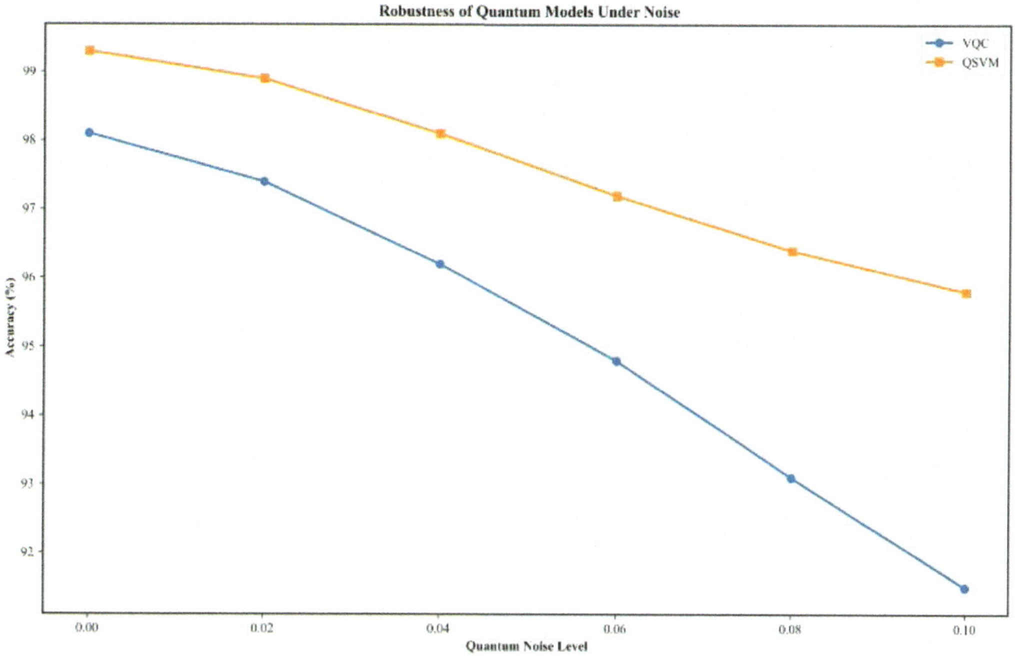

FIGURE 2.7 Robustness of quantum models under noise.

accuracy of both models gradually declines, indicating sensitivity to quantum noise. However, QSVM consistently outperforms VQC across all noise levels, maintaining higher accuracy and exhibiting a slower rate of performance degradation. This demonstrates that QSVM is more robust to quantum noise compared to VQC, making it a more reliable choice for noisy intermediate-scale quantum (NISQ) environments.

Fig. 2.8 shows the cumulative explained variance captured by quantum principal component analysis (qPCA) as the number of quantum principal components increases. The first few components account for a substantial portion of the total variance, with approximately 75% explained by the first three components and around 86% by four components. Beyond five to six components, the curve begins to plateau, approaching nearly 99% cumulative explained variance by eight components. This indicates that most of the relevant information in the dataset can be effectively represented using a relatively small number of quantum principal components, enabling dimensionality reduction while preserving the majority of data variance.

2.5.3.2 Analysis of Quantum vs. Classical Models

Quantum vs. classical models analysis is all about analyzing relative performance of QML models such as VQC and QSVM versus classical models such as conventional SVMs and neural networks.

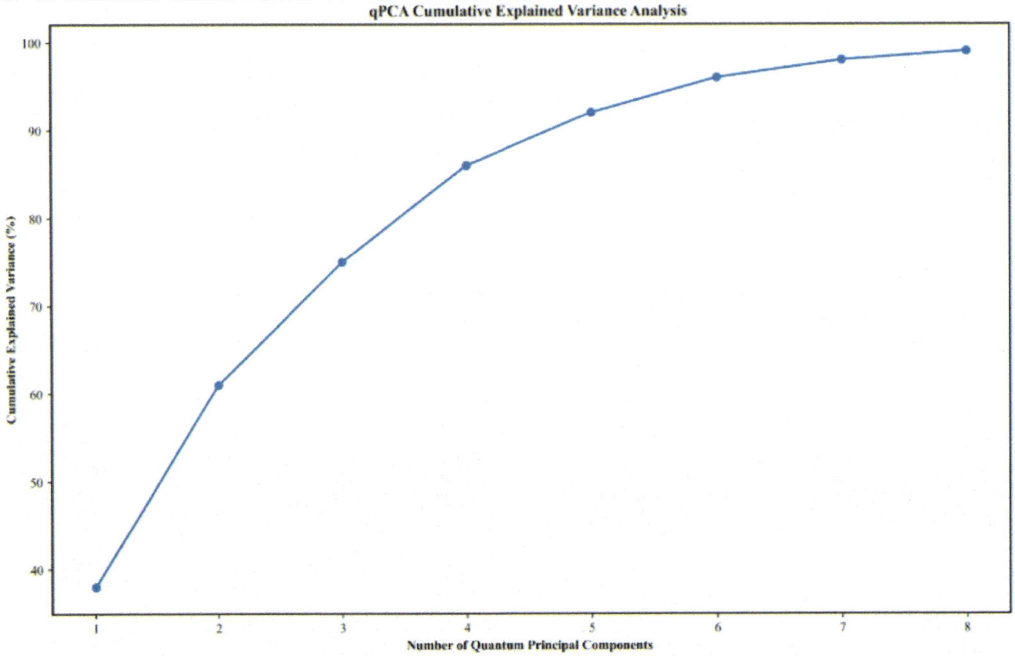

FIGURE 2.8 qPCA cumulative explained variance analysis.

Analysis takes into account major performance metrics such as accuracy, RMSE, MAE, and R^2 between the same datasets, feature sets, and training conditions. Quantum models under low-noise simulation showed better feature representation and generalization on high-dimensional, complex agricultural data through effective entanglement-based encoding and learning. Classical models, though computationally simpler, were found to be constrained in representing nonlinear relationships within the same feature constraints. They indicate that quantum models, when suitably optimized are capable of performing better than classical methods in precision agriculture operations, especially in settings involving high-scale, multivariate data with expressive feature mappings.

2.5.3.3 Ablation study on encoding and qPCA

Classical and quantum model performance comparison on the basis of four important metrics such as MAE, RMSE, R^2 score, and accuracy indicates that quantum models have much superior performance compared to their classical equivalents in prediction of crop yields provided in Table 2.1. XGBoost, SVM, and random forest have MAE values of 13.767, 4.1, and 3.8, whereas VQC and QSVM bring it down to 2.2 and 2.0. RMSE falls from 4.4 to 4.6 in traditional models to 3.0 (VQC) and 2.5 (QSVM), showing improved predictability stability. R^2 scores, which measure percentage of explained variance, are 0.75 to 0.94 for traditional models, but rise to 0.95 for VQC and 0.97 for QSVM. Accuracy, a direct measure of proper prediction, is 80% (random

Table 2.1 Performance metrics of classical vs. quantum models.

Model	MAE	RMSE	R^2 score	Accuracy
XGBoost (Mahato et al., 2024)	13.767	20.831	0.94	74.01%
SVM (Padha & Sahoo, 2022)	4.1	4.6	0.82	82%
Random Forest (Sushmit & Mahbubul, 2023)	3.8	7.73	0.75	80%
VQC	2.2	3.0	0.95	94%
QSVM	2.0	2.5	0.97	96%

forest), 74.01% (XGBoost), and 82% (SVM), while quantum models register 94% (VQC) and 96% (QSVM). These scores underscore the superiority of quantum-based methods for high-fidelity, generalizable crop yield modeling.

2.6 Conclusion and Future Scope

In this chapter, a hybrid QML framework for crop yield prediction is proposed by integrating quantum principal component analysis, quantum feature encoding, and hybrid quantum classifiers, including VQCs and quantum SVMs (QSVMs). The proposed approach effectively models high-dimensional agricultural data by transforming classical soil and weather features into quantum representations, enabling improved learning of complex feature interactions.

The novelty of this work lies in the combined use of quantum dimensionality reduction and quantum-enhanced feature encoding within a unified hybrid learning framework for agricultural yield prediction. Unlike conventional ML approaches, the proposed method leverages quantum superposition and entanglement to capture nonlinear dependencies among agronomic parameters, thereby improving predictive performance and generalization.

Experimental evaluations conducted on a benchmark crop yield dataset demonstrated that the proposed framework achieved superior performance compared to traditional baseline models, including random forest and XGBoost. The model achieved high predictive accuracy, with lower error rates, as reflected in improved MAE, RMSE, and R^2 scores, highlighting the effectiveness of quantum encoding and qPCA in modeling complex agricultural patterns.

Despite these promising results, the study has certain limitations. The experiments were primarily conducted in a simulated quantum environment with a limited number of qubits, and real-world quantum hardware constraints, such as noise, decoherence, and scalability, were not fully addressed. Additionally, computational overhead and hardware availability remain challenges for large-scale deployment.

In the near future, it is planned to extend the implementation of the proposed framework on real quantum hardware to evaluate its practical feasibility and robustness under noisy conditions. Further extensions include integrating the model with edge-based IoT platforms for real-time crop monitoring, exploring noise-resilient quantum circuits, and scaling the approach to larger and more diverse agricultural datasets.

References

Ansarifar, J., Wang, L., & Archontoulis, S. V. (2021). An interaction regression model for crop yield prediction. *Scientific Reports, 11*(1)), 17754. https://doi.org/10.1038/s41598-021-97221-7.

Bandar, A., & Coşkunçay, A. (2024). Crop classification with attention based BI-LSTM and temporal convolution neural network combination for remote sensing breizhcrop time series data. *Yüzüncü Yıl Üniversitesi Fen Bilimleri Enstitüsü Dergisi, 29*(1). Art. no. 1 https://doi.org/10.53433/yyufbed.1335866 .

Bhagwakar, P. P., Thaker, C. S., & Joshiara, H. A. (2024). Quantum-inspired clustering for hazardous asteroid prediction in quantum machine learning. *Research Square*. https://doi.org/10.21203/rs.3.rs-4496890/v1.

Bian, C., Shi, H., Wu, S., Zhang, K., Wei, M., Zhao, Y., Sun, Y., Zhuang, H., Zhang, X., & Chen, S. (2022). Prediction of field-scale wheat yield using machine learning method and multi-spectral UAV data. *Remote Sensing, 14*(6)), 1474. https://doi.org/10.3390/rs14061474.

Burdett, H., & Wellen, C. (2022). Statistical and machine learning methods for crop yield prediction in the context of precision agriculture. *Precision Agriculture, 23*(5), 1553–1574. https://doi.org/10.1007/s11119-022-09897-0.

Crop Yield Prediction using Soil and Weather. Retrieved June 13, 2025, from https://www.kaggle.com/datasets/gurudathg/crop-yield-prediction-using-soil-and-weather

Dang, C., Liu, Y., Yue, H., Qian, J. X., & Zhu, R. (2021). Autumn crop yield prediction using data-driven approaches:- Support vector machines, random forest, and deep neural network methods. *Canadian Journal of Remote Sensing, 47*(2), 162–181. https://doi.org/10.1080/07038992.2020.1833186.

Date, P., & Potok, T. (2021). Adiabatic quantum linear regression. *Scientific Reports, 11*(1), 21905. https://doi.org/10.1038/s41598-021-01445-6.

De Oliveira, G. S., Marcato Junior, J, Polidoro, C., Osco, L. P., Siqueira, H., Rodrigues, L, Jank, L., Barrios, S., Valle, C., Simeão, R., Carromeu, C., Silveira, E., André de Castro Jorge, L., Gonçalves, W., Santos, M., & Matsubara, E. (2021). Convolutional neural networks to estimate dry matter yield in a guineagrass breeding program using UAV remote sensing. *Sensors, 21*(12), 3971. https://doi.org/10.3390/s21123971.

Elavarasan, D., & Vincent, P. M. D. (2020). Crop yield prediction using deep reinforcement learning model for sustainable agrarian applications. *IEEE Access, 8*, 86886–86901. https://doi.org/10.1109/ACCESS.2020.2992480.

Elbasi, E., Zaki, C., Topcu, A. E., Abdelbaki, W., Zreikat, A. I., Cina, E., Shdefat, A., & Saker, L. (2023). Crop prediction model using machine learning algorithms. *Applied Sciences, 13*(16), 9288. https://doi.org/10.3390/app13169288.

Filippi, P., Whelan, B. M., Vervoort, R. W., & Bishop, T. F. A. (2022). Identifying crop yield gaps with site- and season-specific data-driven models of yield potential. *Precision Agriculture, 23*(2), 578–601. https://doi.org/10.1007/s11119-021-09850-7.

Ghosh, S. S., Dey, S., Bhogapurapu, N., Homayouni, S., Bhattacharya, A., & McNairn, H. (2022). Gaussian process regression model for crop biophysical parameter retrieval from multi-polarized C-band SAR data. *Remote Sensing, 14*(4). Art. no. 4 https://doi.org/10.3390/rs14040934 .

Gong, L., Yu, M., Jiang, S., Cutsuridis, V., & Pearson, S. (2021). Deep learning based prediction on greenhouse crop yield combined TCN and RNN. *Sensors, 21*(13). Art. no. 13 https://doi.org/10.3390/s21134537 .

Incudini, M., Grossi, M., Mandarino, A., Vallecorsa, S., Pierro, A. D., & Windridge, D. (2023). The quantum path kernel: A generalized neural tangent kernel for deep quantum machine learning. *IEEE Transactions Quantum Engineering, 4*, 1–16. https://doi.org/10.1109/TQE.2023.3287736.

Jabed, Md. A., & Murad, M. A. A. (2024). Crop yield prediction in agriculture: A comprehensive review of machine learning and deep learning approaches, with insights for future research and sustainability. *Heliyon, 10*(24), e40836. https://doi.org/10.1016/j.heliyon.2024.e40836.

Jin, B., & Xu, X. (2024). Forecasting wholesale prices of yellow corn through the gaussian process regression. *Neural Computing and Applications, 36*(15), 8693–8710. https://doi.org/10.1007/s00521-024-09531-2.

Jin, X., Zhang, T., Wang, L., Luo, Q., Li, S., Zhang, X., Lu, J., & Rao, Y. (2023). Using an optimised neural architecture search for predicting the quantum yield of photosynthesis of winter wheat. *Biosystems Engineering, 230*, 442–457. https://doi.org/10.1016/j.biosystemseng.2023.04.015.

Joshi, A., Pradhan, B., Gite, S., & Chakraborty, S. (2023). Remote-sensing data and deep-learning techniques in crop mapping and yield prediction: A systematic review. *Remote Sensing, 15*(8), 2014. https://doi.org/10.3390/rs15082014.

Karthikeyan, J., & Murugan, A. (2024). A study on enhancing crop yield prediction with accuracy through machine and deep learning models. *SEEJPH*, 1217–1227. https://doi.org/10.70135/seejph.vi.2270.

Keerthana, M., Meghana, K. J. M., Pravallika, S., & Kavitha, M. (2021). An ensemble algorithm for crop yield prediction. In *2021 third international conference on intelligent communication technologies and virtual mobile networks (ICICV)* (pp. 963–970). https://doi.org/10.1109/ICICV50876.2021.9388479.

Kumar, R., Bhanu, M., Mendes-Moreira, J., & Chandra, J. (2024). Spatio-temporal predictive modeling techniques for different domains: A survey. *ACM Computing Surveys, 57*(2), 38:1–38:42. https://doi.org/10.1145/3696661.

Li, B., Qiao, B., Zhao, Q., Yang, D., Zhu, R., Wang, Z., & Yang, Y. (2025). A predictive model of the photosynthetic rate of chili peppers using support vector regression and environmental multi-factor analysis. *Horticulturae, 11*(5). Art. no. 5 https://doi.org/10.3390/horticulturae11050502 .

Logeshwaran, J., Srivastava, D., Kumar, K. S., Rex, M. J., Al-Rasheed, A., Getahun, M., & Soufiene, B. O. (2024). Improving crop production using an agro-deep learning framework in precision agriculture. *BMC Bioinformatics, 25*(1), 341. https://doi.org/10.1186/s12859-024-05970-9.

Lou, Z., Quan, L., Sun, D., Xia, F., Li, H., & Guo, Z. (2024). Multimodal deep fusion model based on Transformer and multi-layer residuals for assessing the competitiveness of weeds in farmland ecosystems. *International Journal of Applied Earth Observation and Geoinformation, 127*, 103681. https://doi.org/10.1016/j.jag.2024.103681.

Lu, M., Gao, P., Li, H., Sun, Z., Yang, N., & Hu, J. (2023). An optimization approach for environmental control using quantum genetic algorithm and support vector regression. *Computers and Electronics in Agriculture, 215*, 108432. https://doi.org/10.1016/j.compag.2023.108432.

Magallanes-Quintanar, R., Galván-Tejada, C. E., Galván-Tejada, J. I., Gamboa-Rosales, H., de J. Méndez-Gallegos, S., & García-Domínguez, A. (2024). Auto-machine-learning models for standardized precipitation index prediction in North–Central Mexico. *Climate, 12*(7). Art. no. 7 https://doi.org/10.3390/cli12070102 .

Mahato, K. D., Kumar Das, S. S. G., Azad, C., & Kumar, U. (2024). Machine learning based hybrid ensemble models for prediction of organic dyes photophysical properties: Absorption wavelengths, emission wavelengths, and quantum yields. *APL Machine Learning, 2*(1), 016101. https://doi.org/10.1063/5.0181294.

Maheshwari, D., Sierra-Sosa, D., & Garcia-Zapirain, B. (2022). Variational quantum classifier for binary classification: Real vs synthetic dataset. *IEEE Access, 10*, 3705–3715. https://doi.org/10.1109/ACCESS.2021.3139323.

Metawei, M. A., Eldeeb, H., Nassar, S. M., & Taher, M. (2022). Quantum computing meets artificial intelligence: Innovations and challenges. In M. Virvou, G. A. Tsihrintzis, N. G. Bourbakis, & L. C. Jain (Eds.), *Handbook on artificial intelligence-empowered applied software engineering: Vol.1: Novel methodologies to engineering smart software systems* (pp. 303–338). Cham: Springer International Publishing. https://doi.org/10.1007/978-3-031-08202-3_12.

Noorunnahar, M., Chowdhury, A. H., & Mila, F. A. (2023). A tree based extreme gradient boosting (XGBoost) machine learning model to forecast the annual rice production in Bangladesh. *PloS One, 18*(3), e0283452. https://doi.org/10.1371/journal.pone.0283452.

Nosratabadi, S., Imre, F., Szell, K., Ardabili, S., Beszedes, B., & Mosavi, A. (2020). Hybrid machine learning models for crop yield prediction. *arXiv preprint arXiv:2005.04155*. https://doi.org/10.48550/arXiv.2005.04155.

Nti, I. K., Zaman, A., Nyarko-Boateng, O., Adekoya, A. F., & Keyeremeh, F. (2023). A predictive analytics model for crop suitability and productivity with tree-based ensemble learning. *Decision Analytics Journal, 8*, 100311. https://doi.org/10.1016/j.dajour.2023.100311.

Padha, A., & Sahoo, A. (2022). Quantum enhanced machine learning for unobtrusive stress monitoring. In *Proceedings of the 2022 fourteenth international conference on contemporary computing, in IC3-2022* (pp. 476–483). New York, NY, USA: Association for Computing Machinery. https://doi.org/10.1145/3549206.3549288.

Rajagopa, A. (2021). A novel approach in prediction of crop production using recurrent cuckoo search optimization neural networks. *Applied Science, 11*(21)). https://doi.org/10.3390/app11219816.

Rashid, M., Bari, B. S., Yusup, Y., Kamaruddin, M. A., & Khan, N. (2021). A comprehensive review of crop yield prediction using machine learning approaches with special emphasis on palm oil yield prediction. *IEEE Access, 9*, 63406–63439. https://doi.org/10.1109/ACCESS.2021.3075159.

Sadenova, M., Beisekenov, N., Varbanov, P. S., & Pan, T. (2023). Application of machine learning and neural networks to predict the yield of cereals, legumes, oilseeds and forage crops in Kazakhstan. *Agriculture, 13*(6), 1195. https://doi.org/10.3390/agriculture13061195.

Servia, H., Pareeth, S., Michailovsky, C. I., de Fraiture, C., & Karimi, P. (2022). Operational framework to predict field level crop biomass using remote sensing and data driven models. *International Journal of Applied Earth Observation and Geoinformation, 108*, 102725. https://doi.org/10.1016/j.jag.2022.102725.

Setiadi, D. R. I. M., Susanto, A., Nugroho, K., Muslikh, A. R., Ojugo, A. A., & Gan, H.-S. (2024). Rice yield forecasting using hybrid quantum deep learning model. *Computers, 13*(8), 191. https://doi.org/10.3390/computers13080191.

Setiadi, D. R. I. M., Susanto, A., Nugroho, K., Muslikh, A. R., Ojugo, A. A., & Gan, H.-S. (2024). Rice yield forecasting using hybrid quantum deep learning model. *Computers, 13*(8), 191. https://doi.org/10.3390/computers13080191.

Shah, B., Thapa, S., Diyali, R. S., S. , H. K., & Maharjan, S. (2020). Rain prediction using polynomial regression for the field of agriculture prediction for Karnatakka. *Social Science Research Network, Rochester, NY*, 3635278. https://doi.org/10.2139/ssrn.3635278.

Sivanantham, V., Sangeetha, V., Alnuaim, A. A., Hatamleh, W. A., Anilkumar, C., Hatamleh, A. A., & Sweidan, D. (2022). Quantile correlative deep feedforward multilayer perceptron for crop yield prediction. *Computers & Electrical Engineering, 98*, 107696. https://doi.org/10.1016/j.compeleceng.2022.107696.

Song, Y. (2024). Neural quantum states in variational monte carlo method: A brief summary. *arXiv preprint arXiv:2406.01017*. https://doi.org/10.48550/ARXIV.2406.01017.

Song, Z., Yin, J., & Zhang, R. (2023). Revisiting quantum algorithms for linear regressions: Quadratic speedups without data-dependent parameters. *arXiv: arXiv:2311.14823*. https://doi.org/10.48550/arXiv.2311.14823.

Sushmit, M. M., & Mahbubul, I. M. (2023). Forecasting solar irradiance with hybrid classical–quantum models: A comprehensive evaluation of deep learning and quantum-enhanced techniques. *Energy Conversion and Management, 294*, 117555. https://doi.org/10.1016/j.enconman.2023.117555.

PART

II

Quantum instrumentation for smart agriculture

3

Quantum sensors in soil health monitoring

Himanshu[a], Ankit[a], Deepika Saxena[a], Rahmatullah Kazimi[b] and Abhishek Tanwer[c]

[a]DEPARTMENT OF HORTICULTURE, SCHOOL OF AGRICULTURE, LOVELY PROFESSIONAL UNIVERSITY, PHAGWARA, PUNJAB, INDIA [b]DEPARTMENT OF HORTICULTURE, DAIKUNDI INSTITUTE OF HIGHER EDUCATION, DAIKONDI UNIVERSITY, DAIKUNDI, AFGHANISTAN [c]LINCOLN INSTITUTE FOR AGRI-FOOD TECHNOLOGY, SCHOOL OF AGRI-FOOD TECHNOLOGY AND MANUFACTURING, UNIVERSITY OF LINCOLN, LINCOLNSHIRE, UNITED KINGDOM

3.1 Introduction to quantum sensing

The agricultural sphere is experiencing an extreme change, as diagnostic and automation technologies are rapidly gaining popularity due to the latest innovative advances. Quantum sensing has proved to be such a revolutionary device that it has the potential to change the way in which we see and evaluate life and environmental systems and monitor them. This technological transition is closely connected to the dynamics of the Fourth Industrial Revolution (Industry 4.0) as a global trend, which is defined by the combination of artificial intelligence (AI), robotics, biotechnology, nanotechnology, the Internet of Things (IoT), and quantum technologies. In agrarian production, that line of convergence even more dominates in areas such as fleeting driverless automobiles, continuous climate sensing, intelligent irrigating mechanisms, and accurate nutrient management, all of which are extending the incongruence-free manure output. In this revolution, quantum sensing can be singled out due to its extreme sensitivity, specificity, and noninvasive detection (Hassoun et al., 2023). Using the concept of quantum mechanics, such as superposition, entanglement, and coherence, quantum sensors can quantify tiny variations in temperature, electric and magnetic fields, or the gradient of gravitational fields down to the nanoscale. This enables online real-time monitoring of the most important soil properties, such as moisture, microbial respiration, nutrient availability, salinity, and subsurface hydrology, even in tough environments. Quantum technologies provide a shift in paradigm in data fidelity and agricultural decision-making as conventional sensing tools start to show weaknesses in data capture in contemporary agroecosystems. The history of quantum sensing proceeds in the direction of understanding quantum physics (early 20th century). The first demonstrations of the potential of quantum effects to achieve ultraprecise measurements were

Quantum Computing and Instrumentation for Smart Agriculture. DOI: https://doi.org/10.1016/B978-0-443-44247-6.00002-6

made with atomic clocks, the quantum Hall effect, and superconducting quantum interference devices (SQUIDs). The past 20 years saw major advances in solid-state quantum physics, laser cooling, and quantum information science, and sensors based on nitrogen-vacancy (NV) centers in diamond, atomic vapor cells, and trapped ions are being miniaturized and deployed in the field (Aarif et al., 2025). Specifically, NV-diamond sensors have become very promising in soil diagnostics because of their stability, biocompatibility, and magnetic and electric fields. These sensors can measure the ionic changes, redox, and subsurface microbial breathing, which are all critical clues of soil health. Also, quantum gravimeters are being sought in groundwater motion mapping, soil compaction strata, and the moisture gradient, with post-2020 pilot schemes being realized in Australia, the UK, and India, and their effectiveness is being tested in varying field circumstances. Likewise, the quantum-enhanced low-temperature atomic biosensors and entangled photon sources are being developed to detect gaseous emitters (e.g., methane and ammonia), pesticide residuals, or nutrient imbalances (Liu et al., 2023). Even though opportunities exist in technology, there is an evident research gap regarding adjusting quantum sensing platforms to large-scale agricultural use. The majority of the current studies are either conducted in laboratories or are more restricted pilot studies. The combination of digital agriculture ecosystems, such as AI-based analytics, Unmanned Aerial Vehicle (UAV) and satellite remote sensing, and cloud computing, with quantum sensors, is in its infancy. Not much field data exists on long-term reliability, cross-sensor calibration, and environmental robustness of these systems in field conditions when dealing with dynamic agricultural settings. In addition, the majority of the current existing agricultural models are based on the classical input data stream and are not yet retrofitted to accommodate the quantum level of spatial and temporal data resolutions.

Objectives of the chapter

The objectives of the chapter are:

- To clarify the basic principles of quantum sensing and its applicability to agricultural diagnostics. This goal is aimed at revealing how the quantum processes of superposition, entanglement, and coherence allow to create ultraprecise sensing in complex soil conditions;
- To specify the presence of indicators of soil health that should be monitored in high resolution and in real time. The essential parameters that the chapter attempts to address are the soil moisture, pH, nutrient availability, salinity, and the activity of the microbes that directly affect productivity and sustainability;
- To find out how and what people can do with big quantum sensor platforms. It highlights the use of such technologies as NV-diamond sensors, quantum gravimeters, and quantum magnetometers, and how these devices can identify the slightest changes in physical and biochemical characteristics of soil;
- To analyze the quantum soil sensing and the digital agriculture system integration. This goal involves insight into the potential of IoT, AI, remote sensing, and cloud analytics to improve the interpretation of quantum sensor data to make decisions in precision farming;

- And, to find issues, constraints, and opportunities of applying quantum sensors in soil health. This chapter also discusses the obstacles of cost, complex calibration, field durability, and scalability and provides the prospects of the broader adoption of it in sustainable agriculture.

Organization of the chapter

The rest of the chapter is organized as: Section 3.2 enlightens soil health parameters and the need for precision monitoring. Section 3.3 focusses on principles of quantum sensors, followed by discussion on quantum technologies in soil sensing in Section 3.4. Section 3.5 highlights types of quantum sensors for agriculture. Section 3.6 discusses case studies and field applications, while Section 3.7 illustrates integration with digital agriculture platforms. Sections 3.8 and 3.9 discuss challenges and limitations cum research directions. And, finally, Section 3.10 concludes the chapter with future scope.

3.2 Soil health parameters and the need for precision monitoring

The health of soil in terms of moisture, pH, nutrient contents, microbial activities, and electrical conductivity (EC) is vital to sustainable agriculture. These indicators are monitored precisely, thus allowing real-time and location-specific management to make the most of resources and crop yields. Given the rising environmental pressures, the need to apply superior sensing technology would be useful in guaranteeing the sustainability of soil fertility, ecological stability, and food security.

3.2.1 Key indicators of substrate quality: Moisture, pH, microbial biomass, electrical conductivity, and nutrients

Sustainable agriculture depends on the concept of soil health since a healthy soil is a living ecosystem capable of supporting livestock, agricultural production, and human beings. It is very important to assess specific biophysical and biochemical indications of the health of the soil. The most relevant indicators (Fig. 3.1) are soil moisture, pH, microbial biomass, EC, and nutrient content. All these are fundamental in maintaining a strong and productive agroecosystem. Soil moisture is critical to the physiology of plants, as their growth, cellular turgor, enzymatic activities, and nutrient uptake are dependent on it. Soil moisture levels are kept at optimum levels that keep the microbial activity very active, and the plant roots are able to absorb nutrients very effectively. However, both scarcity and abundance situations are not good. As the drought stress reduces the availability of nutrients and impedes plant development, waterlogging may create an anaerobic environment and thus deny roots the chance of releasing oxygen. The chemical availabilities of nutrients, which depend on the pH in the soil, are also of critical importance. A value of between 6.0 and 7.5, i.e., slightly acidic to neutral, is the pH most crops favor. Many of the minerals exhibit insolubility or excessive solubility beyond this range, resulting in toxicities or deficiencies that inhibit crop growth. Rich in soil health, microbial biomass is a sensitive

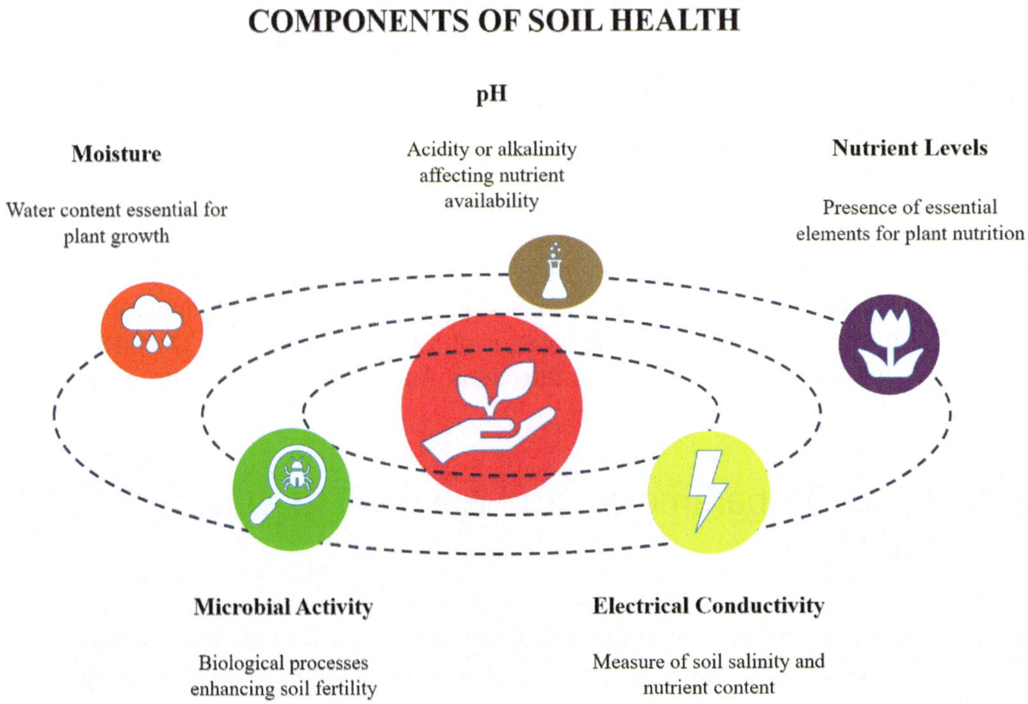

FIGURE 3.1 Important components of soil health.

indicator of the biological part of health. It renders the total living microorganisms in the soil and is important in the fixation of nitrogen, mineralization of nutrients, and the breakdown of organic matter. A decline in microbial biomass is a common symptom of pollution, the stressful conditions of soil, or organic matter loss. On the contrary, EC measures the quantity of water-soluble salts in the soil. It is the key indicator of salinity, which directly affects both the osmotic balance of plants as well as their water uptake. The various crops differ in EC tolerance, and high EC readings may reduce yields, especially in arid and irrigated regions where salinization is a common problem. Finally, the availability of nutrients remains among the important elements of soil fertility (Zhang et al., 2023).

3.2.2 Impact of degraded conditions on crop yield and ecosystem services

Soil degradation is also a major issue of concern in agriculture, and it is determined as a combination of unsustainable land use farming activity, climate-related variabilities, and chemical overuse. This degradation manifests itself in many ways, which include erosion, nutrient mining, organic matter depletion, acidification, and salinization, amongst numerous others. The consequences of such changes are far-reaching, and when specific crops are grown in ways that alter the ecosystem services and food production systems in the long term. Degraded soils have a few signs of degraded functions. The crops cannot infiltrate and capture water, making them more drought-prone. Reduced decomposition of organic matter and inefficient cycling of nutrients are other outcomes of the reduction of soil biological activity. Compaction and other crusting processes ultimately limit gaseous exchange as well as limit root growth in the soil. This degradation of biological, chemical, and physical properties leads to low production. Healthy soils contain carbon dioxide and absorb carbon dioxide contained in the atmosphere by acting as carbon sinks, thus slowing down global warming. This capacity is also diminished because soils degrade, and they can also begin to release carbon. Further, worn-out soils will no longer have the ability to purify water and support favorable faunal and microbiological diversity, both of which are critical in organic matter recycling and pest control (Rodrigues et al., 2023).

3.2.3 Rationale for high-resolution, real-time monitoring in agroecosystems

In conventional soil monitoring methodology, manual sampling is normally accompanied by laboratory testing. Although they are very accurate, the techniques possess certain limitations in terms of time and space resolution. They do not provide an insight into what the soil is at a given period of time, and they are often costly, time-consuming, and rare in occurrence. Conversely, problems besetting the present-day agricultural sector require context-specific, site-specific, and dynamic data so that competent decisions can be made in real-time. Consequently, precision agriculture technologies, i.e., the technologies that allow combining sensors, data analytics, and digital communication platforms to locate high-resolution monitoring capabilities, are getting increasingly more popular. High-resolution soil monitoring enables farmers and academics to detect variation on small spatial scales of soil properties within a field. As an example, the IoT could be used to provide continuous measures of temperature, EC, pH, and moisture as measured by soil sensors. The data are then distributed to the cloud servers to be analyzed to give useful information (Prasanna et al., 2024). The remote sensing technologies based on drones or satellites can also provide nutrient maps, the indices of stress, and observations of vegetation health. This data can later be compared with soil measurements. This multifactor approach allows pinpointing such challenges as water deficiency, salt build-up, and nutrient deficiencies early enough before other factors can bring down yields. Current inquiries are evidence of the applicability of these technologies in real-life scenarios. Real-time monitoring also promotes adaptive management techniques. Interventions can also be deployed where they are most needed and not spread and dumped uniformly onto a field in the form of fertilizers, insecticides, or water. This promotes sustainable utilization of resources, reduces the cost of inputs, and has a reduced adverse impact on the environment. Trends in sensor-based predictive models may also be able to foresee stain trends in soil, and thus interventions that combat it may be implemented in advance, such as crop rotation, the use of organic matter, or phytoremediation (Mallik et al., 2025).

3.3 Principles of quantum sensors

The utilization of quantum sensors is one of the most important advances in the science of measuring today. Utilizing the unusual properties of quantum mechanics, which include entanglement, superposition, coherence, and tunneling, these gadgets can sense and quantify physical quantities with much greater sensitivity than do traditional sensors. Besides transforming scientific studies, quantum sensors are likely to lead to an expanded number of applications in fields as diverse as environmental and health monitoring, imaging, navigation, and metrology.

3.3.1 Fundamental physics: Entanglement, tunneling, coherence, and superposition

Quantum sensors are one of the most important aspects to have occurred in the field of measurement science thus far. These devices exploit the special properties of quantum mechanics, including entanglement, superposition, coherence, and tunneling, to sense and detect physical variables a great deal more delicately than commonplace sensors. One other notable event is the effect of tunneling, whereby particles that under classical mechanics would have crossed potential energy barriers would be intractable. This phenomena find extensive use in the scanning tunneling microscopy and the quantum point contact sensors, which measure the probability of particles being transmitted between electrodes and thus are used in the detection of small forces, currents, or field variations. The atomic- or molecular-level sensing of weak electric or magnetic signals is enhanced by tunneling in sensing applications. Quantum coherence refers to the presence of the relative phase between multiple states in a quantum superposition. Quantum systems are highly controllable in their quantum state (able to make the same measurements over again, repeatedly and reliably), at the same time remaining coherent. Nonetheless, coherence needs delicate safeguarding and error fixing operations because it is very vulnerable to foreign interference. Some of the challenges to the applicability of quantum sensors include ensuring long coherence times of systems, especially under noisy and open applications (Wang et al., 2025). Superposition of quantum systems allows them to be in multiple states simultaneously, until they are measured, at which point they will convert to one outcome. The ability of a quantum system to explore an exponentially greater number of possibilities enhances the informative content significantly. The superposition enhances accuracy in measurements that are used in sensing applications by encoding many routes or states of a signal. As an example, atomic clocks that have superposed atomic states are far more accurate than classical clocks by orders of magnitude (Tomaz et al., 2025).

Recent advances have enabled much more practical implementation of quantum sensing principles by enhancing the control of entanglement, coherence, and tunneling in practice. Due to the recent improvements in the noise-resilient entanglement engineering, it is now possible to sustain multipartite entangled states (spin-squeezed and Greenberger–Horne–Zeilinger (GHZ) states) over a significantly longer time. This has a direct increase in sensitivity over and above the usual quantum limit of magnetometry and inertial sensing applications (Zhou et al., 2025). Also, in open-system and field-deployed environments, dynamical decoupling, quantum error correction, and machine learning-assisted feedback-based error-mitigated quantum coherence preservation methods have demonstrated substantial increases in coherence times. The recent progress in two-dimensional materials and hybrid quantum point contacts has enabled the detection of small, electric, and magnetic perturbations at the nano-scale with a high degree of sensitivity in tunneling-based quantum sensors and the expansion of the utilization of tunneling

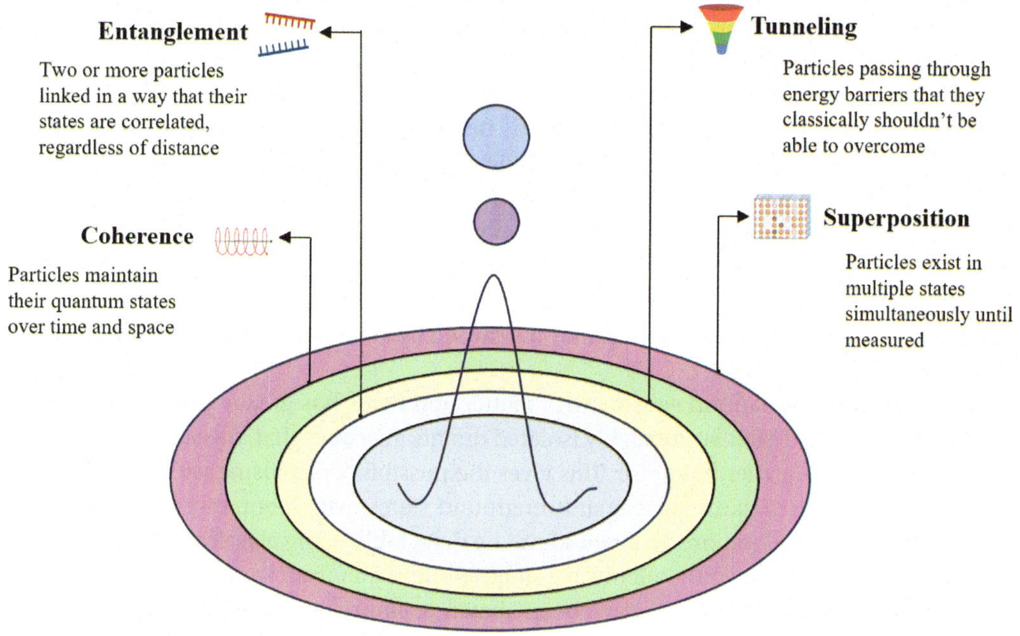

FIGURE 3.2 Fundamental quantum phenomena underpinning quantum sensor operation.

effects in the broader scanning probe systems (Yoo & Maeng, 2025). Also, the superposition-based sensing is assisted by advancements in optical lattice clocks and trapped-ion systems, where the errors in the fractional regime are down to 10^{-19}, which in turn have enhanced their role as fundamental physics tests and ultraprecise metrology (Caldwell et al., 2025). The physics behind the exceptional sensitivity and precision of quantum-based sensors making use of correlated states, wave-like persistence, probabilistic state occupancy, and barrier-penetration effects are shown in (Fig. 3.2).

3.3.2 Interaction with physical matter: Electromagnetic fields, particles, and energy shifts

Quantum sensors interact with the external world by responding to the smallest changes in the particle's immediate environment, energy states, and electromagnetic fields. It is because of this that we can measure physical quantities such as steepness of temperatures, acceleration, pressure, magnetic and electric fields, and even gravitational forces. Detectable and interpretable quantum jumps govern such interactions, and they are discrete changes to the energy states caused by external stimuli.

Magnetic field sensing is one of the most developed applications of quantum interaction with matter. Atomic magnetometers, especially optically pumped magnetometers and NV-diamond magnetometers, can detect changes in the field by monitoring changes in the atomic energy levels. Such shifts may be read out as changes in spin precession, fluorescence, or resonance frequencies using laser or microwave methods, respectively. It is noted that the resolution is very high compared to SQUIDs or classical Hall-effect sensors because such changes occur at the quantum level. Electric fields are measured by devices that include charged particles that are controlled by optical, radio-frequency fields under electromagnetic traps, such as trapped ion sensors. An external electric field disrupts the trapping potential, thereby leading to visible variation in the motional states of the ions or fluorescence signals (Kantsepolsky et al., 2023). One of the main features of numerous quantum sensor processes is the energy shifts. As one example, optical or microwave optical transitions between electronic or hyperfine states of atomic systems (ytterbium or cesium) are monitored with atomic clocks. It is exceptionally sensitive in measuring energy changes, however slight they may be, caused by movement, gravitational potential, or magnetism (Shi et al., 2024).

The interaction of quantum sensors with matter and fields has greatly improved. New levitating particle arrays demonstrate highly isolated quantum probes that are capable of detecting perturbations with low thermal noise. This gives the possibility of sensing with unprecedented sensitivity, even to the regime near quantum ground states, which enhances the detection of minute forces and accelerations that would not be detectable using classical methods. Higher-energy physics superconducting microwave single-photon quantum sensors have already been experimentally demonstrated and have been shown to improve individual particle measurements of collider conditions and time-resolved measurements, which exploit quantum states to identify a particle and measure its energy shift (Fratelli et al., 2025). Moreover, biological quantum bits (qubits) have shown that using fluorescent proteins can be used to create quantum sensing, and now quantum probing of nanoscale biological processes might become feasible by coupling the sample to cellular electromagnetic and electric activities (Yukawa et al., 2025). The controlled interactions between quantum systems and external fields or particles can be converted into measurable energy shifts to allow high-precision applications, such as the discovery of new technologies, navigation, imaging, metrology, and scientific research, (Fig. 3.3) displays such interactions.

3.3.3 Basic design and mechanism of quantum-enabled measurement tools

Quantum-enabled measurement tools are designed based on working with and reading the quantum states at the highest degrees of precision. Such sensors operate (Fig. 3.4) by initializing a quantum system to a specific state and then permitting it to couple to an external perturbation (e.g., the magnetic, electric, or gravitational field housed within), and tracking the subsequent quantum changes. The most common platforms include NV centers in diamond, cold atom interferometers, trapped ions, superconducting circuits, and quantum dots, each of which is optimized for various sensors. NV centers are NV defects in a diamond lattice, wherein control occurs through laser and microwave excitation. They are also very sensitive to nanoscale magnetic and electric fields and can even be operated at room temperature, which makes them a good fit in environmental monitoring, biomedical diagnosis, and material characterization.

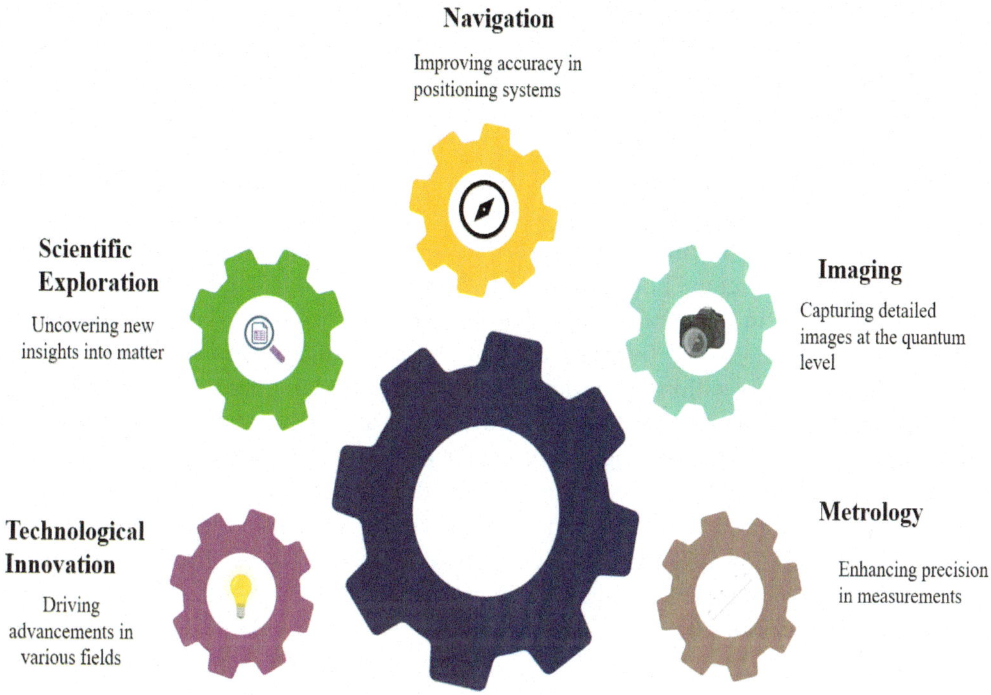

FIGURE 3.3 Application domains enabled by matter–field interactions in quantum sensing.

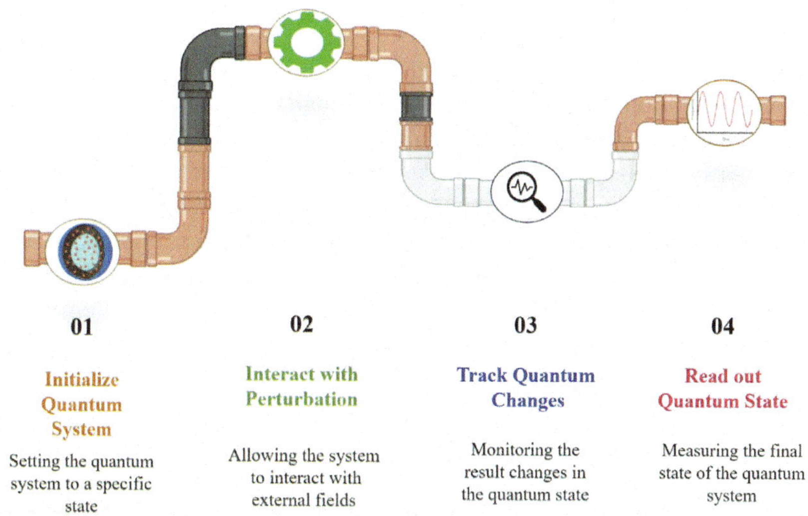

FIGURE 3.4 Working of quantum-enabled measurement tools.

Cold atom interferometers operate in this way: they cool atoms to near absolute zero and then measure the pattern of interference to measure acceleration, rotation, or a gradient of gravity, which at surface pressures can be used to map soil density and subsurface character. Trapped ion systems, entailing the suspension of ions in a vacuum by electromagnetic fields, are regarded as sensitive to electric field and temperature modification (Oh et al., 2024).

Recent developments have increased the sensitivity and the practical applications of quantum sensors greatly. To observe electromagnetic phenomena down to the scale of the atomic scale, scientists are also turning to new quantum materials and defect platforms, including the color centers in hexagonal boron nitride (hBN), whose centers will be improved over the traditional NV centers in the application of ultrafine sensors. Moreover, lattice-scale quantum electrometers developed using the technology of observing minute changes in optical emission of the defect centers within the structure of manufactured diamonds have demonstrated the ability to detect single electrical charges (Pieplow et al., 2025). At nanoscale dimensions, the enhanced processes with entangled NV center pairs have shown a power of magnitude of magnetic sensitivity, pushing the accuracy beyond their previous limits. The combination of quantum sensors and AI-based data analytics, as well as sustained interferometric control, makes it possible to perform resilience functioning in noisy conditions, thereby increasing the extent of applying quantum measurement systems to industrial diagnostics, medical imaging, and navigation (Panyaram, 2025).

3.4 Quantum technologies in soil sensing

Quantum technologies are making a revolution in sustainable agriculture and environmental monitoring, especially in soil sensing. The adultacy, complexity, and explicitness of the traditional soil sensing approaches, attributed to mechanical, chemical, or electromagnetic tenets, are limited. The increasing complexity of agroecosystems is causing an increasing need for tools that can be employed non-destructively to sense, visualize, and characterize the dynamics of the soil in real time with nanometers of resolution. Quantum sensors have an unparalleled advantage in their field because they are based on the core concepts of quantum physics. The technologies have the advantage of measuring with very small resolutions and accuracy compared to classical systems due to their use of quantum effects such as entanglement, coherence, and superposition. Three significant changes that quantum technologies are bringing to soil sensing are using NV centers in diamond substrates to probe subsurface density and moisture through quantum gravimetry, sensing ionic motions, and microbial activity with quantum magnetometry.

3.4.1 Use of nitrogen-vacancy centers in diamond substrates

NV centers in diamond have been made highly flexible, nanoscale-sensing quantum systems. The atomic-scale defects consist of a nitrogen atom and an empty diamond lattice site. NV centers are so useful in sensing because the quantum state of their spin can be initialized, controlled, and read in ambient conditions with laser and microwave excitation. They make an ideal choice in the analysis of complex structures such as soil due to their temperature, pressure, magnetic, and electric force sensitivity. NV diamonds have been incorporated into nanodiamonds and utilized in agricultural studies to follow microenvironments near plant roots, in particular the rhizosphere, where interactions among nutrients, bacteria, and plant roots take place (Adeniji & Stine, 2023; Zulkharnay & May, 2024) demonstrated the use of functionalized nanodiamonds with NV centers to detect the redox differences of microbial respiration within agricultural soils. The sensors have a spatial resolution of several micrometers and could give real-time results on changes that took place in a biochemical manner when using compost to loosen up the soil.

Centers in diamond NV are evolving to be a useful platform in high-resolution environmental and soil-related sensing, and not just a laboratory curiosity. It was also shown that NV-center-containing nanodiamonds could be used as in-vivo probes in plant root zones to detect paramagnetic species such as iron and oxygen radicals in a significant rhizosphere quantum sensing experiment (Ibrahim et al., 2024). This method provides a new perspective on characterizing metal dynamics and radical chemistry in soils in previously unimaginable space. Recent advances in entanglement-enhanced NV sensing have demonstrated that pairs of NV centers can gain more than 3x in sensitivity and higher spatial resolution when operated in ambient conditions (Du et al., 2024). This is an important move toward the identification of small soil chemical gradients. Moreover, manufactured double-layered silica-coated nanodiamonds with NV centers have made it possible to use spin relaxometry techniques to regulate the detection of reactive radicals. This technology can be adjusted to real-time tracking of the soil redox processes (Su et al., 2025). NV centers made in diamond help to detect local changes in magnetic and electric fields with precision, which allows better analysis of the soil, targeting specific sensing methods, and ecologically responsible control of agriculture, as shown in (Fig. 3.5).

3.4.2 Gravimetry for subsurface density and moisture mapping

Quantum gravimetry can sense minute variations in gravity that can be attributed to a shift in the distribution of the mass beneath the surface. Due to this reason, quantum gravimeters are ideal sources of measurement of soil bulk density and moisture, which is a very crucial aspect to the productivity and soil status. The neutron moisture probes and conventional gravimeters are often bulky, require calibration to the soil type, and have low resolution. Quantum gravimeters, especially those using atom interferometry, provide, in contrast, high-resolution, contact-free, scalable measuring platforms. Most recently, there has been the use of quantum gravimeters to monitor the redistribution of soil water due to irrigation and precipitation. Horel and Zsigmond (2023) examined the dynamics of soil moisture in grapevine soils measured at the field with the help of a portable atom interferometer-based gravimeter. The device enabled locating precise irrigation based on mapping the fluctuation in the gravity force due to water content in the root zone, which eliminated waterlogging or deficit stress. Such sensors are particularly effective in determining deep soil moisture profiles that cannot be seen with the help of conventional sensors based on the ground level, since the gravitational responses play out with depth. De Silva et al., 2023 explained that the AI-assisted gravimetric system incorporates the data on

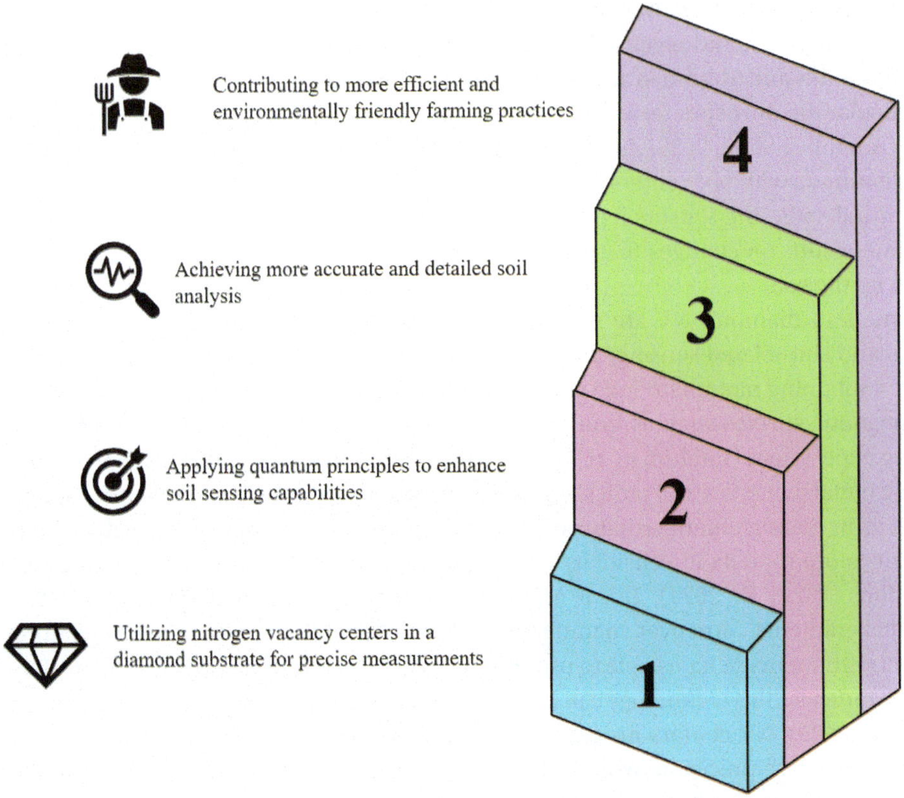

FIGURE 3.5 Conceptual framework for soil sensing using nitrogen-vacancy centers in diamond.

remote sensing with quantum data. Their hybrid model was able to act as a decision support tool on watershed management and land regeneration since it was able to predict subsurface compaction areas, erosion-prone areas, and areas within dry regions in India. Since the changes in the carbon stock in the soil alter the density and, therefore, gravitational reactions, such gravimetric methods can also promote studies of carbon sequestration.

In addition to traditional geophysical surveys, more recent inventions in quantum gravimetry have demonstrated practical applications in subsurface density and moisture mapping. A large-scale aerial experiment with a cold-atom quantum gravimeter (GIRAFE) and classical gravimeters on Greenland and Iceland generated the high-precision gravity measurements (1–2 mGal) over complex topography and showed that quantum sensors can resolve small-scale variations in the mass distribution due to density differences between ice and rock that can directly tell us about subsurface hydrological conditions as well as potential density variations related to moisture (Muccini et al., 2025). A separate hybrid survey on-site gravimetry field survey utilized an atomic gravimeter and a spring gravimeter to measure absolute gravity with an accuracy of approximately 7.7 mGal. This has shown the way a quantum gravimeter can serve the purpose of absolute density references through the mapping of standardized gravity between survey

FIGURE 3.6 Comparative positioning of gravimetric techniques within soil sensing approaches.

points. This offers superior baseline calibration of relative density variations that are crucial to determining subsurface water content variations (Shettell et al., 2024). Besides field tests, a realistic performance estimation of future satellite gravimetry missions based on the concept of cold atom interferometry has been determined through simulation studies of the noise model of quantum gravimetric sensors. These studies established a continual monitoring in the world by highlighting sensitivity improvement and error reduction needed to deduce time-varying gravitational signals through the mass redistribution, including the groundwater movement and changes in soil moisture (Kantsepolsky & Aviv, 2024). Fig. 3.6 shows that the gravimetric methods alone provide the precision of measurements and practical influence in characterizing subsurface density and moisture in a more efficient way as compared to the conventional soil sensing and sampling technique of the product using quantum-enabled solutions.

3.4.3 Magnetometry for detecting ionic movement and microbial activity

Quantum magnetometry aims at applying quantum-based materials and procedures to attain sensitive and spatially resolved measurements of magnetic fields. The most promising technologies in this sector include SQUIDs, atomic vapor cells, and NV-center-based sensors. The sources of magnetic signals in soil research are ionic transport (e.g., nutrient flux), redox reactions, and magnetotactic microbial activity. Traditional sensors have poor sensitivity, or the signal is swamped by soil particles, and are regularly unable to clean up these small, transient signals. Quantum magnetometers overcome these constraints by detecting the difference in spin resonance or fluorescence to a weak magnetic field. One of the uses is monitoring the circulation of such nutrients as phosphates and nitrates in the soil matrix. Yuan et al. (2023) fitted a small magnetometer (based on NV) straight in maize fields to monitor the movement of nutrient ions following fertilization. The device measured the magnetic fingerprints of nutrient pulses in the root zone in order to correlate them to soil texture and plant increment. These discoveries enabled precision fertilization techniques to be possible, leading to lower input costs and the prevention of leaching in the groundwater. The metabolism of microorganisms also generates magnetic signals. As an example, magnetotactic bacteria have the internal magnetite crystals that cause them to align with the magnetic field of the Earth automatically. These organisms are indicators of redox potential and take part in the recycling of nutrients. Schofield et al. (2022) spatially resolved microbial hotspots in organically maintained fields using quantum magnetometers. Based on their study, the biological activity is likely to gather in the neighborhood of decaying residues and root hairs, which are the areas where biofertilizers may be concentrated.

Moreover, magnetometry allows the detection of soil stress indicators at an early stage, such as compaction and salt. Quantum sensors have the potential to detect changes in electromagnetic signatures in the area before the development of plant symptoms, due to ion mobility and mineral solubility changes in response to stress. These real-time diagnostics could be critical to the adaptive process of soil management under the climate stressors. Integration of quantum magnetometers and sequencing of microbial DNA in combination with soil resistivity and hyperspectral imaging would provide a complete image of soil functioning. Khondakar and Kaushik (2024) showed that multiplexed quantum soil sensors can be deployed throughout agricultural lands to offer real-time soil health dashboards that are accessed by mobile applications. As sensors become cheaper and data processing improves, these systems are most likely to become necessities of the next generation of digital agriculture. The outlines core application areas (Table 3.1) where quantum magnetometers offer transformative potential for monitoring soil biophysical processes with high precision.

3.5 Types of quantum sensors for agriculture

Quantum sensors are leading precision agriculture due to the ultrasensitive and noninvasive measurement of physical, chemical, and biological factors in the soil, plants, and atmosphere. Unlike standard sensors, quantum sensors exploit coherence, entanglement, and superposition to enhance sensitivity and spatial resolution. Such capabilities enable real-time monitoring of those variables that are critical to soil management, plant health, and climate adaptation planning. The various forms of quantum sensors (Fig. 3.3) that have been developed are undergoing modifications to be used in the field as the world demands data-driven and reproducible methods of agriculture. The most promising platforms falling under this category are the fiber-optic and solid-state platforms, NV-based magnetometers, and portable gravimeters to conduct in situ diagnostics, mapping of microbial activity, and characterization of terrain (Fig. 3.7).

3.5.1 Portable gravimeters for terrain analysis

Gravimetry is one of the oldest geophysics techniques that has recently been revolutionized once again due to quantum technologies. Quantum gravimeters measure even the minutest variations

Table 3.1 Key application areas of quantum magnetometry in soil health monitoring.

Application area	Function	Quantum magnetometer type	Relevance to soil health and agriculture
Ionic transport and nutrient cycling	Movement of nitrate, phosphate, and potassium ions through the soil profile	NV center magnetometers in diamond	Enables precision fertilization, reduces nutrient leaching, and improves root-zone nutrient availability
Microbial activity mapping	Magnetic response of magnetotactic bacteria and redox-active microbial metabolism	Atomic vapor cell magnetometers; NV-based sensors	Supports targeted biofertilizer application and informs microbial health in organic soil management
Soil stress detection	Changes in ionic mobility and mineral magnetism due to salinity, compaction, and drought	NV-diamond and SQUID magnetometers	Provides an early warning system for abiotic stress, enabling adaptive irrigation or reclamation practices
Redox fluctuations	Oxidation–reduction reactions during microbial respiration and anaerobic processes	Trapped ion magnetometers; miniaturized vapor cells	Helps in monitoring carbon sequestration, denitrification, and anaerobic microbial performance
Integration with digital platforms	Real-time soil health visualization using magnetic data streams	Multiplexed quantum sensors integrated with IoT	Enables field-scale decision-making, improves spatial resolution in digital soil health monitoring systems

in the gravitational acceleration by using laser-cooled atoms in free fall, typically with atom interferometry. Aids are important tools of terrain analysis in agriculture due to their outstanding accuracy in the detection of the variations of the subsurface density and the mass distribution. These sensors provide data on subsurface geologic structures: aquifers and soil moisture gradients, compacted layers and voids, without making physical contact or disturbing the ground, which can be used to conduct terrain classification and slope mapping as well as irrigation design. Quantum gravimeters were recently miniaturized so they became at least portable, field-deployable instruments, and are now perhaps on the verge of merging the worlds of precision in the laboratory and on the farm. The sensor enabled the viticulturists to adjust irrigation schemes to the accuracy of the distribution of the subsurface water at different elevations and different slopes to define water retention patterns related to the topography (Wang et al., 2023). Such applications aid in adjusting land management to drought, water-use efficiency, and reducing over-irrigation. Moreover, this is becoming an increasing challenge to climate-smart agriculture; the need to store carbon in soils, which can be further explained by topography analysis through gravimetry. Greater branched mass production or organic substance decay are commonly factors of more compacted soils. Kumara et al. (2023) modeled geographical changes in soil carbon

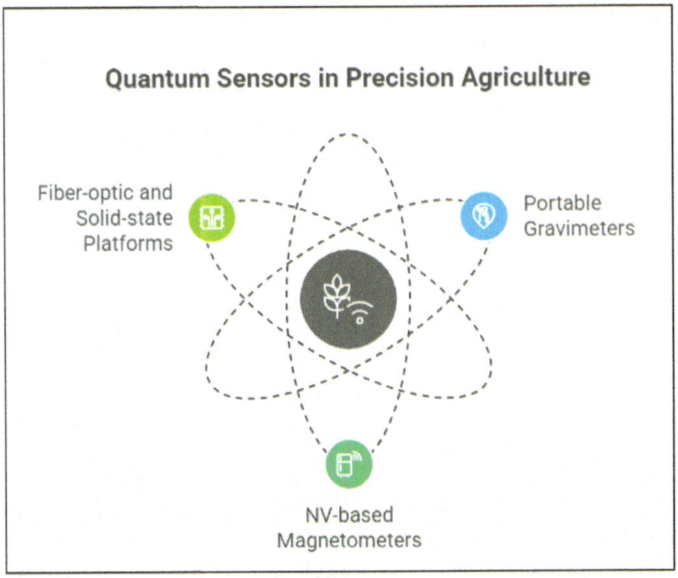

FIGURE 3.7 Types of quantum sensors in precision agriculture.

sequestration in centrally located Indian agroforestry systems using carbon flux data and the gravimetric readings to combine them. Such models are required to analyze the outcomes of regenerative agriculture, carbon trade, and land recovery. Today, such terrain vulnerabilities as erosion, salinization, or groundwater departures can be predicted through the combination of gravimeters and AI-powered mapping solutions. They are also noninvasive, meaning that they are ideal when it comes to routine monitoring in organically managed systems or in fields of conservation agriculture where minimal disturbance of soil is given utmost priority.

3.5.2 NV-based magnetometers for ionic and microbial fluctuations

The diamond NV core is one of the most versatile and stable platforms of quantum sensing. NV center is considered a defect on the diamond lattice and is a nitrogen atom that neighbors a missing carbon. These centers are very temperature sensitive, strain sensitive, and electric and magnetic field sensitive and can be optically read out and optically controlled at ambient temperature. NV-based magnetometers are particularly suitable for detecting soil microbial activity and ionic movements, which are imperative processes in the nutrient cycle and plant health. Ions such as potassium, nitrate, and phosphate, during their transfer through soil water or absorption by roots, generate weak soil-system magnetic signs. Such small magnetic differences are detectable by NV magnetometers, which will allow data on soil fertility gradients, trends of leaching, and rates of nutrient uptake. Ameer et al. (2024) demonstrated that NV sensors established in corn fields could monitor the real-time diffusion of nitrates following fertilizer application. This method was more effective in comparison with the traditional methods (i.e., ion-selective electrodes), slower, and often requires a calibration to the ionic strength or pH of the material. Other areas where NV magnetometers are used in the detection of microbial activity. Several microorganisms involved in nitrogen fixation, degradation, or metal reduction can be detected using magnetic signals due to redox reactions or the formation of magnetic nanoparticles. As an example, the metabolism of magnetotactic bacteria due to intracellular magnetite crystals makes the bacteria align with the magnetic field of the Earth, and the metabolism of the bacteria influences the local magnetic signature. By observing microbial colonization patterns in no-till compared to/traditional till systems using NV-based sensors, Shah et al. (2025) confirmed that microbial hotspots were making their presence felt and more persistent on conservation practices than with traditionally practiced till systems. This impacts the ability to locate disease-suppressive zones or create microbial-inoculant practices. Further, it is now possible to in situ sense the rhizosphere, which is the tiny patch of soil impacted by the roots, due to the introduction of NB centers to the ends of soil probes or within a nanodiamond coating that is biocompatible. This enables the farmers and scientists to be able to monitor the activities of the root–soil–microbe interactions on a scale that has never been observed before. NV-based magnetometer is under development as a diagnostic soil health tool capable of detecting early signs of nutrient deficiencies, salt saturation, or microbe stress due to pesticides in combination with imaging devices and data analytics platforms.

3.5.3 Fiber-optic and solid-state platforms for field applications

The fertile ground under the quantum-enabled measurement in agriculture consists of NV and gravimetric sensors and the translation of their potential concept into a robust, scalable implementation. Two of the more exciting examples of these are solid-state quantum platforms, or quantum sensors using fiber optics, which can achieve scalability in dispersed sensing systems, environmental noise resilience, and form factors. The most common telecommunication fiber-optic examples of a quantum sensor include those based on optical coherence, Bragg gratings, or interferometry, where a photon can be used as the quantum carrier. These sensors can detect thermal, pressure, strain, or electromagnetic variations along the length of a fiber that may be buried in the ground, wound around tree roots, or laid in a field. Xu et al. (2024) developed a fiber-optic quantum soil sensor that could continuously detect soil salinity in the arid coastal soils of China through simultaneously sensing variation in soil temperature and strain caused by the salinity. The sensors had to be inserted in the existing irrigation pipes, be sturdy, and never be affected by electromagnetic interference. Fiber technologies would be especially promising when it comes to the distributed sensing of vast swathes of farmland. The fiber-optic techniques, compared to the point probes, permit continuous measurements up to tens to hundreds of meters, so that it is possible to spatially map the compaction gradient or soil moisture. These discoveries add strength to variable-rate planting, fertilizing, and irrigation methods, which enter precision agricultural platforms. In contrast, the miniature quantum systems utilized as solid-state quantum sensors to detect environmental signals typically include superconducting circuits, quantum dots, or atomic imperfections in silicon carbide. The crop-related sensors are being developed more and more as plug-and-play components of soil analysis kits or as attachments to tractor frames or on drones. Wu et al. (2022) developed practical steps towards solid-state magnetometry, and, with the development of a chip-based magnetic sensor that has been able to detect the ionic changes in saline soils, they have made a step change in the field of solid-state magnetometry. This sensor may be used to diagnose salinity stress of wheat fields at an early stage. Solid-state platforms are also incorporated into AI and IoT ecosystems. Such smart sensors transmit real-time quantum sensor data to cloud applications, which train machine learning models to understand the soil and trigger automated responses, which may adjust watering, apply nutrients in alternative ways, or publish disease risk warnings.

3.6 Case studies and field applications

Due to the recent revolutions in quantum sensing, the manner in which soil and plant health are monitored in real-life farming conditions has radically shifted. Quantum technologies, such as NV centers, gravimeters, and fiber-optic sensors, are widely explored in more and more field tests and pilots to determine their suitability in different agroecological zones. The technologies are providing unparalleled detail about nutrient recycling, water deficit, and soil pollutant risks and are creating real-time, high-resolution information that conventional practices simply cannot compete with. To give an example, NV-based magnetometers have been successfully used in potato fields by the QuantumSoil Project in Germany, leading to a 15% enhanced production and a 12% fall in fertilizer use. Similarly, the QUANTAQUA Lab in India was used to measure changes in subsurface moisture in a field of parched cotton by use of quantum gravimeters and thus schedule irrigation more intelligently (Videgain et al., 2024). The fiber-optic probes proved useful in the early detection of saline saturations in the reclaimed marshland of the UK, highlighting remedial actions before obvious crop damage could be detected. Combined, the experiments demonstrate that quantum sensing technologies have become practical tools under consideration and testing in different soil and climate conditions. Location-based, practical intelligence is already generated through the combination of quantum sensors and digital infrastructure such as drones, LoRaWAN, and artificial intelligence platforms. In Alberta, Canada, the fiber-optic drone sensor was used to map soil compaction regions down to the canopy height level, which will change the implementation of deep tillage practices through the entire vineyard block (Ansari et al., 2025). Also, the California vineyards could irrigate their fields with reduced water volumes as much as 20% due to AI-enabled gravimetric platforms. These examples point to the fact that data collection has been substituted by data-driven decision automation in real-time scenarios in smart agriculture. The most compelling emerging frontier is probably to leverage quantum sensors on a microscale to monitor the dynamics of soilpre-veraison stressplant–environment interactions. To give a scientifically sound realization of sustainable nutrient management, the NV sensors in the Netherlands aided in assessing the mobility of nitrates under different covers of crop residues. NV sensors were capable of detecting water shortage in grapevines in Italy several days before evident signs occurred. In the interim, the endeavors in monitoring pollution in Brazil and Vietnam have indicated how fiber-optic and NV nanodiamond sensors are enabling the early detection of heavy metals and degradation of pesticides, respectively (Bandara et al., 2025).

The case studies highlight in Table 3.2 demonstrate how diverse quantum sensing platforms are being field-tested to address nutrient dynamics, moisture variability, salinity, contamination, and stress detection in real-time agricultural settings.

3.7 Integration with digital agriculture platforms

The growing sophistication of quantum sensors has encouraged a revolution in agricultural surveillance, especially where these sensors form part of wider digital agriculture systems. When combined with AI algorithms, IoT devices, remote sensing systems, and cloud-based analytics, quantum sensing technology will prove to be one of the elements of real-time, intelligent, and adaptive farming systems. Now farmers can not only notice the changes in the soil and plant's health but can also expect them due to this convergence, which will lead to more intelligent treatments, reduce input costs, and ultimately make them more sustainable.

3.7.1 Synergies with AI, IoT, and machine learning models

The natural fusion of the quantum sensors, the IoT, and the AI makes farms dynamic feedback systems. An example that stands out is the SmartRoot Q System released in South Korea in 2023. The technology integrates AI-based algorithms that use historical data on soil chemistry and predict when the ground will have nutrient deficits, many days before they can be detected in the field with NV-based ionic sensors located near root zones. The forecasts are thus transmitted to automated fertigation devices, which means the preemptive correction becomes possible. The gravimetric sensors continuously record the minute changes in the bulk density of the soil in line with water content and root respiration. These data are processed through machine learning models developed to determine water stress in sugarcane fields. Over two years of experimental growing seasons, the platform helped to cut down the amount of irrigation by 18% without any effect on the production (Gan et al., 2023). The AgriSense Grid in Kenya, portable NV sensors are deployed around different field topographies, connected to IoT towers powered by solar energy. These towers transmit the data read by the sensors through sensors to a cloud hub in the region, which is then processed through an ML engine using inputs like soil temperature, salinity, and microbial flux. The algorithm also suggests microbial inoculants dosages and techniques of crop rotation based on these real-time data. They are even bettering the treatment of the disease. McDonald and collaborators in Canada have begun a joint project, PathoTrace Quantum, to forecast outbreaks of *Fusarium* wilt in tomato greenhouses using deep learning with quantum soil temperature and magnetic fluctuation sensors. The algorithm brings about an early warning system through which application of fungicides is significantly reduced because it can learn pathogen dynamics each time it sees small changes in the soil magnetic properties (Dhavale et al., 2025).

3.7.2 Remote sensing and satellite-based data integration

The quantum sensors, in conjunction with remote sensing platforms such as satellites and unmanned aerial vehicles, open the possibility of multiscale monitoring. The use of quantum

Table 3.2 Case studies demonstrating quantum sensor applications in soil monitoring and precision agriculture.

Deployments	Application area	Quantum sensor type	Implementation	Key outcomes	References
Pilot trials in precision farming zones	Nutrient monitoring in root zones	NV-based magnetometers	QuantumSoil Project, Germany, in potato fields using NV sensors to track potassium flux in the rhizosphere.	Enabled 15% yield increase and 12% fertilizer reduction via real-time nutrient tracking	Toselli et al. (2023)
	Subsurface moisture analysis in arid zones	Portable quantum gravimeters	Gravimeters deployed in cotton fields to track root-zone water content	Improved irrigation scheduling and reduced water stress in the dry season	Çetin (2025)
	Salinity detection in coastal soils	Fiber-optic quantum sensors	Sensors installed in reclaimed marshlands for salt stress monitoring	Detected salinity hotspots 2 weeks before symptoms, allowing early gypsum application	Goud et al. (2022)
Data outputs from integrated systems	Multi-layered sensing of moisture and nutrients	NV magnetometers + gravimeters + remote sensing	Integrated LoRaWAN network combined with NDVI and surface temperature sensors	Enabled automated fertigation and irrigation advice based on real-time soil condition maps	Mowla et al. (2023)
	Nitrate leaching detection during rainfall	NV-based sensors + ML dashboard	In maize fields where nitrate surges were tracked post-thunderstorm	Prevented nutrient loss with rapid fertigation adjustments based on a 15-min detection window	Simelane et al. (2024)
	Soil compaction detection from the canopy	Drone-mounted fiber-optic quantum sensors	Sensors mapped subsurface compaction via drone overflights	Informed variable-depth ripping strategy to improve aeration & reduce root restriction	da Silveira Petruci et al. (2025)

(continued on next page)

Insights into nutrient dynamics, water stress, and contamination zones	Nitrate mobility under cover crops	NV magnetometers	Comparison of nitrate retention under vetch vs. rye residues	Demonstrated better NUE under vetch, influencing cover crop selection	Vavlas et al. (2024)
	Early detection of water stress in vines	Gravimeters + temperature-sensitive NV sensors	NV and sap flow data used to predict pre-veraison stress in grapes	Allowed preemptive deficit irrigation, optimizing grape quality and water use.	López-Urrea et al. (2024)
	Arsenic and heavy metal monitoring	Fiber-optic quantum sensors	In situ monitoring of arsenic leaching during irrigation	Issued early SMS alerts to farmers; prompted flushing to reduce uptake	Mutunga et al. (2024)
	Pesticide breakdown and residue monitoring	Functionalized NV nanodiamond sensors	Tracked breakdown of organophosphates in soybeans under the no-till system	Enabled real-time pesticide decay tracking, aiding residue compliance	Banerjee et al. (2024)

sensors allows precision agriculture to operate at a variety of scales due to their mm-level accuracy, and remote sensing allows landscape-level assessments of the farm. Meanwhile, transpiration rate at the canopy level was also obtained with the aid of satellite-based normalized difference water index (NDWI). The two data streams enabled targeted deep tillage and nitrogen amendment because they revealed some blocks of soil where root access was impaired, yet there were good canopy indicators. Quantum sensors and other remote sensing deployed systems, such as satellites and unmanned aerial vehicles, collaborate to offer multiscale surveillance. Precision agriculture will be able to operate on many scales due to the millimeter-accuracy of quantum sensors as well as the landscape-scale assessment of remote sensing (Khalid et al., 2025). The combination of the two data streams allowed targeted deep tillage and nitrogen amendment because specific soil blocks were detected in which root uptake was still impaired in spite of canopy signals being good. Likewise, Argentine farmers participated in the Sat-Soil Sync Project; these farmers monitored porosity in the soil through Gravimetric quantum sensors during the growing season of maize. Meanwhile, the growth of vegetation and the reflectance of the surfaces were being monitored with the help of high-altitude satellite imagery from the Sentinel 2. The combined platform allowed the possibility of real-time diagnosis of subsurface compaction zones that would otherwise not have been visible using the satellite-only systems. In Austria, the AirGround Quantum Integration Trial involved drones equipped with remote-borne, lightweight quantum magnetometers and flew over wheat fields to detect local anomalies in the roots' electrical activity. Fiber-optic sensors on the ground recorded capillary water tension at the same time. The group developed advanced zone maps that guided variable rate irrigation and seed densities alterations by intersecting aerial and in-soil measurements (Manu et al., 2024).

3.7.3 Cloud-based analytics and real-time decision support

Cloud computing is the main nervous system of the quantum-powered agriculture. It enables collection, data storage, processing, and remote access to the explosion of data that quantum sensors produce through easy-to-use dashboards. Through these platforms, the responsiveness and the ability to scale decisions in real time are made possible. One example is the AgroQ-Net Platform in New Zealand, which is considered well known. The interface provides access to a combination of NV magnetometer recordings, gravimetric soil water measurements, as well as microbial flux inputs to farmers, scientists, and agribusiness in order to be in a position to harness water. The dashboard proffers hourly alerts on tendencies of organic matter breakdown, water table drop, and hazard of nitrate leaching. Using only a few mouse clicks, users will be able to add these signals to automated fertilization or irrigation control systems and make field management much simpler. SoilStream Quantum Dashboard enables dryland farmers in Morocco to view moisture retention curves by gravimetric sensors at different soil depths. Combined with local weather predictions and AI-based predictions of the evapotranspiration process, the technique offers early warning indications of stress and proposed adjustive irrigation levels, which are particularly crucial in places with a scarcity of water (Abdulameer et al., 2025).

Another innovative platform that is produced in Germany is called QuantumFieldEdge. It gives current intervention signals as well as trend analysis. As an example, it trains long-term nutrient mobility models with data from the NV sensors that are measured over multiple seasons. In sustainable nutrient budgeting, farmers are able to assess predictions on the accumulation of potassium or loss of phosphorus depending on the crop and the range of rainfall. In the Philippines, a rural cooperative has rolled out the Q-Cloud Smallholder Toolkit that enables community farms to transmit daily gravimetric data back to a central server using cheap smartphones. The ability of the cloud analytics platform to transmit data to understandable and easy-to-interpret visual feedback allows even low-tech users to make data-driven decisions: green means healthy, yellow is a warning, and red calls to take action (Bulacio Fischer et al., 2025).

3.8 Challenges and limitations

Many barriers prevent the practical adoption of quantum sensing technologies in agriculture despite their apparent promise. These challenges include environmental resilience, financial feasibility, and technical constraints. The limitations of the quantum tools have to be taken into consideration as the agricultural systems move to the high-tech precision platforms to ensure sustainability, access, and reliability.

3.8.1 Technical barriers: Calibration, data noise, signal interference

Technically, the calibration can be considered as one of the most persistent shortcomings of the field use of quantum sensors. Many quantum measuring applications, such as NV magnetometers and gravimeters, require ultra-precise alignment and initialization techniques. Due to the changing conditions of the field, e.g., due to changes in temperature, heterogeneity of the soil composition, and electromagnetic variations, it might be necessary to recalibrate quantum devices more often than classic sensors, which are usually factory-calibrated once. Otherwise, uncontrolled, this brings about reading variability and poses a long-term stability. In addition, data noise is a significant issue, particularly in uncontrolled open-field research. Because quantum sensors have their natural sensitivity to background electromagnetic fields, temperature fluctuations, and even mechanical vibrations, they are designed to measure extremely subtle events, including ionic flux, spin resonance, or alteration in atomic energy levels, e.g., as an example, it might include sound pollution, such as noise of nearby power lines, tractor work, or mobile communication signal picking up by NV sensors placed underground. This often demands implementation of complicated signal filtering techniques that may distort the raw data inadvertently or result in slackness in the real-time response.

Other challenges to in situ applications are natural and artificial interference with the signal. Quantum sensors can receive scrambled or garbled signals when they are coupled next to metallic irrigation systems or GPS-controlled gear, more likely to occur when the soil contains magnetic minerals or higher conductivity. Weather-related interference with the gravimetric sensors may provide false readings of the subsurface density or water content, especially when rainfall or exposure to wind pressure or an increase in humidity occurs. In addition to that, communication with digital platforms also leads to an even subtler level of technical dependence. It is possible that unreliable signal transmission infrastructures, i.e., LoRaWAN, Bluetooth, or 5G, lead to lost, delayed, or distorted data from the sensor. Massive decisions made following sensor feedback or continual automation are hard to implement since the reliability is an issue in that case (Bass & Doser, 2024).

3.8.2 Economic factors: Device cost, skill requirements, and market adoption

Quantum sensor cost remains as one of the largest economic factors inhibiting their widespread applications. Unlike more conventional soil moisture probes or electrochemical nutrition meters, quantum devices often demand expensive raw materials, complex fabrication processes, and sensitive calibration equipment (as in the case of NV centers in ultrapure diamonds and superconducting elements in gravimeters). It is due to this fact that smallholder or mid-scale farmers are as yet unable to afford most of the commercially developed quantum sensors. Even high-input research organizations and farms demand the demonstration of the return on investment (ROI) in quantum sensing. The initial cost might not always be compensated by the long-term or indirect payoffs, high fertilizer efficiency rates, or early diagnosis of stress in the situation of competitive agricultural markets. Without either a government or a fully corporate sponsor, such technologies remain relegated to pilot projects and upscale agri-tech projects. Another stumbling block is the need for skills. To utilize quantum sensing tools, a person should not only possess complete knowledge of digital system applications, signal processing, and quantum mechanics, which are not provided in regular farming education programs. Agronomists or farmers not schooled in quantum theory may find it hard to troubleshoot a calibration issue, interpret an anomalous signal, or integrate data into valuable field decisions. A knowledge bottleneck that occurs due to sparsity and the cost of training programs hinders deployment. Moreover, the maintenance and support ecosystem of the market remains in very early stages, as found in the case that something in a quantum technology goes wrong or performs poorly, service and repair capabilities, as well as spares, are often located in remote cities or international centres. This discourages potential users from adopting the technology due to its expensive upkeep as well as its time-consuming nature. The market may see the extensive utilization of quantum agriculture instruments only when the networks of locally available services and user-friendly models are prevalent (Wu et al., 2025).

3.8.3 Environmental and operational durability of quantum devices

Quantum sensors have a very hard time being durable in a real farm scenario. Agrarian environments are dynamic and often harsh, involving direct contact with muck, dust, and periodic wetness, as well as pests, UV light, and mechanical impacts from machinery. Most of the quantum devices, particularly the primitive ones, were designed to operate in controlled laboratories and not extremely lousy external environments. In tropical or monsoonal areas, there is greater stress due to humidity and waterlogging of the soil. Fiber-optic quantum sensors can be impacted by moisture intrusion, condensation, and the formation of microbial biofilm over sensor tips, weakening an otherwise very strong signal so that regular cleaning or recalibration is required. Likewise, magnetic interferences of such soil as ferruginous or saline soil may affect the magnetometer's stability. There are two other major constraints, such as autonomy and power consumption. Quantum sensors regularly require high-energy light sources, e.g., lasers or microwave fields, to manipulate quantum states and generate readable signals. Such energy requirements become prohibitive in remote farms with a scant supply of solar infrastructure or the grid. Although the low-power NV sensor technology is developing, even the current models cannot find a way to provide a collapse in terms of performance and endurance. Finally, software-sensor compatibility is not guaranteed in all cases. Cloud-based platforms and decision support systems often use pre-coded protocols built to work with common electrochemical sensors. The data generated by quantum sensors should be injected and analyzed through special APIs and converters, and this is an impediment to integration and a continued dependence on specialized IT solutions (Haensch et al., 2023).

3.9 Research directions

Quantum sensing for soil health monitoring is a nascent field, but it has an outstanding potential to transform and improve the way soil processes are measured, understood, and controlled under precision agriculture (Naorem et al., 2024). The future research should focus on creating rugged field-deployable quantum sensor devices that would be able to withstand actual farm environments such as changing temperature, salinity, moisture, biofouling, mechanical vibration, and electromagnetic noise without the deterioration of sensitivity and readout stability. It is also crucial that the standardization of calibration procedures and location-specific benchmarking systems should be enhanced in order to enhance cross-location comparability, sensor drift, and long-term monitoring in different agroecological regions (Thingujam et al., 2025). Another significant area of research is to scale quantum sensing applications all the way to the rhizosphere level, with NV-center platforms and spin-based sensing systems capable of providing micro-scale imaging of nutrient fluxes, redox gradients, and microbial hotspots controlling plant resilience and soil fertility (Zalieckas et al., 2024). Moreover, by connecting quantum sensor deliverables to the AI-controlled analytics, digital twins, IoT links, and cloud-based decision-support systems, the high-resolution quantum data can be used to irrigate, fertilize, and restore the health of the soil. Both hybrid sensor networks (quantum devices as an ultraprecise reference tool and classical low-cost sensors as an in situ continuous field-scale coverage) and fusion algorithms (quantum data of in situ devices with satellite and drone-based remote sensing layers) will also be practical (Cengiz & Sama et al., 2025). Lastly, studies should involve reduction of costs, scalable manufacturing, low power operation, and a model of deployment friendly to farmers (with service-based models) so that such technologies can be implemented beyond pilot projects and thereby enhancing climate-intelligent soil management, nutrient use efficiency, soil carbon sensors, and a more robust positive food security profile.

3.10 Conclusion

The quantum sensors offer sensitivity, precision, and real-time feedback that the traditional instruments cannot offer when compared, leading to a revolutionary change in monitoring the health of soil. These tools have the potential to localize ionic fluxes, moisture gradients, and microbiological activity on the order of nanometers through phenomena such as entanglement, coherence, or spin resonance. With its combination of AI, IoT, cloud analytics, and remote sensing, it allows predictive and adaptive decision-making across agroecosystems. In vineyards, cotton fields, and reclaimed lands, case studies have demonstrated conceivable benefits such as lower use of fertilizers, optimized irrigation, and early indication of contamination or saltiness. However, what gets in the way of the widespread acceptance is there. One has to overcome technical challenges like electromagnetic interference, noise on the signal, and sensor calibration. The economic barriers to accessibility, especially to the smallholders, are the costs of their devices, expertise, and inadequate market infrastructure. Given that most quantum machines are yet to come to terms with the logistics of field operation, the problem remains environmental durability. Despite these limitations, the integration of quantum sensing into digital agriculture opens up a new age of more specific soil control. As technology progresses and corresponding enabling regulation and training programs are developed, quantum sensors can become central to long-term, data-driven farming that meets the demands of environmental preservation and agricultural output. The opportunities of quantum soil sensing in the future are miniaturization, a decrease in costs, and ruggedization of sensors to be used in the field over the long term. It is believed that advancements in edge computing, AI-enhanced signal denoising, and quantum-classical hybrid systems would result in more robust and interpretable quantum data in practice. Equipped with quantum gravimeters on drones and satellites, it might also be possible to track the soil moisture and subsurface processes in the regions and globally. A combination of supportive regulatory frameworks, standardized calibration procedures, and focused capacity building efforts can support quantum sensing technologies to become vital components of long-term, data-driven, climate-resilient agricultural systems that balance environmental stewardship and productivity.

References

Aarif KO, M., Alam, A., & Hotak, Y (2025). Smart sensor technologies shaping the future of precision agriculture: Recent advances and future outlooks. *Journal of Sensors, 2025*(1), 2460098. https://doi.org/10.1155/js/2460098.

Abdulameer, L., Al-Khafaji, M. S., Al-Awadi, A. T., Al Maimuri, N. M., Al-Shammari, M., & Al-Dujaili, A. N. (2025). Artificial intelligence in climate-resilient water management: A systematic review of applications, challenges, and future directions. *Water Conservation Science and Engineering, 10*(1), 44. doi:10.1007/s41101-025-00371-2.

Adeniji, T. M., & Stine, K. J. (2023). Nanostructure modified electrodes for electrochemical detection of contaminants of emerging concern. *Coatings, 13*(2), 381. https://doi.org/10.3390/coatings13020381.

Ameer, S., Ibrahim, H., Kulsoom, F. N. U., Ameer, G., & Sher, M. (2024). Real-time detection and measurements of nitrogen, phosphorous & potassium from soil samples: A comprehensive review. *Journal of Soils and Sediments, 24*(7), 2565–2583. doi:10.1007/s11368-024-03827-5.

Ansari, R., Manna, A., Hazra, S., Bose, S., Chatterjee, A., & Sen, P. (2025). Breeding 4.0 vis-à-vis application of artificial intelligence (AI) in crop improvement: An overview. *New Zealand Journal of Crop and Horticultural Science, 53*(4), 812–854. https://doi.org/10.1080/01140671.2024.2409775.

Bandara, R. M. P. N. S., Jayasignhe, A. B., & Retscher, G. (2025). The integration of IoT (Internet of Things) sensors and location-based services for water quality monitoring: A systematic literature review. *Sensors, 25*(6), 1918. https://doi.org/10.3390/s25061918.

Banerjee, A., Jhariya, M. K., Raj, A., & Mechergui, T. (2024). *Ecosystem management: Climate change and sustainability*. John Wiley & Sons.

Bass, S. D., & Doser, M. (2024). Quantum sensing for particle physics. *Nature Reviews Physics, 6*(5), 329–339. doi:10.1038/s42254-024-00714-3.

Bulacio Fischer, P. T., Carella, A., Massenti, R., Fadhilah, R., & Lo Bianco, R. (2025). Advances in monitoring crop and soil nutrient status: Proximal and remote sensing techniques. *Horticulturae, 11*(2), 182. https://doi.org/10.3390/horticulturae11020182.

Caldwell, E. D., Triano, T. M., & Sinclair, L. C. (2025). High-precision optical time and frequency transfer. *Advances in Optics and Photonics, 17*(2), 375–440. https://doi.org/10.1364/AOP.545290.

Cengiz, A., & Sama, A. (2025). Artificial intelligence techniques for quantum-enhanced nanosensor development in precision agriculture and real-time crop monitoring. *Journal of Quantum Nano- Green Environmental Systems*. https://doi.org/10.70023/qnges.251101.

Çetin, Ö. (2025). Irrigation management and innovative approaches in cotton under climate change. In *Agriculture and water management under climate change* (pp. 65–92). Cham: Springer Nature Switzerland.

da Silveira Petruci, J. F., da Silva Sousa, D., & Mizaikoff, B. (2025). Advanced mid-infrared sensors for molecular analysis. *Analytical Chemistry, 97*(13), 6871–6890. doi:10.1021/acs.analchem.4c06799.

Dhavale, M. C., Aiwale, H. N., & Sarnobat, D. H. (2025). Modern diagnostic techniques in plant pathology. *Plant Health and Disease Management: A Modern Approach*, 184.

De Silva, Abhiram, G., Perera, T.A.N.T., Herath, H.M.S.K., Rathnayake, R.M.K.T., Bandara, P.K.G.S.S. (2023). Artificial intelligence (AI) assisted precision agriculture: Applications in Asian context. In de Vries, W.T., Rudiarto, I., & Milinda Piyasena, N. M.P. (Eds.), Geospatial science for smart land management (pp. 353–365). CRC Press.

Du, J., Shi, F., Kong, X., Jelezko, F., & Wrachtrup, J. (2024). Single-molecule scale magnetic resonance spectroscopy using quantum diamond sensors. *Reviews of Modern Physics, 96*(2), 025001. https://doi.org/10.1103/RevModPhys.96.025001.

Fratelli, A., Zaffalon, M. L., Mazzola, E., Dirin, D. N., Cherniukh, I., Otero-Martínez, C., & Brovelli, S. (2025). Size-dependent multiexciton dynamics governs scintillation from perovskite quantum dots. *Advanced Materials, 37*(5), 2413182. https://doi.org/10.1002/adma.202413182.

Gan, H., Roper, W. R., Groffman, P. M., Morris, T. F., & Guillard, K. (2023). Automated sensor-based quantification of soil water retention and microbial respiration across drying conditions. *Soil Biology and Biochemistry, 180*, 108987. https://doi.org/10.1016/j.soilbio.2023.108987.

Goud, B. R., Raghavendra, M., Prasad, P. S., Hatti, V., Halli, H. M., Nayaka, G. V., & Rajpoot, S. K. (2022). Sustainable management and restoration of the fertility of damaged soils. *Agriculture Issues and Policies, 113*.

Haensch, W., Raghunathan, A., Roy, K., Chakrabarti, B., Phatak, C. M., Wang, C., & Guha, S. (2023). Compute in-memory with non-volatile elements for neural networks: A review from a co-design perspective. *Advanced Materials, 35*(37), 2204944. https://doi.org/10.1002/adma.202204944.

Hassoun, A., Aït-Kaddour, A., Abu-Mahfouz, A. M., Rathod, N. B., Bader, F., Barba, F. J., & Regenstein, J. (2023). The fourth industrial revolution in the food industry—Part I: Industry 4.0 technologies. *Critical reviews in food science and nutrition, 63*(23), 6547–6563. https://doi.org/10.1080/10408398.2022.2034735.

Horel, Á., & Zsigmond, T. (2023). Plant growth and soil water content changes under different inter-row soil management methods in a sloping vineyard. *Plants, 12*(7), 1549. https://doi.org/10.3390/plants12071549.

Ibrahim, A.A., Abd-Ellatif, S., Abdel Razik, Hamedo, H.A., & Salem, K.F. (2024). Nanosensors for enhancing plant growth and productivity. In Al-Khayri, J.M., Alnaddaf, L.M., Jain, S.M., & Penna, S. (Eds.), Innovative methods in horticultural crop improvement: Biosensors and nanosensors (pp. 25-50). Switzerland: Springer Nature.

Kantsepolsky, B., & Aviv, I. (2024). Quantum sensing for the cities of the future. *The International Archives of the Photogrammetry, Remote Sensing and Spatial Information Sciences, 48*, 93–100. https://doi.org/10.5194/isprs-archives-XLVIII-4-W10-2024-93-2024.

Kantsepolsky, B., Aviv, I., Weitzfeld, R., & Bordo, E. (2023). Exploring quantum sensing potential for systems applications. *IEEE Access, 11*, 31569–31582. doi:10.1109/ACCESS.2023.3262506.

Khalid, U., Paracha, U. I., Naveed, Z., Duong, T. Q., Win, M. Z., & Shin, H. (2025). Quantum fusion intelligence for integrated satellite-ground remote sensing. *IEEE Wireless Communications, 32*(3), 46–55. https://doi.org/10.1109/MWC.001.2400425.

Khondakar, K. R., & Kaushik, A. K. (2024). *Next-generation smart biosensing: Nano-platforms, Nano-microfluidics interfaces, and emerging applications of quantum sensing*. Academic Press.

Kumara, K., Pal, S., Chand, P., & Kandpal, A. (2023). Carbon sequestration potential of agroforestry systems in Indian agricultural landscape: A meta-analysis. *Ecosystem Services, 62*, 101537. https://doi.org/10.1016/j.ecoser.2023.101537.

Liu, G. Q., Liu, R. B., & Li, Q. (2023). Nanothermometry with enhanced sensitivity and enlarged working range using diamond sensors. *Accounts of Chemical Research, 56*(2), 95–105. doi:10.1021/acs.accounts.2c00576.

López-Urrea, R., Oliveira, C. M., Montoya, F., Paredes, P., & Pereira, L. S. (2024). Single and basal crop coefficients for temperate climate fruit trees, vines and shrubs with consideration of fraction of ground cover, height, and training system. *Irrigation Science, 42*(6), 1099–1135. doi:10.1007/s00271-024-00964-0.

Mallik, S., Chakraborty, A., Podder, K., Talukdar, S., Rahman, A., & Mishra, U. (2025). Enhancing soil moisture prediction with explainable AI: Integrating IoT and multi-sensor remote sensing data through soft computing. *Applied Soft Computing, 180*, 113406. https://doi.org/10.1016/j.asoc.2025.113406.

Manu, A., McDanel, J., Brummel, D., Avornyo, V. K., & Lawler, T. (2024). Using unmanned aerial systems technology to characterize the dynamics of small-scale maize production systems for precision agriculture. *Drones, 8*(11), 633. https://doi.org/10.3390/drones8110633.

Mowla, M. N., Mowla, N., Shah, A. S., Rabie, K. M., & Shongwe, T. (2023). Internet of things and wireless sensor networks for smart agriculture applications: A survey. *IEEe Access, 11*, 145813–145852. https://doi.org/10.1109/ACCESS.2023.3346299.

Muccini, F., Greco, F., Cocchi, L., Marsella, M., Zanutta, A., Borghi, A., & Rivalta, E. (2025). Airborne strapdown gravity survey of Sos enattos area (NE Sardinia, Italy): Insights into geological and geophysical characterization of the italian candidate site for the einstein telescope. *Remote Sensing, 17*(13), 2309. https://doi.org/10.3390/rs17132309.

Mutunga, T., Sinanovic, S., & Harrison, C. S. (2024). Integrating wireless remote sensing and sensors for monitoring pesticide pollution in surface and groundwater. *Sensors, 24*(10), 3191. https://doi.org/10.3390/s24103191.

Naorem, A., Patel, A., Adak, S., Singh, P., & Udayana, S. K. (2024). Precision technologies and digital solutions: Catalyzing agricultural transformation in soil health management. *Digital Agricultural Ecosystem: Revolutionary Advancements in Agriculture*, 175–190. https://doi.org/10.1002/9781394242962.ch10.

Oh, E., Gregoire, M. D., Black, A. T., Hughes, Jeramy, K. , Kunz, P. , D., Larsen, M., & Sackett, C. A (2024). Perspective on quantum sensors from basic research to commercial applications. *AIAA Journal, 62*(11), 4029–4053. https://doi.org/10.2514/1.J062707.

Panyaram, S. (2025). Intelligent manufacturing with quantum sensors and AI a path to smart industry 5.0. *International Journal of Emerging Trends in Computer Science and Information Technology*, 140–147. https://doi.org/10.56472/ICCSAIML25-117.

Pieplow, G., Torun, C. G., Gurr, C., Munns, J. H., Herrmann, F. M., Thies, A., & Schröder, T. (2025). Quantum electrometer for time-resolved material science at the atomic lattice scale. *Nature Communications, 16*(1), 6435. doi:10.1038/s41467-025-61839-2.

Prasanna, R., Saravanan, P., Jenath, M., Saravanan, G., & Rajesh Kanna, R. (2024). IoT based soil moisture monitoring and nutrients suggestion based on the soil PH value. In *2024 International Conference on Power, Energy, Control and Transmission Systems (ICPECTS)* (pp. 1–6). IEEE.

Rodrigues, C. I. D., Brito, L. M., & Nunes, L. J. (2023). Soil carbon sequestration in the context of climate change mitigation: A review. *Soil Systems, 7*(3), 64. https://doi.org/10.3390/soilsystems7030064.

Schofield, H., Boto, E., Shah, V., Hill, R. M., Osborne, J., Rea, M., & Brookes, M. J. (2022). Quantum enabled functional neuroimaging: The why and how of magnetoencephalography using optically pumped magnetometers. *Contemporary Physics, 63*(3), 161–179. https://doi.org/10.1080/00107514.2023.2182950.

Shah, S. S., van Dam, J., Singh, A., Kumar, S., Kumar, S., Bundela, D. S., & Ritsema, C. (2025). Impact of irrigation, fertilizer, and pesticide management practices on groundwater and soil health in the rice-wheat cropping system—a comparison of conventional, resource conservation technologies and conservation agriculture. *Environmental Science and Pollution Research, 32*(2), 533–558. doi:10.1007/s11356-024-35661-0.

Shettell, N., Lee, K. S., Oon, F. E., Maksimova, E., Hufnagel, C., Wei, S., & Dumke, R. (2024). Geophysical survey based on hybrid gravimetry using relative measurements and an atomic gravimeter as an absolute reference. *Scientific Reports, 14*(1), 6511. doi:10.1038/s41598-024-57253-1.

Shi, T., Wei, Q., Qin, X., Liu, Z., Chen, K., Cao, S., & Chen, J. (2024). Dual-frequency optical-microwave atomic clocks based on cesium atoms. *Photonics Research, 12*(9), 1972–1980. https://doi.org/10.1364/PRJ.528942.

Simelane, M. P. Z., Soundy, P., & Maboko, M. M. (2024). Effects of rainfall intensity and slope on infiltration rate, soil losses, runoff and nitrogen leaching from different nitrogen sources with a rainfall simulator. *Sustainability, 16*(11), 4477. https://doi.org/10.3390/su16114477.

Su, J., Kong, Z., Kong, F., Liu, X., Zeng, L., Wang, Z., & Shi, F. (2025). Double-Layered Silica-Engineered Fluorescent Nanodiamonds for Catalytic Generation and Quantum Sensing of Active Radicals. *arXiv preprint arXiv:2512.21934*. https://doi.org/10.48550/arXiv.2512.21934.

Thingujam, U., Prabha, D., Bag, A. G., Thingujam, V., Darshan, N. P., Dutta, S., & Gorain, S. (2025). From point sensing to intelligent systems: A comprehensive review on advanced sensor technologies for soil health monitoring. *Discover Sensors, 1*(1). doi:10.1007/s44397-025-00028-8.

Tomaz, A. A., Mattos, R. S., & Barbatti, M. (2025). The quantum measurement problem: A review of recent trends. *arXiv preprint arXiv:2502.19278*. https://doi.org/10.1080/14786435.2025.2601922.

Toselli, M., Baldi, E., Ferro, F., Rossi, S., & Cillis, D. (2023). Smart farming tool for monitoring nutrients in soil and plants for precise fertilization. *Horticulturae, 9*(9), 1011. https://doi.org/10.3390/horticulturae9091011.

Vavlas, N. C., Seubring, T., Elhakeem, A., Kooistra, L., & De Deyn, G. B. (2024). Remote sensing of cover crop legacies on main crop N-uptake dynamics. *European Journal of Soil Science, 75*(5), e13582. https://doi.org/10.1111/ejss.13582.

Videgain, M., Martínez-Casasnovas, J. A., Vigo-Morancho, A., Vidal, M., & García-Ramos, F. J. (2024). On-farm experimentation of precision agriculture for differential seed and fertilizer management in semi-arid rainfed zones. *Precision Agriculture, 25*(6), 3048–3069. doi:10.1007/s11119-024-10189-y.

Wang, W., Straffelini, E., & Tarolli, P. (2023). Steep-slope viticulture: The effectiveness of micro-water storage in improving the resilience to weather extremes. *Agricultural Water Management, 286*, 108398. https://doi.org/10.1016/j.agwat.2023.108398.

Wang, Z., Guo, Z., Chen, Z., Li, M., Zhou, Z., Zhang, C., & Ma, Z. (2025). Quantum coherence: A fundamental resource for establishing genuine multipartite correlations. *Physical Review A, 111*(3), 032410. https://doi.org/10.1103/PhysRevA.111.032410.

Wu, K., Tonini, D., Liang, S., Saha, R., Chugh, V. K., & Wang, J. P. (2022). Giant magnetoresistance biosensors in biomedical applications. *ACS applied materials & interfaces, 14*(8), 9945–9969. doi:10.1021/acsami.1c20141.

Wu, Y., Nagy, A., Rajnai, Z., Fregan, B., & Takács-György, K. (2025). Quantum machine learning in crop disease monitoring: Opportunities and challenges to practical implementation. In *2025 IEEE 12th international conference on computational cybernetics and cyber-medical systems (ICCC)* (pp. 000059–000064). IEEE.

Xu, J. J., Tang, C. S., Cheng, Q., Vahedifard, F., Zhang, D. Z., Sun, C., & Shi, B. (2024). Soil water potential measurement using fiber-optic sensing technique. *Journal of Hydrology, 636*, 131324. https://doi.org/10.1016/j.jhydrol.2024.131324.

Yoo, H., & Maeng, S. (2025). Investigation of electron tunneling in ultra-precision machine tool for measuring the distance of tool and workpiece. *The International Journal of Advanced Manufacturing Technology, 136*(7), 3239–3248. doi:10.1007/s00170-025-15024-w.

Yuan, Q., Sheng, W., Zhang, Z., Li, H., & Zhang, M. (2023). Recent advances in soil nutrient monitoring: A review. *Sensing Technologies for Field and In-House Crop Production: Technology Review and Case Studies, 7*, 19–38. doi:10.1007/978-981-99-7927-1_2.

Yukawa, H., Kono, H., Ishiwata, H., Igarashi, R., Takakusagi, Y., Arai, S., & Baba, Y. (2025). Quantum life science: Biological nano quantum sensors, quantum technology-based hyperpolarized MRI/NMR, quantum biology, and quantum biotechnology. *Chemical Society Reviews*. https://doi.org/10.1039/D4CS00650J.

Zalieckas, J., Greve, M. M., Bellucci, L., Sacco, G., Håkonsen, V., Tozzini, V., & Nifosì, R. (2024). Quantum sensing of microRNAs with nitrogen-vacancy centers in diamond. *Communications Chemistry, 7*(1), 101. https://doi.org/10.1038/s42004-024-01182-7.

Zhang, X., Bilyera, N., Fan, L., Duddek, P., Ahmed, M. A., Carminati, A., & Razavi, B. S. (2023). The spatial distribution of rhizosphere microbial activities under drought: Water availability is more important than root-hair-controlled exudation. *New Phytologist, 237*(3), 780–792. https://doi.org/10.1111/nph.18409.

Zhou, Z., Zhang, H., Huang, R., He, F., Xu, D., Yan, X., & Li, L. (2025). Advancements in research and current applications of "quantum sensor: Superconducting quantum interference device. *The European Physical Journal Special Topics, 234*(30), 9985–10013. https://doi.org/10.1140/epjs/s11734-025-01492-4.

Zulkharnay, R., & May, P. W. (2024). Applications of diamond films: A review. *Functional Diamond, 4*(1), 2410160. https://doi.org/10.1080/26941112.2024.2410160.

4

Quantum cryptography for agri-data security quantum sensors for soil health monitoring

N Sumalatha[a], Elavarasi Gunasekaran[b] and Shahul Hameed[c]

[a]ICT, ALAGAPPA CHETTIAR GOVERNMENT COLLEGE OF ENGINEERING AND TECHNOLOGY, KARAIKUDI, TAMIL NADU, INDIA [b]DEPARTMENT OF CIVIL ENGINEERING, ALAGAPPA CHETTIAR GOVERNMENT COLLEGE OF ENGINEERING AND TECHNOLOGY, KARAIKUDI, TAMIL NADU, INDIA [c]DEPARTMENT OF CIVIL ENGINEERING, TERRA FIRMA GROUND ENGINEERING, KINGDOM OF SAUDI ARABIA

4.1 Introduction of soil health monitoring

As the basis for agricultural production, ecological balance, and environmental sustainability, soil is one of the most essential natural resources supporting life on Earth. Food production, water management, carbon sequestration, and biodiversity conservation all depend on healthy soils. Crop productivity, water retention, nutrient cycling, and resistance to erosion and degradation are all directly impacted by soil health in agricultural systems. Monitoring, managing, and maintaining soil health has become crucial for long-term food security and sustainable development as land-use demands, climatic variability, and population growth increase globally.

The ability of soil to continue operating as a living system that sustains biological productivity, preserves environmental quality, and supports human, animal, and plant health is commonly referred to as soil health (Adak et al., 2023). Soil health highlights the dynamic and biological aspects of soil systems, in contrast to soil quality, which is frequently assessed for a particular land use. It acknowledges soil as a living, dynamic ecosystem made up of biological activity, chemical composition, and physical structure that constantly interact over time and space. Climate-resilient agriculture and sustainable land management depend on maintaining this equilibrium.

4.1.1 Background and importance of soil health monitoring

Traditional farming methods frequently ignored the long-term effects on soil structure, organic matter, and biological diversity in favor of optimizing short-term crop productivity. Widespread

Quantum Computing and Instrumentation for Smart Agriculture. DOI: https://doi.org/10.1016/B978-0-443-44247-6.00007-5

soil degradation has been caused by intensive tillage, overuse of chemical pesticides and fertilizers, monocropping, deforestation, and inadequate irrigation management. Reduced soil fertility, elevated salinity, nutritional imbalance, compaction, erosion, and a decrease in microbial activity are all consequences of these activities. Recent research indicates that food security, water quality, ecosystem resilience, and greenhouse gas emissions have become major concerns due to soil degradation (Adamchuk et al., 2004; Ahmed et al., 2018).

Soil health monitoring is essential in this situation because it makes it possible to evaluate soil conditions, spot early indicators of deterioration, and assist in making well-informed management decisions. Farmers, agronomists, and legislators can assess the effects of agricultural practices, maximize the use of inputs, and put conservation plans that improve soil resilience into action with the help of effective monitoring. However, evaluating soil health is a difficult and complex undertaking because it varies naturally across terrain and changes over time.

4.1.2 Conventional soil health assessment approaches

In the past, laboratory-based examination of physical, chemical, and biological indicators and regular soil sampling have been the mainstays of soil health evaluation. Bulk density, texture, pH, electrical conductivity (EC), organic carbon, cation exchange capacity, microbial biomass, and nutrient content (nitrogen, phosphorus, and potassium) are often measured parameters. These markers offer important information about soil fertility, structure, and biological activity, and laboratory techniques are typically thought to be very precise and trustworthy. Despite their accuracy, standard soil testing methodologies suffer from various limitations. They are expensive, time-consuming, and labor-intensive, especially when frequent monitoring is needed. Additionally, it is difficult to capture soil heterogeneity and dynamic changes brought on by weather events, crop growth cycles, and management methods because soil sampling is frequently geographically sparse and temporally rare (Akhter and Sofi, 2022). Therefore, particularly in large-scale or resource-constrained agricultural systems, standard methods might not be sufficient to enable real-time decision-making or precision agriculture applications.

4.1.3 Emerging challenges and the need for advanced monitoring systems

The shortcomings of static and reactive soil management techniques have been shown by the growing complexity of contemporary agricultural systems. Changes in precipitation patterns, an increase in the frequency of extreme weather events, and rising temperatures are some of the extra uncertainties brought about by climate change. These factors have a substantial impact on soil moisture dynamics, nutrient availability, and biological processes. Simultaneously, the increasing need for sustainable farming methods calls for the economical use of energy, water, and fertilizers while reducing negative effects on the environment. The need for automated, scalable, high-resolution soil health monitoring systems that can deliver ongoing, site-specific data has increased as a result of these difficulties. In order to support prompt interventions and adaptable management techniques, modern soil monitoring must transition from periodic evaluation to dynamic, data-driven methods.

4.1.4 Technological advancements in soil health monitoring

Soil health monitoring has undergone a substantial transformation thanks to recent developments in sensing technologies, remote sensing platforms, machine learning (ML), and data analytics. Soil characteristics, including EC, moisture content, temperature, and nutrient levels, can be quickly measured thanks to proximal soil sensing technologies, which include portable and in-situ sensors. Particularly, it has been demonstrated that soil EC correlates with texture, salt, wetness, and organic matter concentration, making it a useful stand-in for a number of soil characteristics (Ali et al., 2020). By offering extensive, noninvasive observations of soil and land characteristics, remote sensing technologies have further increased monitoring possibilities. Vegetation indices, surface soil moisture, changes in land cover, and agricultural stress indicators are increasingly being estimated using satellite-based sensors and unmanned aerial vehicles (UAVs) with multispectral and hyperspectral cameras. These metrics are useful but indirect markers of soil health and geographical variability (Akhter and Sofi, 2022). Particularly, UAVs have excellent spatial and temporal resolution, which makes them appropriate for precision agriculture and field-level monitoring.

4.1.5 Role of machine learning and data fusion

A potent method for improving soil health evaluation is the combination of artificial intelligence (AI) and ML. To find patterns, estimate degradation trends, and predict soil attributes, ML algorithms may examine vast amounts of diverse data gathered from sensors, remote sensing platforms, and historical records. Regression models, artificial neural networks, support vector machines, and deep learning frameworks are some of the techniques that have shown great promise for increasing the precision and scalability of estimating soil properties (Atatüre et al., 2018). The reliability of soil health assessments is further improved by data fusion techniques that integrate proximal sensing data, remote sensing observations, and geographic information systems (GIS). These approaches transcend the drawbacks of individual techniques and offer a more thorough depiction of soil conditions across various geographical and temporal dimensions by integrating several data sources. The shift from reactive soil management to proactive, predictive, and adaptive decision-making frameworks is made possible by such integrated systems.

4.1.6 Problem definition and research gaps

Despite tremendous advancements in technology, there are still a number of obstacles to the widespread use of sophisticated soil health monitoring systems. Practical application is still hampered by sensor calibration, data ambiguity, model transferability across areas, and expensive initial investment costs, especially in developing nations. Furthermore, instead of using comprehensive soil health indices that take into account physical, chemical, and biological aspects, a large number of current studies concentrate on specific soil parameters. Converting complicated data outputs into useful insights for farmers and land managers is another significant difficulty. Although sophisticated models are capable of producing precise forecasts, their efficacy ultimately hinges on their accessibility, interpretability, and interaction with decision-support systems. Interdisciplinary research bridging soil science, agronomy, data science, and environmental engineering is necessary to address these issues.

4.1.6.1 Scope

Given these factors, the goal of this chapter is to present a thorough overview of soil health monitoring techniques, including both traditional procedures and cutting-edge technological solutions. The scope encompasses ML applications, proximal and remote sensing methods, laboratory-based evaluations, and integrated monitoring systems. The chapter addresses upcoming developments in soil health monitoring, emphasizes practical applications, and critically assesses the benefits and drawbacks of each strategy. This chapter aims to aid in the creation of effective, scalable, and sustainable soil monitoring systems by summarizing existing knowledge and pointing out important research gaps. In the end, increasing soil health monitoring is crucial for sustainable land-use planning, climate resilience, environmental protection, and agricultural output. From traditional sensing techniques to new quantum sensor technologies in smart agriculture, this chapter offers soil health monitoring in a methodical manner. Before addressing the ideas of quantum sensing, it covers the types, principles, and limitations of classical soil sensors. The use of quantum sensors for subsurface investigation, pollution, nutrients, and soil moisture is emphasized. Lastly, the relevance of hybrid classical-quantum systems in sustainable agriculture, future research areas, and obstacles are discussed.

4.2 Sensors in agriculture

A sensor is a simple, compact device that measures or detects elements of the natural environment, including motion, heat, or light, and transforms them into an analog or digital representation. Sensors are also used to keep updated on soil characteristics. Numerous kinds of sensors are employed in agriculture to implement smart agriculture. Using variable-rate technologies (VRT), farmers use precision agriculture to make sure insecticides, fertilizers, and herbicides are delivered to crops precisely, at the appropriate time, and at the right rate.

4.2.1 Role and importance of sensors in agriculture and broad

Agricultural techniques have changed dramatically as a result of the quick development of sensing technology, making it possible to switch from traditional farming methods to precision and smart agriculture systems. Because they enable real-time monitoring, data collection, and automated decision-making, sensors are essential parts of contemporary agricultural ecosystems. Their use of data analytics, AI, and Internet of Things (IoT) platforms has increased agricultural operations' sustainability, productivity, and resource efficiency.

4.2.1.1 Soil and environmental monitoring

To keep an eye on important variables, including temperature, pH, moisture content, and nutrient concentration, soil-based sensors are frequently employed. Water stress, nitrogen leaching, and soil deterioration can be avoided by precisely scheduling irrigation and fertilization through ongoing monitoring of these variables. Environmental sensors that detect temperature, humidity, rainfall, wind speed, and sun radiation give important microclimatic information that helps make well-informed decisions about crop protection, harvesting, and sowing.

4.2.1.2 Crop health and growth assessment

Crop health and growth conditions are evaluated using advanced sensing technologies, such as optical, multispectral, and infrared sensors. These sensors allow for the early identification of plant stress brought on by disease, pest infestation, nutritional imbalance, or water scarcity. Early detection of such circumstances enables focused responses that minimize overuse of agrochemicals and reduce crop losses. Drones and satellite-based sensors are examples of remote sensing systems that improve yield estimation and large-scale agricultural monitoring.

4.2.1.3 Precision irrigation and nutrient management

Precision irrigation and fertigation systems rely heavily on sensors to provide real-time data on crop and soil conditions. Based on actual crop demand, automated irrigation systems that are connected to sensor networks provide water and nutrients in the ideal amounts. Particularly in areas with limited water resources, this strategy greatly increases agricultural output, lowers operating costs, and improves water usage efficiency.

4.2.1.4 Livestock monitoring and farm automation

Sensors are being used more and more in livestock management to track environmental conditions, animal behavior, and health in addition to crop yield. Early disease identification and better animal welfare are made possible by wearable and ambient sensors that monitor variables including body temperature, activity, and feeding habits. Additionally, sensor technologies reduce labor dependency and boost operational efficiency by supporting farm automation through applications in robotic harvesting, autonomous machinery, and smart greenhouses.

4.2.1.5 Importance of sensors for sustainable agriculture

The development of sustainable agriculture is greatly aided by the use of sensor-based technologies. Sensors help reduce environmental effects while maintaining high output levels by maximizing resource consumption and minimizing overuse of water, fertilizer, and pesticides. Additionally, by offering early alerts for drought conditions, insect outbreaks, and extreme weather occurrences, sensor-driven data analytics improves climate resilience. In order to ensure long-term agricultural sustainability and reduce the dangers connected with climate change, such competencies are crucial.

4.2.1.6 Overall impact and future outlook

In general, sensors enable data-driven and site-specific farm management techniques, which make them the foundation of smart agriculture systems. In the face of rising global food demand and environmental limits, their role in boosting productivity, increasing resource efficiency, and promoting sustainable food production is becoming more and more crucial. Future agricultural systems are anticipated to benefit even more from ongoing developments in wireless communication, data analytics, and sensor precision.

4.2.2 Sensor types in smart agriculture

In order to gather real-time data on soil conditions, crop health, environmental parameters, and livestock behavior, smart agriculture uses a wide variety of sensors. Precision farming systems rely on these sensors for data-driven decision-making, automated control, and continuous monitoring. Soil, environmental, crop, water and irrigation, and livestock sensors are the general categories into which agricultural sensors can be divided based on their use and measured data.

4.2.2.1 Optical sensors in agriculture

In agriculture, optical sensors are sophisticated instruments that employ light-based technologies to track ambient conditions, soil characteristics, and crop health, allowing for more accurate and effective agricultural methods. By detecting changes in light reflectance, absorbance, or fluorescence from plants and soil, these sensors aid in the early detection of illnesses, pest infestations, water stress, and nutrient deficiencies. Common uses include yield prediction, irrigation management, and precision farming, where decisions about crop treatment, fertilizer application, and watering schedules are guided by data from multispectral, hyperspectral, thermal, or LIDAR sensors. Optical sensors save input costs, increase productivity, and support sustainable farming methods by offering site-specific, real-time insights. It uses different light frequencies to measure the characteristics of the soil. These sensors, which are mounted on cars or drones, enable the collection and processing of data on plant color and soil reflectance. Clay, nutrients, and soil levels of moisture can all be detected using optical sensors.

4.2.2.1.1 Working principle

The basis for optical sensors' operation is the fact that the physical and biochemical characteristics of soil and vegetation affect how they reflect, absorb, and emit electromagnetic radiation. Information regarding surface conditions is carried by the reflected or emitted signal that results from the interaction of crops or soil with incident radiation from either an artificial source (such as LiDAR) or a natural source (such as sunshine). This signal is picked up by optical sensors, which then transform it into digital data that is processed to produce agronomically significant values.

Due to chlorophyll activity, healthy vegetation usually shows considerable absorption in the red region and high reflectance in the near-infrared range. Stressed plants, on the other hand, exhibit different spectral signatures, making it possible to identify pest infestation, illness, water stress, and nutrient deficiencies early on. Similar to this, optical sensors can determine soil

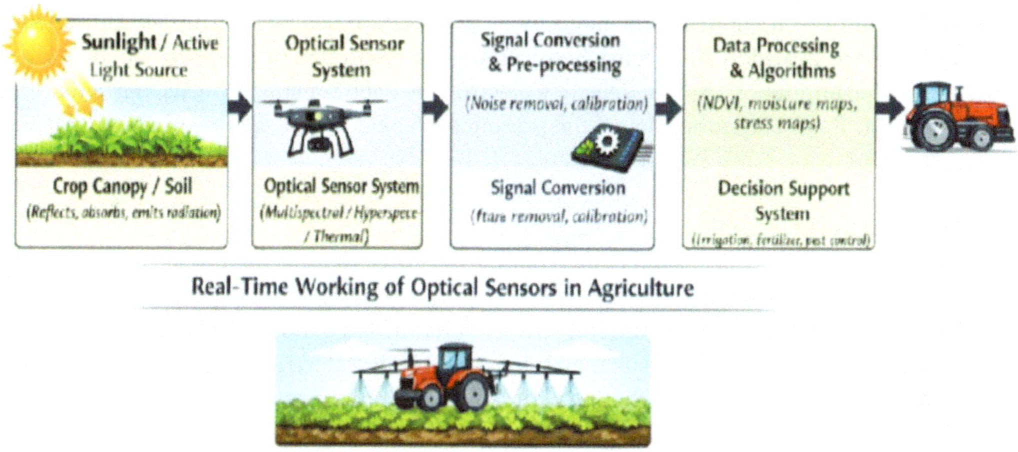

FIGURE 4.1 Optical sensors in agriculture.

moisture, clay content, and nutrient levels since soil reflectance patterns change with moisture content, texture, organic matter, and mineral composition.

Types of Optical Sensor Used:

- Multispectral sensors are frequently employed for vegetation indices like NDVI since they record data in a few distinct wavelength bands (such as red, green, blue, and NIR).
- Crop stress, nutrient deficits, and disease signs can be precisely identified thanks to hyperspectral sensors, which gather data over hundreds of small spectral bands.
- By measuring surface temperature, thermal sensors can identify irrigation inefficiencies and water stress.
- LiDAR sensors create 3D data on crop height, canopy structure, and biomass estimation using laser pulses.

4.2.2.1.2 Real-time data acquisition and processing

Optical sensors can be installed on ground-based platforms, tractors, satellites, and UAVs. Sensor data is wirelessly transferred to cloud-based or edge-computing systems in real-time agricultural applications, where preprocessing, feature extraction, and analytics are carried out. Decision support systems generate and incorporate stress indicators, vegetation indices, and soil reflectance maps. Site-specific farm management is made possible by these outputs, which direct focused pest control, fertilizer application, and variable-rate irrigation (Fig. 4.1).

Applications in agriculture

- Monitoring crop health (early pest and disease detection)
- Evaluation of soil moisture and texture
- Identification of nutrient deficiencies, forecasting, and estimating yield
- Accurate fertilization and watering
- Sustainable management of inputs

4.2.2.2 Electrochemical sensor

Data collection on soil chemicals is aided by it. Information sensors for the measurement of soil nutrients are provided by electrochemical sensors. Samples of soil have been sent to a laboratory for examination. An ion-selective electrode is used for specific tests, particularly pH determination. These electrodes detect the presence of specific ions, like hydrogen, potassium, or nitrate.

4.2.2.2.1 Working principle

Electrochemical sensors have widely been reported as efficient tools for assessing soil chemical content because of their high sensitivity, selectivity and potential of real-time measurement. The sensors allow rapid assessment of the soil fertility and nutrient status through the conversion of chemical processes occurring in a soil matrix into measurable electrical signals. The mechanism of operation uses ionic species interacting with selected membranes or electrode surfaces across the interface between soil pore solution and the sensor via electrochemical rates 2,21,21,1.

Ions move in the direction of an ion-selective membrane at the sensor–soil interface, which is made to interact preferentially with a particular target ion. Due to variations in ion activity on either side, this selective contact creates an electrochemical potential difference across the membrane. The Nernst equation, which creates a logarithmic connection between electrode potential and ion activity 333, is one of the traditional electrochemical laws that are used to understand the resulting potential when it is measured in relation to a stable reference electrode. The quantitative measurement of ion concentration in electrochemical sensing systems is based on this theoretical framework.

Ion-selective electrodes (ISEs) have become the most widely utilized sensors for soil chemical analysis among the several electrochemical devices. Key soil parameters such as hydrogen ions (pH), nitrate (NO_3^-), potassium (K^+), and ammonium (NH_4^+) can be selectively detected using ISEs. These parameters are essential for soil fertility, plant nutrition, and biogeochemical cycling 4,54,54,5. ISEs are especially well-suited for in-situ monitoring applications, such as environmental assessment and precision agriculture, because of their quick response time and low sample preparation needs (666). Improving membrane selectivity, lowering signal drift, and boosting long-term stability under changing soil conditions have been the main goals of recent developments in electrochemical sensor technology. Research has shown that solid-state reference electrodes, polymeric membranes, and nanostructured materials can greatly improve sensor durability and performance 7,87,87,8. Additionally, the deployment of electrochemical sensors into distributed sensor networks for continuous soil monitoring has been made easier by integration with wireless communication modules and low-power electronics 999.

When used in heterogeneous soil conditions, electrochemical sensors have certain drawbacks despite their benefits. Variability in soil moisture, temperature changes, ionic interference, and membrane fouling can all have an impact on sensor accuracy 101010. Therefore, to guarantee dependable long-term functioning 111111, suitable calibration processes, temperature compensation strategies, and routine sensor maintenance are crucial. To address these issues and enhance measurement robustness in field settings, a number of academics have suggested data-driven correction techniques, such as ML-based calibration models 121212.

FIGURE 4.2 Schematic diagram of an electrochemical sensor.

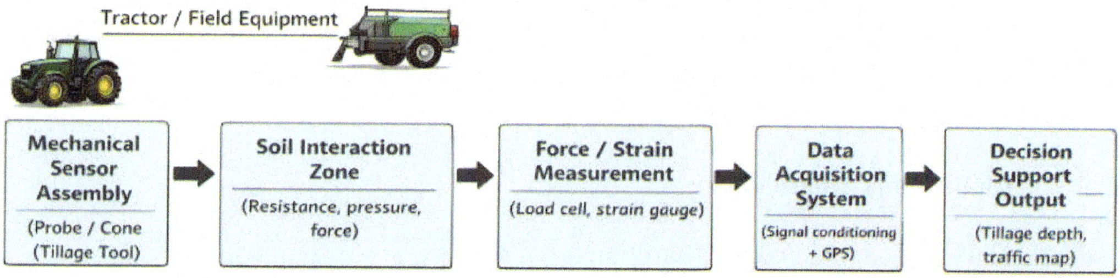

FIGURE 4.3 Working of mechanical soil sensors.

4.2.2.2.2 Schematic representation of an electrochemical sensor

Fig. 4.2 shows the steps involved in operating an electrochemical soil sensor. Ions from the soil pore water diffuse in the direction of the ion-selective membrane, where selective ion binding or exchange takes place. The produced electrochemical potential is transformed into an electrical signal, which is then conditioned and processed to produce quantitative data on pH 131313 or soil nutrient concentration.

4.2.2.3 Mechanical soil sensors

These sensors record the force as determined by load cells or pressure scales using a device that pierces the earth. The holding forces caused by the soils cutting, breaking, and displacement are recorded by a sensor as it passes through the soil. The force needed to penetrate the soil medium divided by the frontal region of the tool in contact with the soil is known as the soil mechanical resistance, and it is expressed in units of pressure.

Working Principle

The dirt resists the mechanical sensor element when it enters or travels through the soil. Using load cells, strain gauges, or pressure transducers, this resistance is quantified in terms of force, torque, pressure, or deformation. The bulk density, moisture content, compaction level, and texture of the soil all affect how much resistance there is (Fig.4.3).

4.2.2.3.1 Types of mechanical soil sensors

- **Cone penetrometers:** Determine root restrictiveness and compaction by measuring soil penetration resistance.
- **Draft force sensors:** Determine soil strength and tillage suitability by measuring the force needed to pull tillage tools.

- **Shear vane sensors:** These are particularly helpful in cohesive soils as they measure the shear Strength of the soil.
- **Compression sensors and pressure plates:** Used to calculate the bearing capacity and compressibility of soil.

4.2.2.3.2 Real-time field operation

Mechanical soil sensors are frequently installed on handheld probes, tillage tools, and tractors. Continuous measurements are taken and sent to GPS-based systems or onboard data loggers as the sensor travels throughout the field. Real-time variable-depth tillage plans and soil compaction maps are then created using these data.

4.2.2.3.3 Applications in agriculture

- Finding layers of compacted soil (hardpans)
- Improving the depth and intensity of tillage
- Enhancement of aeration in the root zone
- Evaluation of soil degradation caused by traffic
- Encouragement of site-specific soil management

Benefits

- Measuring the mechanical behavior of soil directly
- Excellent compaction assessment accuracy
- Efficient in field settings
- Beneficial for making decisions in real time

Restrictions

- Wear and tear are caused by physical contact
- Soil moisture affects performance
- Restricted spatial coverage in contrast to distant sensors

4.2.2.4 Dielectric soil moisture sensor

A dielectric soil moisture sensor is an electronic device that measures the dielectric permittivity of soil to calculate its volumetric water content. The idea is based on the stark difference between the dielectric constants of air (~1) and dry soil (~3–5) and water (~80). The sensor detects an increase in the soil-water-air mixture's total dielectric constant when soil moisture levels rise.

Tractors, tillage tools, and handheld probes are frequently equipped with mechanical soil sensors. Continuous readings are taken as the sensor travels throughout the field and sent to GPS-based systems or on board data loggers. Real-time soil compaction maps and variable-depth tillage plans are then created using these data. It measures the amount of moisture in the soil. The moisture sensors are linked to the farm's rain checkpoints. This allows the soil moisture conditions to be observed when there is less vegetation.

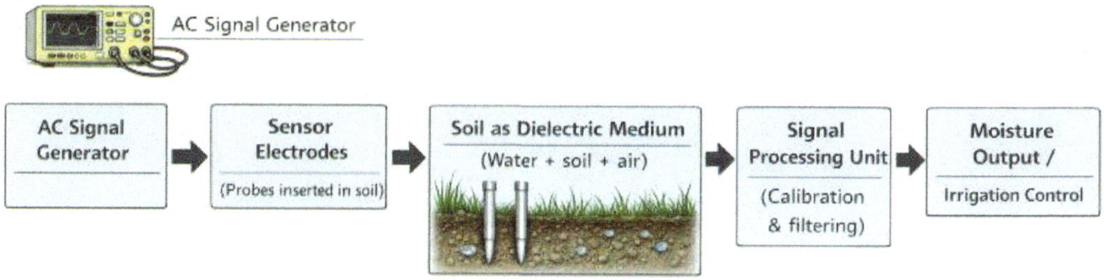

FIGURE 4.4 Working of dielectric soil moisture sensor.

4.2.2.4.1 Working Principle

The sensor is made up of metal electrodes or probes that are put into the ground to create a capacitor. The soil functions as a dielectric medium when an alternating electromagnetic signal is applied. The moisture content of the soil affects the capacitance or signal propagation velocity. The electrical signal that results from this shift is calibrated to determine the volumetric water content (Fig. 4.4).

Types of dielectric moisture sensors:

- **Capacitance sensors:** Determine how variations in soil moisture affect capacitance.
- **Frequency Domain Reflectometry (FDR):** Measure frequency shifts in an oscillating circuit caused by changes in soil moisture.
- **Time Domain Reflectometry (TDR):** Determine how long electromagnetic pulses take to travel along probes inserted into the soil.
- **Amplitude Domain Reflectometry (ADR):** Determine moisture-related signal attenuation.

Real-time measurement process:

Dielectric sensors send data to wireless networks or data loggers while continuously monitoring soil moisture. They enable real-time control of watering schedules depending on crop needs because they are frequently coupled with IoT platforms and automated irrigation systems.

Utilization in agriculture:

- Accurate irrigation control
- Crop water stress detection
- Profiling soil moisture at various depths
- Monitoring in greenhouses and open fields
- Water conservation and drought management

Benefits:

- Quick and nondestructive measurement
- Extremely sensitive to changes in moisture
- Fit for monitoring in real time

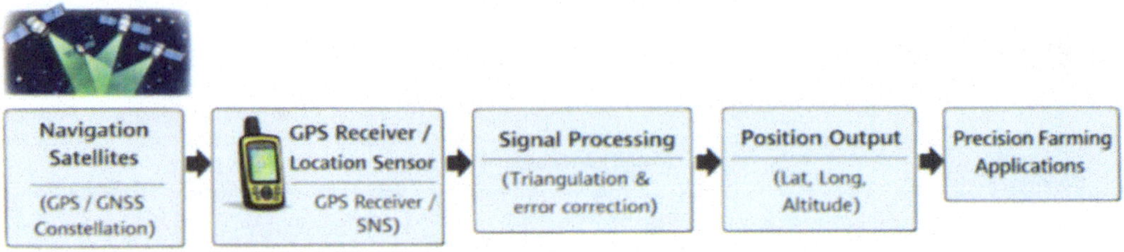

FIGURE 4.5 Working of location survey.

- Suitable for automation systems

Limitations:

- Temperature, salinity, and soil texture all affect accuracy
- Needs to be calibrated for various soil types
- Restricted sensing volume in the vicinity of probes

4.2.2.5 Location sensors

In agricultural applications, location sensors employ GPS (Global Positioning System) technology to pinpoint the exact geographic location of machinery, fields, or individual crops. By giving farmers information on latitude, longitude, and altitude, these sensors help them use site-specific management techniques. Farmers can map fields precisely, observe machinery operations in real time, apply inputs (such as pesticides or fertilizers) in diverse ways, and assess crop performance across zones by combining geographic information with other sensor inputs. In addition to increasing production and decreasing expenses, this also promotes precision agriculture methods like automated guiding systems and variable rate application. Location sensors are tools that track an object's movement, spatial orientation, and geographic location in real time. Location sensors are essential to agriculture and precision farming because they make site-specific field operations, navigation, and spatial mapping possible. Drones, farm equipment, and sensor systems may all function with extreme precision because to the precise positioning data these sensors offer.

4.2.2.5.1 Working Principle

The GPS and other satellite navigation systems are the foundation for the majority of location sensors. By timing the transmission of signals from several satellites, a GPS receiver determines its location. The sensor calculates latitude, longitude, altitude, speed, and time by triangulating information from a minimum of four satellites. To get centimeter-level precision, advanced systems include real-time kinematic (RTK) corrections or differential GPS (DGPS) (Fig. 4.5).

Types of location sensor:

- **Global positioning system (GPS):** Accurate positioning at the meter level.
- **Differential GPS (DGPS):** To increase accuracy, it uses base station correction signals.

- **Real-Time Kinematic Global Positioning System (RTK-GPS):** Offers real-time adjustments for applications requiring high precision.
- **Global navigation satellite systems (GNSS):** This system integrates several satellite systems, including GPS, GLONASS, Galileo, and BeiDou.
- **Inertial navigation sensors (INS):** These devices, which are frequently combined with GPS, track motion using accelerometers and gyroscopes.

4.2.2.5.2 Agriculture's real-time operation

Drones, autonomous cars, harvesters, and tractors all have location sensors built in. Continuous positional data is recorded and synchronized with various sensor outputs, including yield monitors, soil sensors, and optical sensors, while equipment moves over the field. This makes it possible to produce georeferenced maps of crop health, soil characteristics, and yield variability.

Utilization in agriculture:

- Accurate sowing and planting
- Variable-rate application of pesticides and fertilizers
- Farm equipment that is guided and autonomous
- Field zoning and yield mapping
- Crop scouting and drone navigation

Benefits:

- Makes site-specific crop management possible
- Lessens input waste and overlap
- Increases accuracy and operational efficiency—Encourages smart farming and automation

Limitations:

- Signal obstructions (trees, buildings) impair accuracy
- RTK systems are expensive
- Signal loss in isolated or steep locations (Fig. 4.6)

4.2.3 Electronic sensors

Electronic sensors are gadgets that measure, monitor, and control changes in physical, chemical, or biological characteristics by converting them into electrical impulses. Precision farming systems in agriculture rely on electronic sensors to continuously monitor soil, crop, ambient, and equipment characteristics for data-driven decision-making. To verify equipment operations, it was put on vehicles and other field machinery. The data was then sent instantly to computers or sent to individuals via email using satellite and cellphone communication networks. After that, the field director can access the data on their mobile device or office computer.

Working Principle

A change in a measurable characteristic, such as temperature, moisture, pressure, gas concentration, or EC, is how electronic sensors work.

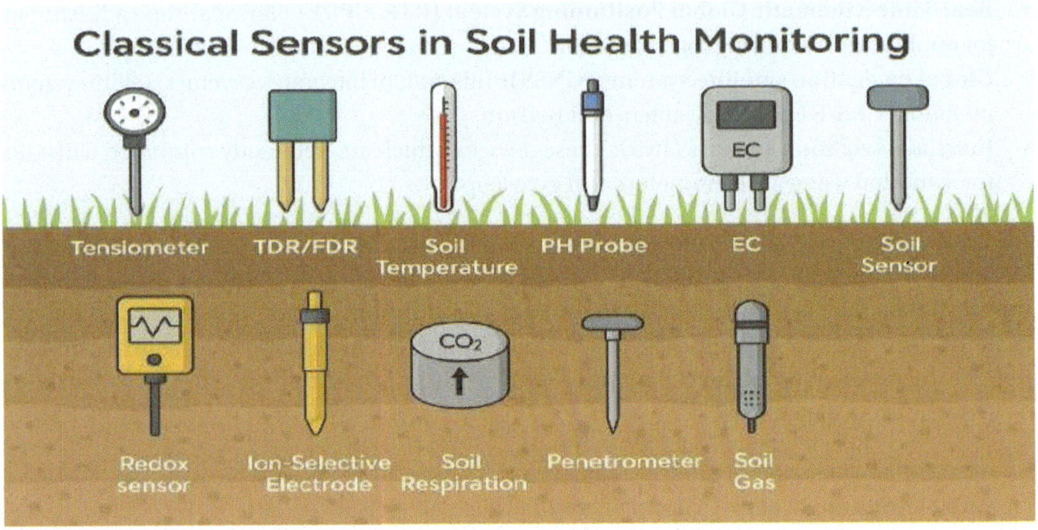

FIGURE 4.6 Types of classical sensors used in soil health monitoring.

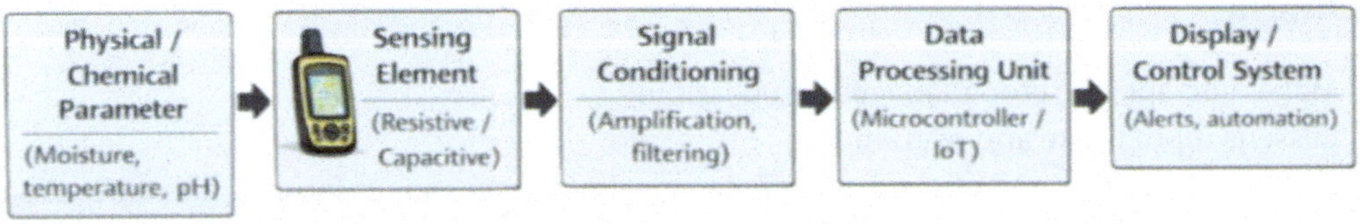

FIGURE 4.7 Working of electronic sensor in agriculture.

This modification modifies a sensor element's electrical characteristic, such as resistance, capacitance, voltage, or current, which is subsequently amplified, conditioned, and processed to generate a readable output (Fig. 4.7).

Types of electronic sensors in agriculture

- **Temperature sensors (thermistors, RTDs):** Track the temperature of the soil and surrounding air.
- **Humidity sensors:** Determine the amount of moisture in the air to analyze crop growth.
- **Dielectric or capacitance-based soil moisture sensors:** Determine the volumetric water content of soil by measuring changes in its electrical capacitance, which fluctuates significantly depending on the amount of moisture present.
- **EC sensors:** Determine the salinity and nutrient availability of the soil.
- **Gas sensors:** They can identify ethylene, CO2, or NH3 in greenhouses.
- **pH sensors:** Determine the acidity and alkalinity of soil and water.

- **Pressure sensors:** Determine fluid pressure within irrigation networks to ensure efficient water distribution, and measure soil water tension to assess how easily plants can extract moisture.

Real-time data acquisition

Microcontrollers, data recorders, or IoT platforms are linked to electronic sensors. Continuous data collection and transmission are done by wired or wireless connectivity (Wi-Fi, LoRa, GSM). This allows for automatic control of climate, fertilization, and irrigation systems as well as real-time monitoring and alarms.

Utilization in agriculture:

- Accurate fertigation and irrigation
- Climate control in greenhouses
- Monitoring the quality of water and soil
- Evaluation of crop growth
- Farm machinery automation
- IoT-based technologies and smart farming benefits
- Excellent sensitivity and quick reaction
- Digital system integration is simple
- Facilitates remote monitoring and automation
- Scalable and economical constraints

Limitations:

- Needs upkeep and a power source
- Vulnerable to environmental harm and electrical noise
- Accuracy requires calibration

4.2.4 Air flow sensors

While on the go, it can take measurements at specific spots. The pressure required to force a predetermined volume of atmospheric air into the earth's surface at a specified depth is the intended output. Different identifying signs are produced by different soil characteristics, such as compaction, structure, type, and moisture content. The rate, velocity, or volume of air movement in a particular environment can be measured with air flow sensors. These sensors are particularly crucial for agriculture in greenhouses, livestock housing, grain storage facilities, and ventilation systems, where regulated air flow is necessary to ensure ideal humidity, temperature, gas concentration, and crop or animal health.

4.2.4.1 Working principle

Air flow sensors work by identifying variations in temperature, pressure, or mechanical displacement brought on by air movement. An electrical signal corresponding to air velocity or flow rate is created from the measured change. Thermal, mechanical, or pressure-based techniques can be used to measure airflow, depending on the sensor device (Fig. 4.8).

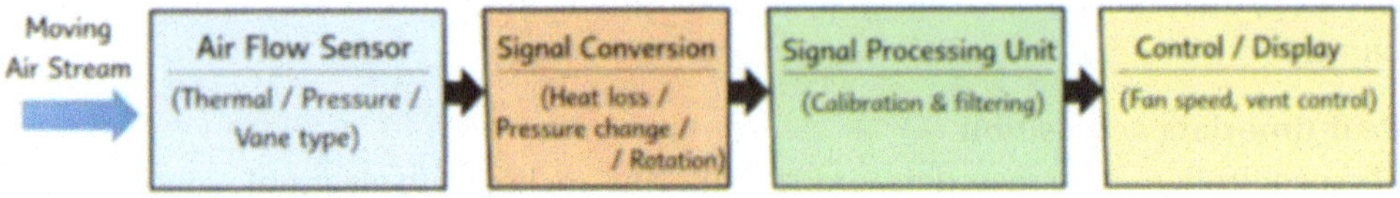

FIGURE 4.8 Working of air flow sensor.

Air flow sensor types:

- **Thermal (hot-wire or hot-film) sensors:** Often employed for low-speed airflow, these sensors measure heat loss from a heated element caused by air movement.
- **Differential pressure sensors:** To estimate airflow, measure the pressure drop across a venturi tube or opening.
- **Vane or anemometer-based sensors:** These employ revolving blades whose speed is directly correlated with the velocity of airflow.
- **Ultrasonic air flow sensors:** Determine how long sound waves take to travel through flowing air.

4.2.4.2 Real-time operation

Air flow sensors are included in HVAC systems, greenhouse control systems, and ventilation ducts. Controllers automatically operate fans, vents, heaters, and cooling systems by continuously monitoring and processing data, guaranteeing stable environmental conditions.

Utilization in agriculture

- Control of greenhouse ventilation
- Monitoring the air quality in livestock housing
- Ventilation for grain drying and storage
- Climate control in dairy and poultry farms
- Wind monitoring and spray drift

Benefits:

- Makes accurate environmental control possible
- Enhances animal welfare and agricultural growth
- Lowers the amount of energy used
- Encourages smart farming and automation

Limitations:

- Humidity and dust have an impact on performance
- Needs routine cleaning and calibrating
- The airflow profile affects accuracy (Fig. 4.9)

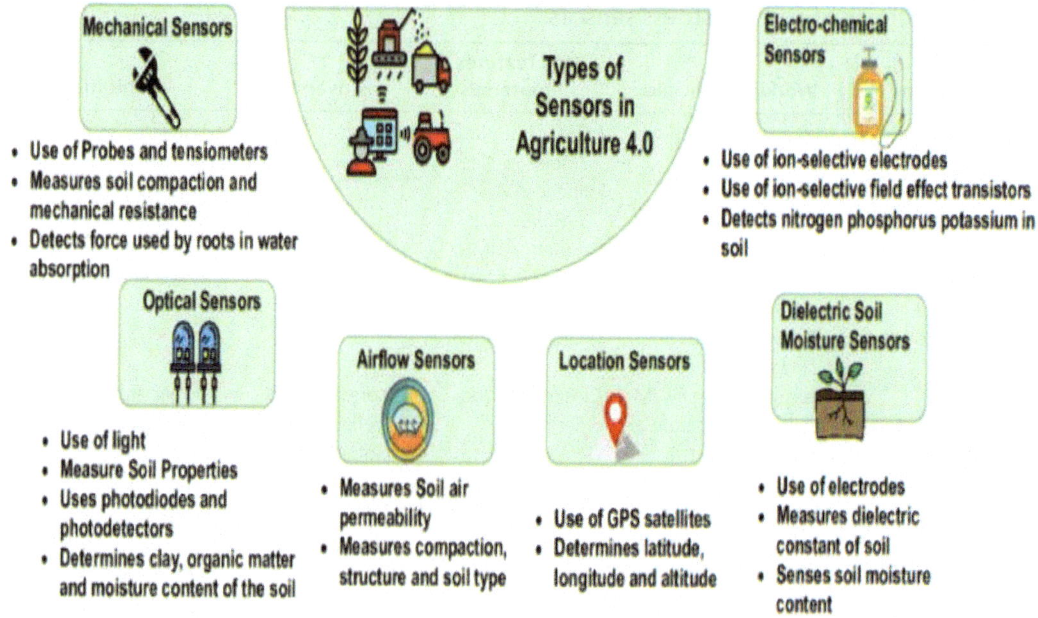

FIGURE 4.9 Types of sensors and their applications in smart agriculture.

4.2.4.3 Types of soil sensors

4.2.4.3.1 Soil moistness sensors

Plant development is significantly influenced by soil moistness, which is a crucial parameter for assessing soil health. The amount of moistness in the soil has a major impact on its physico-chemical characteristics, which in turn affects microbial activity and salt dissolution. Farmers may determine the right quantity and timing of irrigation for plant development by using soil moistness sensors to measure the water content of the soil (Robinson et al., 2008; Yin et al., 2021) (Table 4.1).

4.2.4.3.2 Soil temperature sensor

A crucial factor in agriculture, soil temperature, which ranges from -10°C to 50°C (Bauer & Aschenbruck, 2018), affects a number of plant growth processes, including germination, blooming, and compositing. It also has a major impact on the physical, chemical, and microbiological processes in soils that are essential to plant growth. (Maroli et al., 2021) Specific heat capacity, bulk density, texture, water content, thermal conductivity, and surface covering materials are among the characteristics of soil that have a significant impact on temperature(Bhoyar et al., 2019) (Table 4.2).

4.2.4.3.3 Soil pH sensors

The acidity of soil is measured by its pH, which reflects the combined impacts of the parent material, organisms, and climate that make up the soil. The pH of the soil influences microbial

Table 4.1 Types of soil moistness sensors.

Sensor type	Working principle	Key features/ materials	Advantages	Limitations
Capacitive (Bersani et al., 2022)	Measures changes in soil dielectric permittivity	MEMS, PCB design, Zigbee wireless, linear VWC 1–80%	Low-cost, simple, widely used	Shallow penetration, sensitive to geometry
TDR (Time-domain reflectometry) (Sadowski & Spachos, 2020)	Measures pulse Velocity in a transmission line to Derive dielectric constant	Frequency 1 MHz–1 GHz; Unaffected by texture/salinity/temp	High accuracy, suitable for continuous monitoring	Calibration and cost complexity
TDT (Transmissometry) (Braakman & Poggio, 2019)	Measures one-way EM pulse propagation time	Closed-loop rod design	More robust vs. reflections, improved accuracy	Specialized probes needed
FDR (Frequency-Domain Reflectometry) (Torky & Hassanein, 2020)	Measures frequency difference in EM signal reflection	Coaxial transmission line	Good accuracy, frequency domain analysis	Bulky, affected by signal attenuation
GPR (Ground penetrating Radar) (Thakur et al., 2019)	Sends high-frequency EM waves into subsurface and measures reflection	Operates at ~1 GHz; complex signal interpretation	Deep sensing, subsurface profiling	Expensive, complex, requires signal correction
EMI (Electromagnetic Induction) (Ahmed et al., 2018; Shadrin et al., 2019)	Detects soil conductivity changes at low frequency (~38 kHz)	Coil-based system (e.g., EM38), large separation	Deep sensing, fast, noninvasive	High cost, interference-prone
Impedance-Based (Jo et al., 2020)	Measures voltage drop across a resistive–capacitive network	Uses stainless-steel wire probes; cubic voltage–moisture relation	Simple electronics	Sensitive to temp/salinity
Resistive (GQD-based) (Filgueiras et al., 2020)	Conductivity change due to water adsorption on graphene quantum dots (GQDs)	Au interdigitated electrodes, GQD sensing layer	High sensitivity, good stability	Slower response, material synthesis
Ultrasonic (Jo et al., 2020)	Measures time difference in acoustic wave reflections	40 kHz wave, stainless/brass pipe, dummy pair for temperature correction	Robust, Multi-point capability	Expensive, temperature-sensitive

(*continued on next page*)

Table 4.1 Types of soil moistness sensors—cont'd

Sensor type	Working principle	Key features/ materials	Advantages	Limitations
Optical (VIS–NIR) (Bacco et al., 2018)	Monitors NIR light absorption at water-related bands (1450, 1940, 2950 nm)	Fiber optics, reflection spectrum variation	Nondestructive, In situ measurement	Calibration required, ambient light sensitive
Micro-Nanoporous Ceramic (Triantafyllou et al., 2019)	Diaphragm deflection due to water potential via nanoporous ceramic membrane	SiO_2–Si diaphragm, high-resolution (0.04 kPa)	High sensitivity, Direct water potential	Complex fabrication
Heat Capacity Sensor (Rudolph et al., 2019)	Measures the temperature rise after pulse heating to derive volumetric water content	Dual-probe with heater and sensor	Good accuracy, Heat–moisture correlation	Slow response, hard to install
Gravimetric Quantum Sensor (Zhong et al., 2022)	Measures gravitational field shifts via atomic interferometry to infer the moisture level	Matter-wave interference, quantum coherence, cold-atom interferometers	High sensitivity, Noninvasive, Deep subsurface detection	Expensive, complex, requires skilled handling

Table 4.2 Types of soil temperature sensors.

Sensor type	Working principle	Materials/Structure	Advantages	Limitations
Thermocouples (Deep & Elarabi, 2017)	Voltage generated by two thermoelectrically dissimilar metals	Iron/Constantan, Copper/Consta ntan	Fast response, easy automation, wide temperature range	Needs long cables and calibration for accuracy over distance
RTDs (Resistance Temperature Detectors) (Bacco et al., 2018)	Resistance change of the metal wire with temperature	Platinum wire on ceramic/glass core	High accuracy, stability, repeatability	Fragile, requires protection, thermal lag
Thermistors (Sadowski & Spachos, 2020)	Resistance change in ceramic/polymer with temperature	Ceramic semiconductors, polymers	High sensitivity, good resolution	Nonlinear response, complex calibration
Inkjet-Printed Biocompatible Sensors (Hu et al., 2023)	Temperature-induced resistance change in the printed metal layer	Silver-based inkjet-printed on a flexible substrate	Biocompatible, flexible, substrate-independent sensitivity (0.246 Ω/°C), low-cost fabrication	Durability and long-term environmental stability

activity as well as the chemical forms of various nutrients, which in turn impacts plant nutrition availability. For the majority of plants, soil with a pH between 5.5 and 7.5 is ideal (Soheli et al., 2022). In smart and precision agriculture, the pH value of the soil can offer useful information for managing the soil's health within a range that is appropriate for particular agricultural crops (Table 4.3).

4.2.4.3.4 Soil nutrient sensors

The three most crucial nutrients for crop productivity among the soil organic materials (SOMs) are potassium (K), phosphorus (P), and nitrogen (N) .N, P, and K fertilizers are used in agricultural engineering to increase crop yields. However, there are serious negative economic and environmental impacts of excessive artificial fertilizer inputs (Guignard et al., 2017). The development of precision agriculture and environmental sustainability depends on the effective and precise detection of nutrient molecules in soil, which allows plants to adapt to nutrient-deficient situations (Schachtman & Shin, 2007; Zhu et al., 2019) (Table 4.4).

Comparison between sensors and quantum sensors: Monitoring soil health has historically depended on classical sensors, which detect soil characteristics including moisture, pH, and nutrient levels using standard physical, chemical, and electrical principles. Using the concepts of quantum physics, quantum sensors have become a state-of-the-art substitute in recent years (Table 4.5).

4.2.4.3.5 Introduction to quantum sensors

Quantum Information Science (QIS) integrates the conceptual foundations of information science and quantum mechanics as well as aspects of computer science, engineering, mathematics, and the physical sciences (Benioff, 1980; Kumar, 2023). QIS has emerged as one of the fastest-growing study topics over the past three decades. In order to spur the development of QIS, three U.S. agencies—the National Institute of Standards and Technology (NIST), the National Science Foundation (NSF), and the Department of Energy (DOE)—are directed by the U.S. National Quantum Initiative Act, which was signed into law in late 2018.

For quantum sensing, there are two main approaches that may be used. The first, photonic quantum sensing, uses the quantum nature of light to sense objects remotely and read out memory, among other uses (Pirandola et al., 2018). Nonphotonic quantum sensors have also been created for use in magnetometry, thermometry, and other fields. These sensors rely on materials with well-defined quantum states, such as spin qubits and trapped ions, atomic vapors, superconducting circuits, and solid defects (such as nitrogen-vacancy centers in diamonds), which are all common components of quantum sensors (Atatüre et al., 2018). The performance of conventional sensor systems may also be improved by using either type of quantum sensor, which promises lower noise levels than the so-called "shot limit" (Degen et al., 2017; Pirandola et al., 2018). By measuring physical quantities using quantum systems or events, quantum sensing frequently outperforms traditional sensors in terms of responsiveness and accuracy. To improve measuring capabilities beyond the conventional quantum limit, methods frequently take advantage of quantum entanglement, superposition, and squeezed states.

(DiVincenzo, 2000) and (Degen et al., 2017) listed three prerequisites that should be met to build a functional quantum sensor using any potential material:

Table 4.3 Types of soil pH sensors.

Sensor type	Working principle	Materials/ Components	Advantages	Limitations
Colorimetric methods (Soheli et al., 2022)	pH-sensitive dyes change color, compared against the reference chart	Organic pigments, indicator paper, or strips	Simple, low-cost	Low precision, subjective reading, not suitable for in situ automation
Photometric/Optical sensors (Torky & Hassanein, 2020)	Optical signal change due to pH-sensitive dye or coating (e.g., absorption, fluorescence, and holography)	Optical fiber, fluorophores, and CCD cameras	High resolution, noncontact	Bulky, expensive, not suitable for in-soil measurement
Conductometric sensors (Thakur et al., 2019)	Conductivity changes in a pH-sensitive hydrogel or polymer	Interdigitated electrodes, hydrogel, polyaniline, SnO_2, and TiO_2	Low power, reusable, with good material design	Requires soil solution in most cases; hydrogel stability issues
Potentiometric (ISE) sensors (Soheli et al., 2022)	Potential difference across a pH-sensitive membrane and reference electrode	Glass electrode, Ag/AgCl reference, HCl internal solution	Good accuracy, direct measurement via Nernst equation	Fragile, sensitive to damage, requires maintenance
ISFET (Ion-sensitive FET) (Anisi et al., 2015)	Current change in the transistor due to the gate voltage shift from pH Interaction	SiO_2, Si_3N_4, and Al_2O_3 gate materials	Highly integrable, miniaturized, sensitive	Complex packaging is needed for soil environment
Microcantilever sensors (Filgueiras et al., 2020)	Mechanical deflection due to hydrogel/polymer swelling/shrinking with pH	Piezoresistive layers, SiO_2/Si_3N_4, Au coatings	High sensitivity, suitable for MEMS integration	Sensitive to mechanical noise, complex setup
Magnetoelastic Sensors (Onojeg huo et al., 2018)	Frequency shift due to mass change of pH-responsive material under a magnetic field	Magnetostrictive substrate, pH-sensitive coating, excitation/detection coils	Wireless readout possible, robust response	Newer tech, calibration, and stability are still under evaluation

Table 4.4 Types of soil nutrient sensors.

Sensor type	Nutrient	Working principle	Key materials/ Methods	Advantages	Limitations
NIR reflectance spectroscopy (Ali et al., 2018)	N, P	Measures NIR reflectance correlated to nutrient content	NIR detectors, optical filters, calibration models	Nondestructive, fast, *spatially mapped*	Surface-focused, *affected by soil* moisture
Optical colorimetric Sensors (Chen et al., 2019)	N, P	Detects color change via reaction with soil nutrients	Reagents + LEDs, photodiodes, filters	Simple, portable	Low sensitivity, short lifespan of reagents
Electrochemical ISE/SC-ISE Sensors (Akhter & Sofi, 2022)	NH_4^+, NO_3^-, PO_4^{3-}	The ion-selective membrane generates a potential based on the target ion concentration	PVC membranes, ionophores (e.g., nonactin), LIG electrodes, Ag/AgCl reference	Good sensitivity and selectivity, suitable for multi-depth monitoring	Sensor drift requires calibration
ISFET-based Sensors (Laraoui et al., 2015)	NH_4^+, NO_3^-	Field-effect transistor gate voltage shifts due to ion interaction	$SiO_2/Si_3N_4/Al_2O_3$ gate layers, ionophores, IC integration	Miniaturizable, integratable with electronics, real-t i m e sensing	Soil packaging and robustness issues

(*continued on next page*)

Microcantilever & MEMS Sensor (Filgueiraset al., 2020; Mukherjee et al., 2019)	NO_3^-, PO_4^{3-}	Detects deflection due to material swelling/shrinking upon ion exposure	SiO_2/Si_3N_4 cantilevers, MoS_2, polymers, gold layers	High sensitivity (down to pM), low power	Complex fabrication, environmental sensitivity
Chemiresistive SWCNT Sensors (Han et al., 2020)	NH_3 (indirect N)	Resistance change of carbon nanotube upon gas exposure	Single-walled carbon nanotubes (SWCNT), functional coatings	Ultra-sensitive (ppb), compact	Affected by temperature, humidity, and gas cross-sensitivity
Raman spectroscopy sensors (Gupta & Quan, 2018)	P	Inelastic scattering of laser light identifies chemical bonds	Raman laser (785 nm), fiber optics, portable spectrometer	Fast, high resolution, specific to chemical forms	Expensive, surface-limited, affected by fluorescence
Lab-on-chip systems (Imam et al., 2019; Ocheltree et al., 2020)	N, P	Integrated microfluidics + chem-ical/electrical detection	ISFETs, ISEs, microchannels, MEMS materials	Small, integrable, low sample/reagent use	Limited field durability, expensive to mass-produce
Indirect sensing (e.g., NDVI, RONS) (Wang et al., 2017)	N, P	Measures plant physiological response to nutrient status	Chlorophyll meters, NDVI sensors, ROS analyzers	Useful for mapping deficiency indirectly	Delayed feedback, indirect inference

Table 4.5 Difference between the conventional and quantum sensors.

Aspect	Conventional sensors	Quantum sensors
Working principle	Based on classical physics (e.g., electrical resistance, capacitance, optics)	Based on quantum phenomena such as quantum entanglement, tunnelling, superposition, or atomic transitions
Examples	Soil moisture (TDR, FDR), pH sensors, thermocouples, EC sensors	Quantum gravimeters, atomic magnetometers, nitrogen-vacancy (NV) center sensors
Accuracy	Moderate to high	Extremely high (can detect minute changes in gravity, fields, etc.)
Sensitivity	Limited to sensor resolution and noise floor	Ultra-sensitive, capable of detecting variations at the quantum scale
Environmental impact	Generally robust, but can be affected by temperature, pressure, etc.	Highly stable and less susceptible to classical environmental noise
Size and portability	Often compact and field-deployable	Often bulkier or require specialized lab/field setups (but miniaturization is improving)
Power consumption	Low to moderate	Varies; some require cryogenic cooling or laser systems
Cost	Affordable and widely available	High cost due to complex setup and calibration requirements
Ease of use	User-friendly, plug-and-play designs common	Requires technical expertise and calibration
Applications in agriculture	Soil moisture, temperature, light, and gas detection	Soil mapping (gravity), underground water table sensing, nutrient content mapping (emerging)
Examples of use	Irrigation control, precision fertilization, greenhouse monitoring	Mapping subsoil structures, detecting underground roots or water without physical contact
Technological maturity	Well-established, commercially widespread	Emerging technology, limited but growing field of deployments
Future potential	Incremental improvements	Transformational—new capabilities in sensing and diagnostics

1. The quantum system must have distinct resolvable levels of energy that are distinguished by a finite transition energy.
2. The quantum sensor must also be able to be adjusted coherently, typically using time-dependent fields, and
3. It should initialize the sensor into a known state and read out its state.

Using quantum processes, a promising range of sensing materials (Schirhagl et al., 2014) have been produced that test many environmental parameters, such as temperature (Kraus et al., 2014; Plakhotnik et al., 2014), electric (Tabassum et al., 2024) and magnetic (Cooper et al., 2014; Niethammer et al., 2016) fields, pH (Fujiwara et al., 2019; Rendler et al., 2017), thermal conductivity (Laraoui et al., 2015), strain (Ovartchaiyapong et al., 2014; Trusheim & Englund, 2016), and force (Braakman & Poggio, 2019).

The DiVincenzo and Degen requirements mentioned above must be met by materials that can be used as quantum sensors. A change in the material's quantum states, such as a shift in its energy levels, must also be predictable as a result of the sensor's selective interaction with factors of interest, such as temperature, pH, and magnetic field (Degen et al., 2017). Quantum sensors come in a wide variety of forms, each intended to monitor a certain physical quantity, such as temperature, location, magnetic fields, or time. Atomic clocks, magnetometers, gravimeters, gyroscopes, interferometers, and SQUIDs (Superconducting Quantum Interference Devices) are a few examples of typical quantum sensor types (Seskir & Seskir, 2022). Quantum sensors have a wide range of uses, from geophysical and medical imaging to navigation and timekeeping. Quantum sensors are anticipated to become increasingly important as technology develops in a variety of domains, including oil exploration, modeling, and cryptography.

There are several different types of soil sensors that monitor vital soil characteristics, including temperature, pH, moisture content, and nutrient levels (nitrogen, potassium, and phosphorus) (Soussi et al., 2024). To prevent resource waste and guarantee the best possible nutrient availability for crops, these sensors continuously assess soil conditions to direct precise irrigation and fertilization techniques (Reka et al., 2019).

Digital soil health monitoring uses data analytics techniques and digital technology to assess and monitor soil quality. Decisions regarding crop selection, soil management, and agricultural practices are informed by this priceless data, which increases crop yields, lowers input costs, optimizes resource allocation, promotes environmental sustainability, and fosters accuracy and efficiency in soil care (Adak et al., 2023).

4.2.4.3.6 Importance of quantum sensors in smart agriculture

Among the wide range of sensors used in smart agriculture—such as soil moisture, pH, temperature, and nutrient sensors. Table 4.6 specifically focuses on quantum sensors and their role in advanced agricultural monitoring. Quantum sensors operate based on quantum phenomena like superposition and entanglement, enabling extremely high sensitivity and precision compared to conventional sensors. These sensors can detect minute changes in environmental parameters, including soil composition, magnetic fields, and temperature variations, which are critical for precision farming. Their enhanced accuracy supports real-time decision-making for irrigation, fertilization, and crop health management. In addition, quantum sensors improve early stress and disease detection in crops. Overall, the integration of quantum sensors significantly strengthens data reliability and efficiency in smart agriculture systems.

4.2.4.3.7 Applications, working principle, and advantages of QS

Quantum superposition (Rubino et al., 2021) is defined as the process of combining two legitimate quantum states to create an additional permitted state. Quantum superposition is the ability of two or more quantum states to merge or superimpose to create a new, legitimate state. According to Gauglitz (2021), the quantum superposition notion makes it possible to create more sensitive quantum sensors, such as improved accelerometers, gyroscopes, and atomic clocks. With a 50% chance, the qubits can simultaneously take a 0 or a 1 state. In classical computing, on the other hand, the bits can adopt a state of either 0 or 1 (100 percent probability), shown in Fig. 4.10. One characteristic of quantum physics is quantum entanglement, which allows quantum particles to communicate and exchange information remotely.

Table 4.6 Various types of sensors in agriculture.

Quantum sensor	Quantum principle	Working principle	Use in agriculture	Example applications
Magnetometers	Electron/nuclear spin and	Measures magnetic fields via changes in spin states	Used to detect changes in soil mineral content,	Soil fertility mapping,
Gravimeters	Matter-wave interference and quantum coherence	Measures gravitational changes via atomic interferometry or quantum free-fall experiments	Used to measure soil moisture and underground water distribution	Irrigation management, groundwater monitoring
Gyroscopes	Sagnac effect and matter–wave interference	Detects rotation through the phase shift in interfering matter waves	Used in autonomous agricultural machinery for precise navigation and orientation	Smart tractors, drones, and robotic sprayers
Acoustic sensors	Quantum oscillations and phonon interactions	Detects sound via quantum vibration or oscillation of atoms/molecules	Monitor soil compaction and root growth through sound wave propagation	Soil quality analysis, root system health monitoring
Interferometers	Wave-particle duality and quantum interference	Measures phase difference between two or more wave paths (light or matter waves)	Employed in atmospheric monitoring to analyze greenhouse gases and crop respiration	Precision climate control in greenhouses, carbon flux analysis
Quantum imaging	Quantum entanglement and photon correlations	Uses entangled photons or quantum illumination to form high-resolution or low-light images	Helps in plant health imaging, photosynthesis monitoring, and early disease detection	Hyperspectral imaging, quantum-enhanced remote sensing

According to Khushboo, 2021, quantum entanglement connects two particles in such a way that, regardless of their distance from one another, their states cannot be represented separately, as shown in Fig. 4.11. The photons in these states become indistinguishable, and when quantum particles are entangled, they are inseparably connected regardless of their temporal or geographical dispersion. The measurement of one entangled particle would instantly impact the other, regardless of their distance from one another, which cannot be specified independently for each particle. High resolution, solid-state spins, and spatial sensitivity are among the fundamental ideas of quantum mechanics that served as the foundation for these sensors (Bonato et al., 2016).

Table 4.7 highlights various quantum sensors for soil monitoring.

4.2.4.3.8 Working principle of quantum sensors

1. Carbon Quantum Dots (CQDs) (Dong et al., 2024)–Plant Growth and Stress Monitoring Working Principle: Fluorescence Emission and Biological Interaction

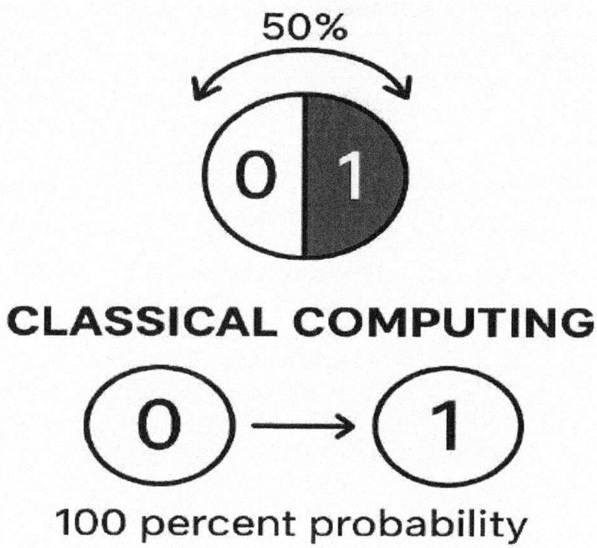

FIGURE 4.10 Quantum superposition.

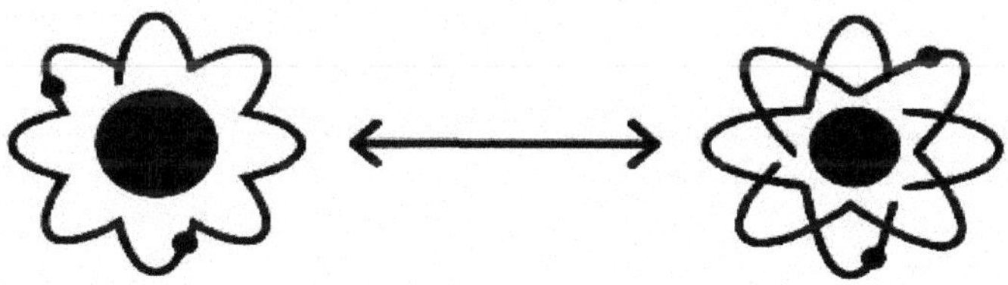

FIGURE 4.11 Quantum entanglement.

CQDs are nanoscale (typically <10 nm) carbon-based fluorescent nanomaterials characterized by excellent photoluminescence, biocompatibility, water solubility, and chemical stability. These properties make them highly suitable for biological and agricultural sensing applications. CQDs absorb ultraviolet or visible radiation and re-emit light at longer wavelengths through fluorescence emission, a process highly sensitive to changes in their surrounding biochemical environment.

In smart agriculture, CQDs are employed as optical nanosensors for monitoring plant physiological conditions and stress responses. Plants subjected to abiotic stresses (such as drought, salinity, temperature extremes) or biotic stresses (such as pathogens and pests) generate elevated levels of Reactive Oxygen Species (ROS).

CQD are introduced or injected in plant tissue by spraying, they responds to physiological differences in tissue and detected by light detectors. The output of photo detectors are signals with decreased strength and in analog form. Signal processing or Micro controllers are used to process the signal and also convert it into digital form for transmission. Wireless communication, including ZigBee, wi-fi, and Bluetooth protocols, is used to store the data in the cloud to perform

Table 4.7 Quantum sensors for soil health monitoring.

Application domain	Types of QD Sensors	Advantages	Key Challenges
Plant growth prozmotion (Ali et al., 2020; Zheng, 2021)	CQDs, GQDs	Enhanced germination and seedling growth via nano-priming. Improved crop stress tolerance. Increased shoot and root lengths	Limited to small-scale trials. High cost of CQD synthesis. Complex integration into large-scale farming systems
Pollutant detection in food & water (Khoshnoud et al., 2020; Nsibande & Forbes, 2016)	CQDs, GQDs, CdSe QDs, CdTe/ZnSe QDs, QB-ICA	Accurate detection of Pb^{2+}, pesticides, additives, and residues. Prevents contamination of food and water. High sensitivity in fluorescence-based detection	Enzyme activity is affected by heavy metals (e.g., Pb). Expensive and fragile fluorescence sensors. Difficult calibration and environmental sensitivity
Soil moisture sensing (Kalita et al., 2016; Kayad et al., 2020)	GQDs	Real-time moisture monitoring. Supports precision irrigation. Sensitive to a wide moisture range	Sensor performance varies across soil types. Small-scale trials dominate current research. Temperature and humidity impact readings
Veterinary drug & pesticide detection (Rehman et al., 2014; Degen et al., 2024)	CdSe QDs, CQDs	Detection of toxic residues in food. Supports food quality and safety. Applicable in water and produce screening	Cost of fabrication-Integration with food surveillance tools is still emerging. Need for user-friendly interpretation tools

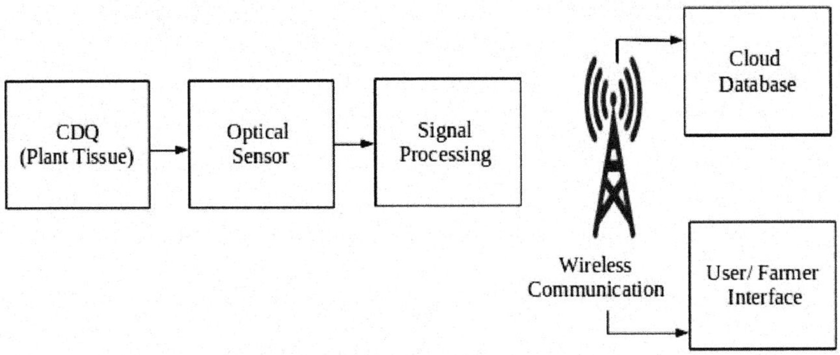

FIGURE 4.12 Real-time connectivity of CQD in smart agriculture.

analytics utilizing ML and AI. The data from the cloud, after doing analytics, preprocessings they send it to farmers through mobiles and provides the direction for pest recommendations. This real-time representation of CQD sensors in smart agriculture is shown in Fig. 4.12.

This integration enables early stress detection, reduced crop loss, optimized resource utilization, and improved yield prediction.

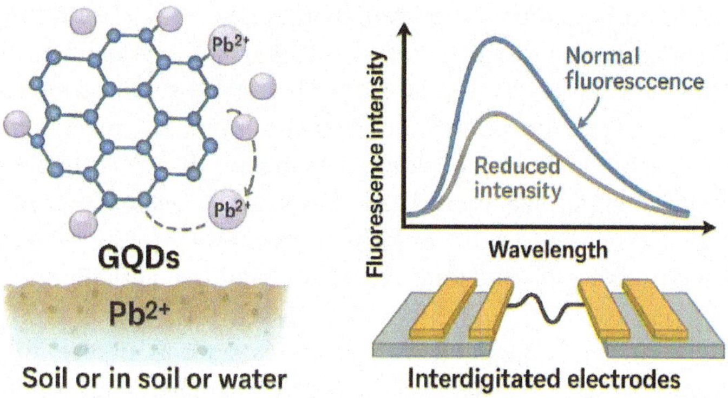

FIGURE 4.13 Graphene quantum dots working principle with outcome graphical representation, real-time smart agriculture integration.

2. Graphene Quantum Dots (GQDs) (Kalita et al., 2016)–Heavy Metal Detection Working Principle: Fluorescence Quenching and EC modulation

GQDs are nanoscale fragments of graphene exhibiting quantum confinement and edge effects, resulting in superior electronic, optical, and electrochemical properties. GQDs display stable fluorescence emission, high surface area, excellent electron mobility, and strong affinity toward metal ions, making them highly effective for environmental and agricultural contaminant sensing.

In contaminated agricultural soils and irrigation water, heavy metals such as lead (PbZ^+), cadmium (CdZ^+), and mercury (HgZ^+) pose serious risks to crop health and food safety. When GQDs interact with these ions, electron transfer mechanisms induce fluorescence quenching, leading to a reduction in emission intensity. This optical change provides a sensitive and selective means for heavy metal detection.

Beyond optical sensing, GQDs can be deposited onto interdigitated electrodes (IDEs) to form electrochemical sensors. In such configurations, soil moisture and ionic concentration variations influence EC. Increased ionic mobility enhances charge transfer, producing measurable changes in current or resistance. This dual sensing capability allows simultaneous monitoring of soil contamination and moisture levels (Fig. 4.13).

1. GQDs continuously sense contaminants and soil ionic variations.
2. Optical and electrical signals are processed by embedded systems.
3. Sensor nodes transmit data through IoT gateways.
4. Cloud platforms perform contamination risk analysis and trend prediction.
5. Decision-support systems generate actionable insights, such as soil remediation alerts, irrigation adjustments, and contamination warnings (Fig. 4.13).

3. CdSe Quantum Dots (She & Wang, 2019)–Enzyme-Assisted Fluorescence Modulation

CdSe QD–urease biosensors offer several advantages for agricultural pollutant monitoring, including high sensitivity, enzyme-assisted signal amplification, rapid response time, and suitability for in situ measurements. Their ability to detect heavy metal contamination in real time makes them valuable tools for safeguarding soil health, water quality, and crop productivity.

Cadmium Selenide (CdSe) quantum dots (QDs) are II–VI semiconductor nanocrystals that exhibit size-dependent optical properties, high quantum yield, narrow emission bandwidth, and excellent photostability. These characteristics make CdSe QDs highly suitable as optical transducers in biosensing applications. Their fluorescence response is particularly sensitive to changes in surface chemistry and local environmental conditions, such as pH and ionic composition, which can be exploited for pollutant detection in agricultural and environmental systems.

When CdSe QDs are bioconjugated with enzymes such as urease, they form functional biosensing complexes capable of selectively detecting agrochemical contaminants and heavy metals in soil and water matrices. Urease catalyzes the hydrolysis of urea into ammonia and carbon dioxide according to the reaction:

$$\text{Urea} + 2H_2O \xrightarrow{\text{Urease}} 2NH_3 + CO_2$$

The release of ammonia increases the local pH in the vicinity of the CdSe QDs. This pH elevation modifies the surface charge distribution and electronic states of the quantum dots, leading to an enhancement in fluorescence intensity. Under normal conditions, the enzymatic activity of urease results in a strong and stable fluorescence signal from the CdSe QDs, which serves as a baseline response of the biosensor.

However, the presence of heavy metal ions such as PbZ^+, CdZ^+, and HgZ^+ significantly affects the enzymatic function of urease. These metal ions exhibit strong affinity toward functional groups within the enzyme, including sulfhydryl, amino, and imidazole residues, causing structural deformation or blockage of the enzyme's active site. As a result, urease activity is inhibited, suppressing the hydrolysis of urea and preventing the associated pH increase. This inhibition leads to a marked reduction in CdSe QD fluorescence intensity.

The decrease in fluorescence emission serves as a quantitative indicator of heavy metal contamination. The extent of fluorescence quenching is directly correlated with the concentration of inhibitory metal ions, enabling sensitive and selective detection of pollutants at trace levels. This fluorescence modulation mechanism allows CdSe QD–urease biosensors to function as reliable indicators of soil and water quality in agricultural environments (Fig. 4.14).

For real-time environmental monitoring, the optical response of the CdSe QD–urease biosensor can be integrated with embedded sensing and communication systems. The biosensor is excited using a suitable light source, typically a UV or blue light-emitting diode, and the emitted fluorescence is captured using a photodetector such as a photodiode or CMOS sensor. The optical signal is converted into an electrical signal, amplified, filtered, and digitized using an embedded processing unit.

The processed data are transmitted through wireless communication modules to a centralized or distributed data platform, where continuous analysis is performed. This enables real-time tracking of contamination levels, temporal trend analysis, and early warning generation for agricultural management and environmental protection (Fig. 4.15).

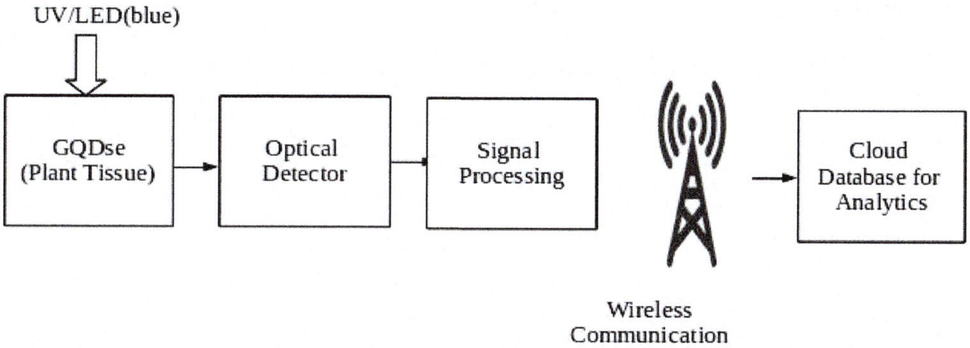

FIGURE 4.14 Real-time connectivity of GQD in smart agriculture. This real-time integration enhances soil safety monitoring, water quality assessment, environmental sustainability, and food security.

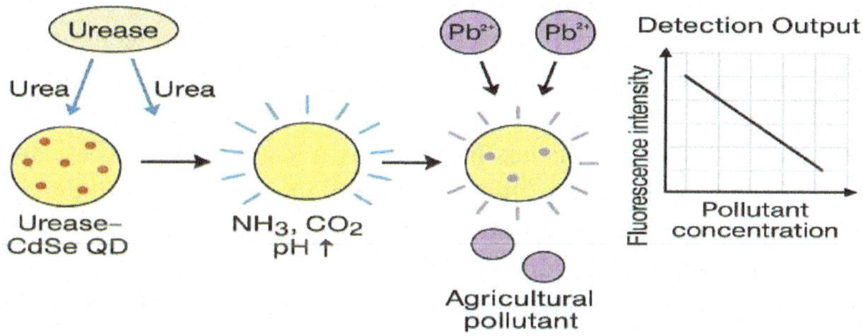

FIGURE 4.15 CdSe quantum dots working principle with outcome graphical representation.

4. CdTe/ZnSe Core (Sevgül et al., 2025)–Shell Quantum Dots–Ratiometric Ion Sensing Working Principle:

Dual-Wavelength Ratiometric Fluorescence Core-shell CdTe/ZnSe quantum dots offer a powerful ratiometric fluorescence-based sensing approach for lead ion detection, enabling accurate, stable, and real-time monitoring of water quality in agricultural environments. Colloidal semiconductor nanocrystals based on cadmium telluride and zinc selenide (CdTe/ZnSe) quantum dots (QDs) have emerged as highly effective optical probes for heavy metal detection in agricultural and environmental applications. These nanocrystals are engineered in a core-shell configuration, wherein the CdTe core serves as the primary emissive center, while the ZnSe shell provides enhanced chemical stability, surface passivation, and improved selectivity toward target analytes. The shell layer effectively reduces surface trap states and suppresses nonradiative recombination, resulting in improved fluorescence intensity and long-term photostability.

The CdTe/ZnSe QDs are specifically designed for ratiometric fluorescence sensing, a technique that relies on the simultaneous generation of two emission bands under a single excitation wavelength. One emission peak is predominantly associated with the CdTe core, whereas the second arises from shell-related or interfacial electronic states. The ratio of these two emission

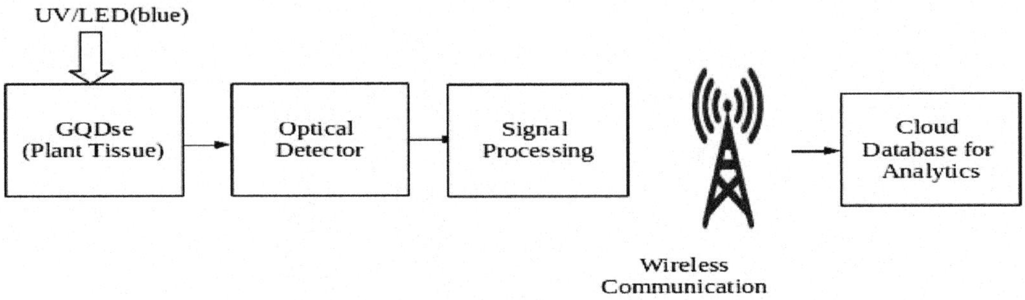

FIGURE 4.16 Real-time monitoring architecture incorporating the CdSe QD.

intensities serves as an internally referenced sensing parameter, enabling high analytical reliability (Fig. 4.16).

Ratiometric sensing offers a significant advantage over single-emission fluorescence techniques by compensating for variations in excitation intensity, probe concentration, optical path length, and environmental conditions. In CdTe/ZnSe QDs, this approach is realized through differential modulation of the two emission peaks upon interaction with lead ions (Pb^{2+}). PbZ^{+} ions selectively bind to functional ligands present on the ZnSe shell, such as carboxyl, amine, or thiol groups. This interaction alters the local electronic environment of the quantum dots, affecting charge transfer and exciton recombination processes. Consequently, one fluorescence emission band undergoes quenching or enhancement, while the second emission band remains relatively unaffected. The resulting change in the fluorescence intensity ratio (I_1/I_2) correlates directly with PbZ^{+} concentration.

For continuous environmental surveillance, CdTe/ZnSe ratiometric QDs can be integrated into real-time optical sensing platforms combined with embedded electronics and wireless communication modules. The QDs are immobilized on solid substrates or embedded within microfluidic channels through which water or soil extracts flow continuously. A single UV or blue light-emitting diode excites the quantum dots, while the emitted fluorescence at two distinct wavelengths is separated using optical filters and detected by photodiodes or CMOS sensors.

The detected optical signals are converted into electrical signals, amplified, filtered, and digitized by an embedded processing unit (Fig. 4.17).

Ratiometric computation is performed in real time, enabling continuous estimation of PbZ^{+} concentration. The processed data are then transmitted through wireless communication protocols such as LoRa, Wi-Fi, or Bluetooth Low Energy to centralized or edge-based platforms for further analysis.

5. QB-ICA (Quantum Dot-Based Immunochromatographic Assay) (Qianzhao, 2024)–Pesticide and Drug Residue Detection

Working Principle: Lateral Flow Immunoassay Enhanced with Quantum Dot Fluorescence

The Quantum Dot-Based Immunochromatographic Assay (QB-ICA) is an advanced form of lateral flow immunoassay that utilizes semiconductor quantum dots as fluorescent signal labels instead of conventional colorimetric reporters. Owing to their high quantum yield, narrow

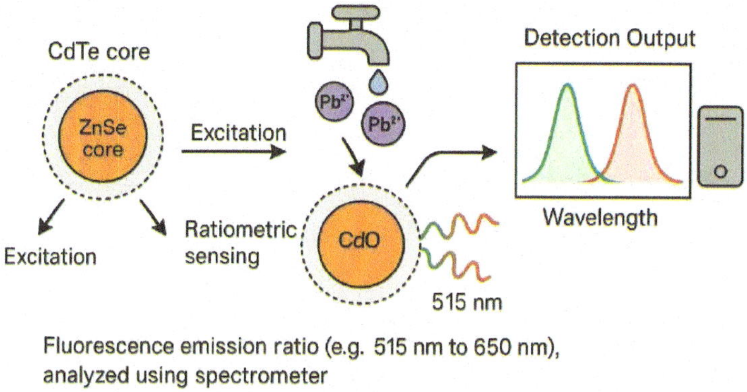

FIGURE 4.17 CdTe/ZnSe Quantum dots working principle with outcome graphical representation.

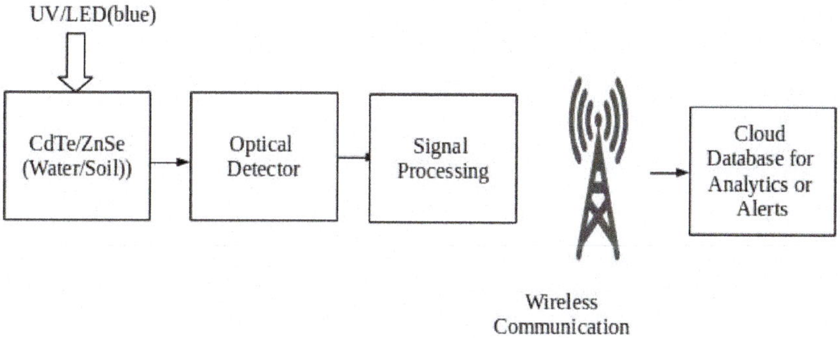

FIGURE 4.18 Real-time monitoring architecture incorporating the CdTe/ZnSe QD.

emission bandwidth, size-tunable fluorescence, and excellent photostability, quantum dots significantly improve the sensitivity and reliability of immunochromatographic detection. In QB-ICA, quantum dots are covalently conjugated with specific antibodies to form QD–antibody probes that selectively recognize target analytes such as pesticide residues, veterinary drugs, mycotoxins, and food additives.

When a liquid sample is applied to the test strip, it migrates through the strip by capillary action, passing sequentially through the sample pad, conjugate pad, nitrocellulose membrane, and absorbent pad. The conjugate pad contains preloaded QD–antibody complexes that are released upon contact with the sample. If the target analyte is present, it binds with the QD–antibody probes to form immune complexes. These complexes continue to migrate along the nitrocellulose membrane and are selectively captured at the test line by immobilized antigens or antibodies, while excess probes are captured at the control line to confirm proper assay performance (Fig. 4.18).

Depending on the molecular size of the analyte, QB-ICA can operate in either competitive or sandwich assay formats. For small-molecule analytes such as pesticides and veterinary drugs, a competitive format is typically employed, in which the analyte in the sample competes

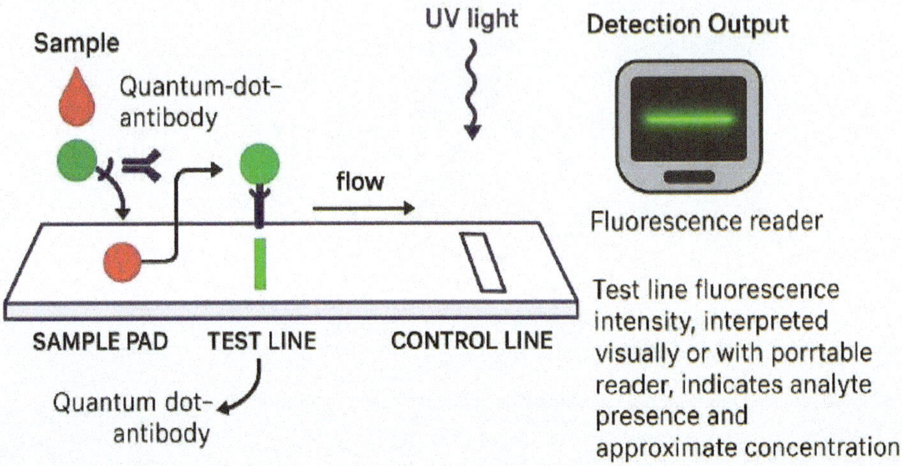

FIGURE 4.19 Quantum dot-based immunochromatographic assay working principle with outcome graphical representation.

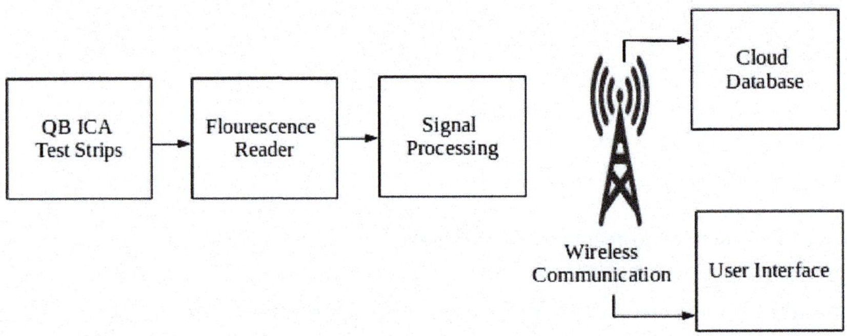

FIGURE 4.20 Real-time monitoring architecture incorporating the QB-ICA QD.

with immobilized antigen at the test line for binding to the QD–antibody probes. In this case, higher analyte concentrations result in weaker fluorescence signals at the test line. For larger biomolecules, a sandwich format may be used, where the target is simultaneously captured by the test line antibody and the QD-labeled detection antibody, producing fluorescence intensity proportional to the analyte concentration.

The presence and concentration of the target analyte are determined by measuring the fluorescence emitted from the test line upon excitation with ultraviolet or blue light. The fluorescence signal can be visually observed using a handheld UV lamp for qualitative analysis or quantitatively measured using a portable fluorescence reader. Compared with traditional gold nanoparticle-based immunochromatographic assays, QB-ICA offers significantly lower detection limits, higher signal-to-noise ratios, improved quantification capability, and the potential for multiplex detection through the use of multicolor quantum dots (Figs. 4.19 and 4.20).

When integrated with compact fluorescence readers and digital electronics, QB-ICA enables real-time data acquisition and transmission. Fluorescence intensities from the test and control lines are converted into digital signals, processed using calibration algorithms, and transmitted wirelessly through Bluetooth, Wi-Fi, or other communication modules to external devices or cloud-based platforms. This allows instant data storage, threshold-based decision-making, and remote monitoring, making QB-ICA particularly suitable for on-site applications in food safety inspection, agricultural monitoring, and regulatory compliance. Overall, QB-ICA represents a robust, rapid, and highly sensitive analytical platform that combines the simplicity of lateral flow assays with the superior optical properties of quantum dots and the capabilities of real-time digital monitoring systems.

4.2.4.3.9 Applications of QS in different areas

In a variety of fields, quantum sensors are enabling previously unheard-of levels of sensitivity and precision in measurements by taking advantage of quantum mechanical concepts like superposition and entanglement. Applications in aircraft, submarines, and autonomous vehicles depend on high-precision data from quantum accelerometers and gravimeters in geophysics and navigation for GPS-independent navigation and subterranean mapping (Bongs et al., 2019). Diamond's nitrogen-vacancy (NV) centers are utilized in the medical field for biomagnetic imaging methods such as magneto encephalography (MEG), which provide high-resolution, noninvasive brain activity monitoring (Barry et al., 2016). Additionally, these sensors are used in nanoscale diagnostics and single-molecule detection, greatly boosting biomedical research (Schirhagl et al., 2014). The use of quantum-enhanced magnetic and gas sensors in environmental monitoring helps identify contaminants, volcanic activity, and geomagnetic abnormalities (Degen et al., 2024).

Situational awareness is improved in security and defense by using quantum radar and magnetometers to detect submerged submarines and stealth aircraft (Khabiboulline et al., 2019). In astronomy and space science, atomic clocks help with accurate satellite navigation, while quantum-enhanced interferometry is employed in gravitational wave detectors like LIGO (Tino et al., 2019). Additionally, these sensors aid in basic physics studies, such as redefining SI units and testing general relativity and dark matter (Safronova et al., 2018). Quantum sensors are used in industry for structural health monitoring and nondestructive testing (NDT) of infrastructure, where they may identify subtle changes in stress and subsurface damage. By using more sensitive detection of magnetic and gravitational anomalies than traditional instruments, they improve subsurface exploration in the oil and gas sector. Thus, in the fields of science, engineering, and technology, quantum sensing is quickly becoming a vital instrument (Table 4.8).

4.2.4.3.10 Challenges in digital soil health monitoring

Sustainable land management techniques and agriculture could undergo a transformation thanks to digital soil health monitoring. But it also presents a unique set of difficulties. To continually evaluate the physical, chemical, and biological characteristics of soil, digital soil health monitoring combines in-situ sensors, remote sensing, IoT platforms, and data analytics. Although this paradigm encourages sustainable land management and precision agriculture, its operational dependability and scalability are constrained by a number of technological issues.

Table 4.8 Types of sensors and their applications in soil health monitoring.

Quantum sensor type	Soil parameter monitored	Application
Quantum gravimeter	Soil density, porosity, water content	Soil compaction, erosion detection
Quantum magnetometer	Magnetic minerals, contaminants	Soil fertility mapping, pollution detection
Quantum moisture sensor	Water content	Irrigation planning, drought assessment
Quantum Ion/pH sensor	pH, nutrient ions	Nutrient management, soil chemistry monitoring
Quantum spectroscopic sensor	Organic content, trace gases	Soil health, microbial activity, carbon monitoring

4.3 Data reliability and standardization

Heterogeneous sensing technologies (dielectric, optical, electrochemical, and biosensors) and laboratory-based validation techniques are used to gather data on soil health. Data inconsistency frequently results from variations in sensor design, measurement depth, temporal resolution, and calibration procedures. Cross-sensitivity effects are also introduced by soil characteristics, including moisture, salinity, and texture, especially in sensors that measure EC and nutrients. Data interoperability between systems and countries is hampered by a lack of common data formats, metadata descriptors, and benchmarking methods. To guarantee the repeatability and comparability of soil health assessments at regional and global scales, consistent sensing techniques, reference soils, and harmonized calibration frameworks must be established.

4.4 Data management, storage, and processing volume

Digital soil monitoring systems produce multi-dimensional, high-frequency datasets that include temperature gradients, microbial activity indicators, soil moisture dynamics, nutrient fluxes, and plant-soil interactions. Scalable cloud infrastructure, effective data compression, and real-time data streaming architectures are necessary for handling these massive volumes. Moreover, preprocessing procedures, including sensor drift correction, spatial interpolation, noise filtering, and missing data imputation, greatly increase computing complexity. Real-time soil monitoring in the field requires more sophisticated big-data frameworks and edge computing solutions.

4.5 Cost, accessibility, and infrastructure constraints

Costs related to sensors, communication modules, power supplies (such as solar units), data transmission, and software platforms are involved in the implementation of digital soil monitoring systems. Adoption is frequently further hampered by ongoing expenses for sensor replacement, recalibration, and subscription-based analytics services. Real-time monitoring is not feasible in remote or resource-constrained areas due to poor internet connectivity and a lack of digital infrastructure. To increase acceptance among smallholder farmers and developing countries, a low-cost, energy-efficient, and modular sensor network design is essential.

4.6 Sensor reliability, stability, and calibration

Soil sensors operate in harsh environments characterized by fluctuating moisture content, temperature extremes, biological activity, and chemical aggressiveness. These conditions can cause sensor drift, fouling, corrosion, and signal degradation over time. For example, ISEs used for nutrient sensing often exhibit reduced selectivity and lifespan under variable soil pH and salinity. Continuous calibration against laboratory reference measurements is necessary but operationally challenging in large-scale deployments. Research is ongoing into self-calibrating sensors, adaptive algorithms, and robust sensor materials to improve long-term measurement stability.

4.7 Data interpretation and multiscale integration

Physical structure, chemical composition, and biological activity interact intricately to produce the emergent attribute of soil health. Advanced analytical methods like ML, data fusion, and mechanistic modeling are needed to convert unprocessed sensor outputs into useful indicators of soil health. Another level of complexity is added when in-situ sensor data is combined with weather information, remote sensing observations, and crop growth models. Crop-specific reactions to soil conditions, temporal variability, and spatial heterogeneity must all be taken into consideration by decision-support systems.

4.8 Capacity building, skill gaps, and knowledge transfer

The ability of end users to employ digital soil monitoring technology effectively depends on their technical capabilities. Land managers and farmers are frequently untrained in data interpretation, sensor deployment, system upkeep, and technology-driven decision-making. Targeted training initiatives, user-friendly interfaces, and interactive strategies that combine farmer expertise with digital information are necessary to close this gap. The advantages of digital soil health monitoring are still underutilized without sufficient capacity building.

4.9 Conclusion

Monitoring soil health is crucial for climate resilience, environmental preservation, and sustainable agriculture. For the measurement of variables including moisture content, pH, temperature, and nutrient levels, traditional sensors like mechanical, optical, and electrochemical probes have been frequently used. These technologies have proven to be reliable, affordable, and well-suited to current farming methods. They frequently have issues with sensitivity, selectivity, and long-term stability in shifting field conditions, though. New developments in quantum sensor technology present a revolutionary opportunity to address these issues. Quantum sensors offer previously unheard-of levels of sensitivity, precision, and miniaturization by utilizing quantum phenomena, including entanglement, superposition, and tunneling. Methods like quantum gravimetry, NV-center-based magnetometry, and quantum-enhanced spectroscopy allow for the more accurate and timely observation of soil properties at the molecular and subsurface levels. Combining quantum and classical sensors can result in a hybrid monitoring system that takes advantage of both the advanced features of quantum sensing and the reliability of conventional techniques. Such collaboration can help with real-time data collection, greatly enhance soil diagnostics, and boost precision agriculture by enabling more intelligent decision-making. In order to fully realize the potential of quantum-enhanced soil health monitoring, interdisciplinary research and developments in quantum hardware, data analytics, and sensor network integration will prove essential.

4.10 Future Scope

The combined use of quantum cryptography for agri-data security and quantum sensors for soil health monitoring, which enables safe and intelligent precision farming systems, represents the future potential of quantum technology in agriculture. In order to protect sensitive data pertaining to crop health, irrigation, yield prediction, and agri-supply chains from cyber threats and potential quantum attacks, quantum cryptography can offer ultra-secure communication for data exchanged between soil sensors, drones, satellites, and cloud-based decision platforms. In addition, quantum sensors provide previously unheard-of sensitivity for real-time monitoring of soil parameters like salinity, pH, moisture, nutrient content, and root-zone dynamics, enabling better resource utilization and early detection of soil degradation. Climate-resilient and sustainable agriculture can be greatly improved by integrating quantum sensing with IoT, AI-driven analytics, and digital soil twins, supported by quantum-secure data transfer. This will open the door for the development of next-generation smart farming ecosystems.

References

Adak, S., Mandal, N., Maity, P. P., Mukhopadhyay, A., & Gunturi, A. (2023). Digital soil health: Unveiling the potential of soil health monitoring. *Food and Scientific Reports, 4*(9), 96–103.

Adamchuk, V. I., Hummel, J. W., Morgan, M. T., & Upadhyaya, S. K. (2004). On-the-go soil sensors for precision agriculture. *Computers and Electronics in Agriculture, 44*(1), 71–91. https://doi.org/10.1016/j.compag.2004.03.002.

Ahmed, N., De, D., & Hussain, I. (2018). Internet of things (IoT) for smart precision agriculture and farming in rural areas. *IEEE Internet of Things Journal, 5*(6), 4890–4899. https://doi.org/10.1109/JIOT.2018.2879579.

Akhter, R., & Sofi, S. A. (2022). Precision agriculture using IoT data analytics and machine learning. *Journal of King Saud University - Computer and Information Sciences, 34*(8), 5602–5618. https://doi.org/10.1016/j.jksuci.2021.05.013.

Ali, M. A., Dong, L., Dhau, J., Khosla, A., & Kaushik, A. (2020). Perspective—electrochemical sensors for soil quality assessment. *Journal of The Electrochemical Society, 167*(3), 037550. https://doi.org/10.1149/1945-7111/ab69fe.

Ali, T. A. A., Choksi, V., & Potdar, M. B. (2018). Precision agriculture monitoring system using green internet of things (G-IoT). In *Proceedings of the 2nd International Conference on Trends in Electronics and Informatics, ICOEI 2018* (pp. 481–487). *May 2018*.https://doi.org/10.1109/ICOEI.2018.8553866.

Anisi, M. H., Abdul-Salaam, G., & Abdullah, A. H. (2015). A survey of wireless sensor network approaches and their energy consumption for monitoring farm fields in precision agriculture. *Precision Agriculture, 16*(2), 216–238. https://doi.org/10.1007/s11119-014-9371-8.

Atatüre, M., Englund, D., Vamivakas, N., Lee, S. Y., & Wrachtrup, J. (2018). Material platforms for spin-based photonic quantum technologies. *Nature Reviews Materials, 3*(5), 38–51. https://doi.org/10.1038/s41578-018-0008-9.

Bacco, M., Berton, A., Gotta, A., & Caviglione, L. (2018). IEEE 802.15.4 Air-ground UAV communications in smart farming scenarios. *IEEE Communications Letters, 22*(9), 1910–1913. https://doi.org/10.1109/LCOMM.2018.2855211.

Barry, J. F., Turner, M. J., Schloss, J. M., Glenn, D. R., Song, Y., Lukin, M. D., Park, H., & Walsworth, R. L. (2016). Optical magnetic detection of single-neuron action potentials using quantum defects in diamond. *Proceedings of the National Academy of Sciences of the United States of America, 113*(49), 14133–14138. https://doi.org/10.1073/pnas.1601513113.

Bauer, J., & Aschenbruck, N. (2018). Design and implementation of an agricultural monitoring system for smart farming. *2018 IoT Vertical and Topical Summit on Agriculture - Tuscany, IOT Tuscany 2018*, 1–6. https://doi.org/10.1109/IOT-TUSCANY.2018.8373022.

Benioff, P. (1980). The computer as a physical system: A microscopic quantum mechanical Hamiltonian model of computers as represented by Turing machines. *Journal of Statistical Physics, 22*(5), 563–591. https://doi.org/10.1007/BF01011339.

Bersani, C., Ruggiero, C., Sacile, R., Soussi, A., & Zero, E. (2022). Internet of things approaches for monitoring and control of smart greenhouses in industry 4.0. *Energies*, (10), 15. https://doi.org/10.3390/en15103834.

Bhoyar, P., Sahare, P., Dhok, S. B., & Deshmukh, R. B. (2019). Communication technologies and security challenges for internet of things: A comprehensive review. *AEU - International Journal of Electronics and Communications, 99*, 81–99. https://doi.org/10.1016/j.aeue.2018.11.031.

Bonato, C., Blok, M. S., Dinani, H. T., Berry, D. W., Markham, M. L., Twitchen, D. J., & Hanson, R. (2016). Optimized quantum sensing with a single electron spin using real-time adaptivemeasurements. *Nature Nanotechnology, 11*(3), 247–252. https://doi.org/10.1038/nnano.2015.261.

Bongs, K., Holynski, M., Vovrosh, J., Bouyer, P., Condon, G., Rasel, E., Schubert, C., Schleich, W. P., & Roura, A. (2019). Taking atom interferometric quantum sensors from the laboratory to real-world applications. *Nature Reviews Physics, 1*(12), 731–739. https://doi.org/10.1038/s42254-019-0117-4.

Braakman, F. R., & Poggio, M. (2019). Force sensing with nanowire cantilevers. *Nanotechnology*, (33), 30. https://doi.org/10.1088/1361-6528/ab19cf.

Chen, W. L., Lin, Y. B., Lin, Y. W., Chen, R., Liao, J. K., Ng, F. L., Chan, Y. Y., Liu, Y. C., Wang, C. C., Chiu, C. H., & Yen, T. H. (2019). AgriTalk: IoT for precision soil farming of turmeric cultivation. *IEEE Internet of Things Journal, 6*(3), 5209–5223. https://doi.org/10.1109/JIOT.2019.2899128.

Cooper, A., Magesan, E., Yum, H. N., & Cappellaro, P. (2014). Time-resolved magnetic sensing with electronic spins in diamond. *Nature Communications, 5*, 1–7. https://doi.org/10.1038/ncomms4141.

Deep, V., & Elarabi, T. (2017). Efficient IEEE 802.15.4 ZigBee standard hardware design for IoT applications. In Proceedings - International Conference on Signals and Systems, ICSigSys *2017* (pp. 261–265). https://doi.org/10.1109/ICSIGSYS.2017.7967053.

Degen, C. L., Reinhard, F., & Cappellaro, P. (2017). Quantum sensing. 1–45.Review of modern physics,89(3), article 035002.

Degen, C. L., Reinhard, F., Cappellaro, P., Malia, B. K., Wu, Y. Y., Martínez-Rincón, J., Kasevich, M. A., Zhang, Z., Zhuang, Q., Aslam, N., Zhou, H., Urbach, E. K., Turner, M. J., Walsworth, R. L., Lukin, M. D., Park, H., Rodzoń, I., Zhang, X., Ivády, V., … Feature, E. (2024). PH- response quantum dots with orange-red emission for monitoring the residue, distribution, and variation of an organophosphorus pesticide in an agricultural crop. *Science, 6*(1), 1–10. https://doi.org/10.1038/s41566-018-0301-6.

DiVincenzo, D. P. (2000). The physical implementation of quantum computation. *Fortschritte Der Physik, 48*(9–11), 771–783. https://doi.org/10.1002/1521-3978(200009)48:9/11⟨771::AID-PROP771⟩3.0.CO;2-E.

Dong, Z., Qi, J., Yue, L., Zhou, H., Chen, L., Gu, J., He, Y., & Wu, H. (2024). Biomass-based carbon quantum dots and their agricultural applications. *Plant Stress, 11*(February), 100411. https://doi.org/10.1016/j.stress.2024.100411.

Filgueiras, R., Almeida, T. S., Mantovani, E. C., Dias, S. H. B., Fernandes-Filho, E. I., da Cunha, F. F., & Venancio, L. P. (2020). Soil water content and actual evapotranspiration predictionsusing regression algorithms and remote sensing data. *Agricultural Water Management, 241*(June), 106346. https://doi.org/10.1016/j.agwat.2020.106346.

Fujiwara, M., Tsukahara, R., Sera, Y., Yukawa, H., Baba, Y., Shikata, S., & Hashimoto, H. (2019). Monitoring spin coherence of single nitrogen-vacancy centers in nanodiamonds during pH changes in aqueous buffer solutions. *RSC Advances, 9*(22), 12606–12614. https://doi.org/10.1039/c9ra02282a.

Gauglitz, G. (2021). The new generation: Quantum sensors. *Analytical and Bioanalytical Chemistry, 413*(23), 5679–5680. https://doi.org/10.1007/s00216-021-03554-7.

Guignard, M. S., Leitch, A. R., Acquisti, C., Eizaguirre, C., Elser, J. J., Hessen, D. O., Jeyasingh, P. D., Neiman, M., Richardson, A. E., Soltis, P. S., Soltis, D. E., Stevens, C. J., Trimmer, M., Weider, L. J., Woodward, G., & Leitch, I. J. (2017). Impacts of nitrogen and phosphorus: From genomes to natural ecosystems and agriculture. *Frontiers in Ecology and Evolution, 5*(JUL). https://doi.org/10.3389/fevo.2017.00070.

Gupta, G. S., & Quan, V. M. (2018). Multi-sensor integrated system for wireless monitoring of greenhouse environment. In 2018 IEEE Sensors Applications Symposium, SAS 2018 - Proceedings, 2018-January (pp. 1–6). https://doi.org/10.1109/SAS.2018.8336723.

Han, Y., Bai, X., Shao, W., & Wang, J. (2020). Retrieval of soil moisture by integrating Sentinel- 1A and MODIS data over agricultural fields. *Water (Switzerland)*, (6), 12. https://doi.org/10.3390/w12061726.

Hu, J. G., Mei, W. Z., Wu, J., Li, J. W., & Wang, D. M. (2023). A fully integrated RFID reader SoC. *Micromachines, 14*(9), 1–15. https://doi.org/10.3390/mi14091691.

Imam, M. Y., Jannat, N., Bibi, F., & Khan, G. S. (2019). Effective study of home plants in purity of territory by utilizing wireless sensor system. In *Proceedings of 2019 16th International Bhurban Conference on Applied Sciences and Technology, IBCAST 2019* (pp. 322–327). https://doi.org/10.1109/IBCAST.2019.8667206.

Jo, H. S., Park, C., Lee, E., Choi, H. K., & Park, J. (2020). Path loss prediction based on machine learning techniques: Principal component analysis, artificial neural network and gaussian process. *Sensors (Switzerland), 20*(7). https://doi.org/10.3390/s20071927.

Kalita, H., Palaparthy, V. S., Baghini, M. S., & Aslam, M. (2016). Graphene quantum dot soil moisture sensor. *Sensors and Actuators, B: Chemical, 233*, 582–590. https://doi.org/10.1016/j.snb.2016.04.131.

Kayad, A., Paraforos, D. S., Marinello, F., & Fountas, S. (2020). Latest advances in sensor applications in agriculture. *Agriculture (Switzerland), 10*(8), 1–8. https://doi.org/10.3390/agriculture10080362.

Khabiboulline, E. T., Borregaard, J., De Greve, K., & Lukin, M. D (2019). Optical interferometry with quantum networks. *Physical Review Letters, 123*(7), 1–13. https://doi.org/10.1103/PhysRevLett.123.070504.

Khoshnoud, F., Quadrelli, M. B., Esat, I. I., & Robinson, D. (2020). Quantum cooperative robotics and autonomy. http://arxiv.org/abs/2008.12230

Khushboo (2021). Entanglement and superposition in quantum computing and communication. *International Journal of Advances in Engineering and Management, 3*(9), 544–552. https://doi.org/10.35629/5252-0309544552.

Kraus, H., Soltamov, V. A., Fuchs, F., Simin, D., Sperlich, A., Baranov, P. G., Astakhov, G. V., & Dyakonov, V. (2014). Magnetic field and temperature sensing with atomic-scale spin defects in silicon carbide. *Scientific Reports, 4*, 1–8. https://doi.org/10.1038/srep05303.

Kumar, P. (2023). Quantum computing – next generation computing technology–A gentle introduction. *Futuristic Trends in Information Technology Volume 2 Book 20, February*, 108–120. https://doi.org/10.58532/v2bs20p1ch8.

Laraoui, A., Aycock-Rizzo, H., Gao, Y., Lu, X., Riedo, E., & Meriles, C. A. (2015). Imaging thermal conductivity with nanoscale resolution using a scanning spin probe. *Nature Communications, 6*(May). https://doi.org/10.1038/ncomms9954.

Maroli, A., Narwane, V. S., & Gardas, B. B. (2021). Applications of IoT for achieving sustainability in agricultural sector: A comprehensive review. *Journal of Environmental Management, 298*(August), 113488. https://doi.org/10.1016/j.jenvman.2021.113488.

Mukherjee, A., Misra, S., & Raghuwanshi, N. S. (2019). A survey of unmanned aerial sensing solutions in precision agriculture. *Journal of Network and Computer Applications, 148*, 102461. https://doi.org/10.1016/j.jnca.2019.102461.

Niethammer, M., Widmann, M., Lee, S. Y., Stenberg, P., Kordina, O., Ohshima, T., Son, N. T., Janzén, E., & Wrachtrup, J. (2016). Vector magnetometry using silicon vacancies in 4H -SiC under ambient conditions. *Physical Review Applied, 6*(3), 1–8. https://doi.org/10.1103/PhysRevApplied.6.034001.

Nsibande, S. A., & Forbes, P. B. C. (2016). Fluorescence detection of pesticides using quantum dot materials – A review. *Analytica Chimica Acta, 945*, 9–22. https://doi.org/10.1016/j.aca.2016.10.002.

Ocheltree, T. W., Mueller, K. M., Chesus, K., LeCain, D. R., Kray, J. A., & Blumenthal, D. M. (2020). Identification of suites of traits that explains drought resistance and phenologicalpatterns of plants in a semi-arid grassland community. *Oecologia, 192*(1), 55–66. https://doi.org/10.1007/s00442-019-04567-x.

Onojeghuo, A. O., Blackburn, G. A., Huang, J., Kindred, D., & Huang, W. (2018). Applications of satellite 'hyper-sensing' in Chinese agriculture: Challenges and opportunities. *International Journal of Applied Earth Observation and Geoinformation, 64*, 62–86. September 2017 https://doi.org/10.1016/j.jag.2017.09.005.

Ovartchaiyapong, P., Lee, K. W., Myers, B. A., & Jayich, A. C. B. (2014). Dynamic strain- mediated coupling of a single diamond spin to a mechanical resonator. *Nature Communications, 5*, 1–6. https://doi.org/10.1038/ncomms5429.

Pirandola, S., Bardhan, B. R., Gehring, T., Weedbrook, C., & Lloyd, S. (2018). Advances in photonic quantum sensing. *Nature Photonics, 12*(12), 724–733. https://doi.org/10.1038/s41566-018-0301-6.

Plakhotnik, T., Doherty, M. W., Cole, J. H., Chapman, R., & Manson, N. B. (2014). All-optical thermometry and thermal properties of the optically detected spin resonances of the NV- center in nanodiamond. *Nano Letters, 14*(9), 4989–4996. https://doi.org/10.1021/nl501841d.

Qianzhao, S. (2024). 量子点免疫层析技术在兽药残留快速检测中的应用. 173–180.

Rehman, A., Abbasi, A. Z., Islam, N., & Shaikh, Z. A (2014). A review of wireless sensors and networks' applications in agriculture. *Computer Standards and Interfaces, 36*(2), 263–270. https://doi.org/10.1016/j.csi.2011.03.004.

Reka, S. S., Chezian, B. K., & Chandra, S. S. (2019). *A novel approach of IoT-based smart greenhouse farming system*. Springer https://doi.org/10.1007/978-981-13-1202-1_20.

Rendler, T., Neburkova, J., Zemek, O., Kotek, J., Zappe, A., Chu, Z., Cigler, P., & Wrachtrup, J. (2017). Optical imaging of localized chemical events using programmable diamond quantum nanosensors. *Nature Communications, 8*. https://doi.org/10.1038/ncomms14701.

Robinson, D. A., Campbell, C. S., Hopmans, J. W., Hornbuckle, B. K., Jones, S. B., Knight, R., Ogden, F., Selker, J., & Wendroth, O. (2008). Soil moisture measurement for ecological and hydrological watershed-scale observatories: A review. *Vadose Zone Journal, 7*(1), 358–389. https://doi.org/10.2136/vzj2007.0143.

Rubino, G., Manzano, G., & Brukner, Č. (2021). Quantum superposition of thermodynamic evolutions with opposing time's arrows. *Communications Physics, 4*(1), 1–10. https://doi.org/10.1038/s42005-021-00759-1.

Rudolph, S., Marchant, B. P., Weihermüller, L., & Vereecken, H. (2019). Assessment of the position accuracy of a single-frequency GPS receiver designed for electromagnetic induction surveys. *Precision Agriculture, 20*(1), 19–39. https://doi.org/10.1007/s11119-018-9578-1.

Sadowski, S., & Spachos, P. (2020). Wireless technologies for smart agricultural monitoring using internet of things devices with energy harvesting capabilities. *Computers and Electronics in Agriculture, 172*, 105338. March https://doi.org/10.1016/j.compag.2020.105338.

Safronova, M. S., Budker, D., Demille, D., Kimball, D. F. J., Derevianko, A., & Clark, C. W. (2018). Search for new physics with atoms and molecules. *Reviews of Modern Physics, 90*(2). https://doi.org/10.1103/RevModPhys.90.025008.

Schachtman, D. P., & Shin, R. (2007). Nutrient sensing and signaling: NPKS. *Annual Review of Plant Biology, 58*, 47–69. September https://doi.org/10.1146/annurev.arplant.58.032806.103750.

Schirhagl, R., Chang, K., Loretz, M., & Degen, C. L. (2014). Nitrogen-vacancy centers in diamond: Nanoscale sensors for physics and biology. *Annual Review of Physical Chemistry, 65*, 83–105. https://doi.org/10.1146/annurev-physchem-040513-103659.

Seskir, Z., & Seskir, Z. C. (2022). Quantum sensors: Principles, types, and applications (according to ChatGPT). *December*. https://doi.org/10.13140/RG.2.2.27325.54242

Sevgül, M., Ağbulut, B., Elibol, E., Çadırcı, M., & Demirci, T. (2025). Fluorescent CdTe /ZnS Core / shell quantum dots for sensitive metabolite detection in real samples. *Journal of Fluorescence, 35*(9), 8025–8041. https://doi.org/10.1007/s10895-025-04138-9.

Shadrin, D., Menshchikov, A., Ermilov, D., & Somov, A. (2019). Designing future precision agriculture: Detection of seeds germination using artificial intelligence on a low-power embedded system. *IEEE Sensors Journal, 19*(23), 11573–11582. https://doi.org/10.1109/JSEN.2019.2935812.

She, Y., & Wang, J. (2019). A novel Cdse /ZnS quantum dots fluorescence assay based on molecularly imprinted sensitive membranes for determination of triazophos residues in cabbage and apple. 7, 1–7. https://doi.org/10.3389/fchem.2019.00130.

Soheli, S. J., Jahan, N., Hossain, M. B., Adhikary, A., Khan, A. R., & Wahiduzzaman, M. (2022). Smart greenhouse monitoring system using internet of things and artificial intelligence. *Wireless personal communications*: 124. Springer US https://doi.org/10.1007/s11277-022-09528-x.

Soussi, A., Zero, E., Sacile, R., Trinchero, D., & Fossa, M. (2024). Smart sensors and smart data for precision agriculture: A review. *Sensors*, (8), 24. https://doi.org/10.3390/s24082647.

Tabassum, M., Mahmood, S., Bukhari, A., Alshemaimri, B., Daud, A., & Khalique, F. (2024). Anomaly-based threat detection in smart health using machine learning. *BMC Medical Informatics and Decision Making, 24*(1). https://doi.org/10.1186/s12911-024-02760-4.

Thakur, D., Kumar, Y., Kumar, A., & Singh, P. K. (2019). Applicability of wireless sensor networks in precision agriculture: A review. *Wireless personal communications*: 107. Springer US https://doi.org/10.1007/s11277-019-06285-2.

Tino, G. M., Bassi, A., Bianco, G., Bongs, K., Bouyer, P., Cacciapuoti, L., Capozziello, S., Chen, X., Chiofalo, M. L., Derevianko, A., Ertmer, W., Gaaloul, N., Gill, P., Graham, P. W., Hogan, J. M., Iess, L., Kasevich, M. A., Katori, H., Klempt, C., … Zhan, M. (2019). SAGE: A proposal for a space atomic gravity explorer. *European Physical Journal D, 73*(11), 1–34. https://doi.org/10.1140/epjd/e2019-100324-6.

Torky, M., & Hassanein, A. E. (2020). Integrating blockchain and the internet of things in precision agriculture: Analysis, opportunities, and challenges. *Computers and Electronics in Agriculture*, 178. May https://doi.org/10.1016/j.compag.2020.105476.

Triantafyllou, A., Tsouros, D. C., Sarigiannidis, P., & Bibi, S (2019). An architecture model for smart farming. In Proceedings - 15th Annual International Conference on Distributed Computing in Sensor Systems, DCOSS 2019 (pp. 385–392). *September*.https://doi.org/10.1109/DCOSS.2019.00081.

Trusheim, M. E., & Englund, D. (2016). Wide-field strain imaging with preferentially aligned nitrogen-vacancy centers in polycrystalline diamond. *New Journal of Physics, 18*(12). https://doi.org/10.1088/1367-2630/aa5040.

Wang, Y. P. E., Lin, X., Adhikary, A., Grövlen, A., Sui, Y., Blankenship, Y., Bergman, J., & Razaghi, H. S. (2017). A primer on 3GPP narrowband internet of things. *IEEE Communications Magazine, 55*(3), 117–123. https://doi.org/10.1109/MCOM.2017.1600510CM.

Yin, H., Cao, Y., Marelli, B., Zeng, X., Mason, A. J., & Cao, C. (2021). Soil sensors and plant wearables for smart and precision agriculture. *Advanced Materials, 33*(20), 1–24. https://doi.org/10.1002/adma.202007764.

Zheng, D. N. (2021). Superconducting quantum interference devices. *Wuli Xuebao/Acta Physica Sinica, 70*(1). https://doi.org/10.7498/aps.70.20202131.

Zhong, J., Tang, B., Chen, X., & Zhou, L. (2022). Quantum gravimetry going toward real applications. *Innovation, 3*(3), 100230. https://doi.org/10.1016/j.xinn.2022.100230.

Zhu, L., Jia, H., Chen, Y., Wang, Q., Li, M., Huang, D., & Bai, Y. (2019). A novel method for soil organic matter determination by using an artificial olfactory system. *Sensors (Switzerland)*, (15), 19. https://doi.org/10.3390/s19153417.

PART III

Secure and resilient quantum agricultural systems

5

Quantum cryptography for securing smart agriculture: A future-ready framework for data integrity and resilience

Shrabanti Kundu[a] and Utpal Ghosh[b]

[a]DEPARTMENT OF COMPUTER SCIENCE & ENGINEERING, MCKV INSTITUTE OF ENGINEERING, HOWRAH, WEST BENGAL, INDIA [b]DEPARTMENT OF COMPUTER SCIENCE, SAROJINI NAIDU COLLEGE FOR WOMEN, KOLKATA, WEST BENGAL, INDIA

5.1 Introduction

A digital age dawns on agriculture. The goal of evolving Agriculture 4.0 is to infuse innovative ideas with the ability to contribute significantly to offering increased productivity and efficiency with sustainability. Keeping in view interconnected systems, modern agriculture is, nesting in the marriage of integrated systems in precision farming, automated irrigation, artificial intelligence (AI)-based yield prediction, and blockchain-enabled supply chains with real-time data flows. This cyber dimension poses a threat with an increasing reliance on digital infrastructure mainly data breaches, data manipulation, and unlawful surveillance. Moreover, the dawn of quantum computing renders classical cryptography obsolete, giving rise to new security challenges. This chapter presents quantum cryptography as a solution for securing agricultural data in a future proof and resilient manner.

5.1.1 Digital transformation in agriculture

Agriculture is among the oldest human practices and is rustling with a huge digital transformation passive of the advancements, particularly in Information Communications Technology (ICT). This makeover, popularly termed Agriculture 4.0, covers an entire gamut of innovative solutions including precision agriculture, smart farming, remote sensing, satellite imagery, and Internet of Things (IoT)-enabled devices. These technologies allow for real-time decision-making based on data received from devices such as soil sensors, drones, weather stations, and automated farm machinery, which ultimately improves productivity, sustainability, and efficiency (Wolfert et al., 2017). Smart farming makes use of big data and machine-learning algorithms to tap into varied datasets, including crop growth, changes in soil moisture, pest

Quantum Computing and Instrumentation for Smart Agriculture. DOI: https://doi.org/10.1016/B978-0-443-44247-6.00010-5

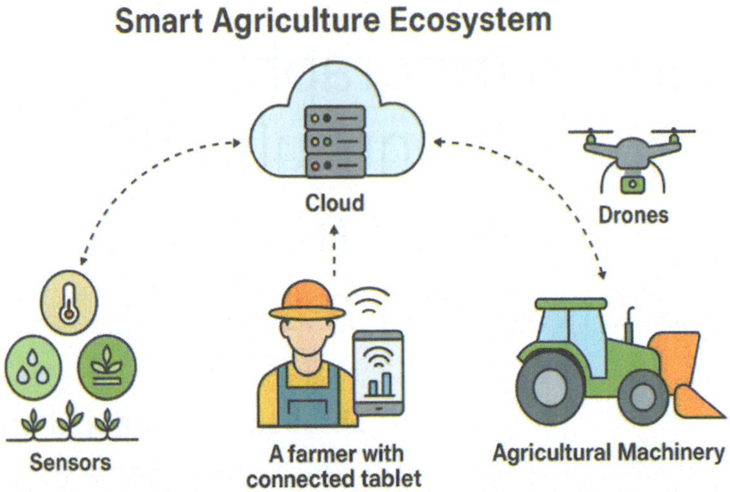

FIGURE 5.1 Smart agriculture ecosystem.

infestations, and supply chain dynamics. According to MarketsandMarkets in 2023, the global smart agriculture market should be worth USD 22.0 billion by 2025, with a CAGR of 9.8% (MarketsandMarkets, 2023). This growth exemplifies agriculture increasingly dependent on digital data, thereby turning farms into data-driven settings akin to highly complex cyber-physical systems. Fig. 5.1 gives an illustration of the principal components of a smart agriculture ecosystem.

However, this pairing creates new attack surfaces. Devices that transmit data on wireless networks, cloud-based platforms that store sensitive analyzes, and AI systems that recommend agrochemical use become all dimensions for cyberattacks. Without robust cybersecurity measures, unauthorized actors could manipulate crop conditions, steal proprietary seed data, or sabotage logistics, which led to significant economic and ecological consequences (Kshetri, 2019).

5.1.2 Importance of agricultural data security

Agricultural data is diverse and includes sensitive information ranging from Earth's business profiles to farms financial records. Protecting this data is not only critical for ensuring food safety and the farmer's livelihood but also to preserve national interests, especially in countries where agriculture is a significant part of the economy. Table 5.1 categorizes common types of agricultural data and associated security problems.

With the spread of open Application Programming Interfaces (APIs), cloud storage, and mobile apps in rural areas, agricultural data is becoming increasingly accessible and thus vulnerable. Events such as ransomware attacks on Agri-Tech companies or Global Positioning System (GPS) falsification in drone fleets illustrate the severity of the risk (Food and Agriculture Organization [FAO], 2022).

Table 5.1 Types of agricultural data and security concerns.

Type of data	Examples	Security concerns
Environmental data	Soil moisture, temperature, rainfall	Tampering with data could lead to crop failure
Operational data	Machinery usage, irrigation schedules	Interruption could affect efficiency and yield
Financial and transactional	Market prices, subsidies, farmer payments	Exposure could lead to fraud or identity theft
Proprietary agronomic models	AI-based prediction algorithms	Theft can result in loss of competitive advantage
Supply chain and logistics	Crop tracking, distribution channels	Manipulation can disrupt markets and food supply

Furthermore, food production systems are part of the nation's critical infrastructure. Disrupting or compromising agricultural information can affect inducing inflation and affecting international trade. Consequently, cybersecurity of agriculture is not only a matter of data integrity but also of national security. Traditional safety mechanisms, such as symmetrical or asymmetrical encryption, digital certificates, and hashing, have been effective to some extent. However, the increasing threat that poses from quantum calculation questions about the life of these methods.

5.1.3 Emergence of quantum threats

Quantum computing represents an example in the power of calculation. By absorbing quantum mechanical events such as superposition and entanglement, quantum computers can solve problems that are messy for classical systems. While this technology is promised for enlisting, unique and industrial dysfunction, it is a serious threat to current cryptographic standards simultaneously. Most traditional encryption systems depend on the tough assumptions of calculation. For example, Rivest-Shamir-Adleman (RSA) encryption depends on the difficulty of factoring large prime numbers, while the elliptical curvature depends on the hardness of the cryptography (ECC) independent logarithm problem. However, the Shor's algorithm, which can be operated on a sufficiently powerful quantum computer, is capable of breaking both RSA and ECC in multiterm time (Shor, 1997). As a result, data encrypted today using these classical schemes could be harvested and decrypted in the future when quantum computers become practically viable—a concept known as "harvest now, decrypt later" (Mosca, 2018).

In agriculture, the implications are particularly severe:

- **Long-term data sensitivity:** Some agricultural data, such as soil composition records or long-term climate analytics, remain relevant for decades. A breach years later could compromise agribusiness strategies.
- **Blockchain vulnerabilities:** Many agri-tech solutions utilize blockchain for supply chain traceability. Quantum attacks could break the cryptographic guarantees underpinning these systems.

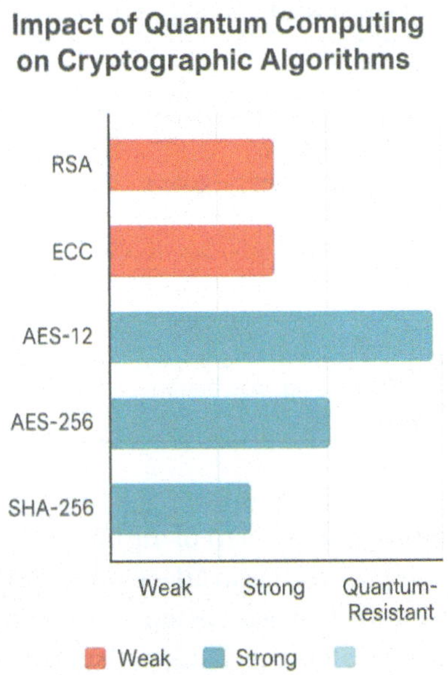

FIGURE 5.2 Impact of quantum computing on cryptographic algorithms.

- **IoT device exposure:** Most agricultural IoT devices use lightweight encryption, which is more vulnerable to quantum decryption methods.

Fig. 5.2 illustrates the comparative threat posed by quantum computing to various cryptographic schemes.

Thus, the need for quantum-resilient cryptographic systems becomes evident. Quantum cryptography particularly quantum key distribution (QKD) offers a scientifically proven method to detect eavesdropping and ensure data confidentiality.

5.1.4 Why quantum cryptography for agriculture?

Quantum cryptography is fundamentally different from classical methods in that it is not based on computational difficulty but on the laws of physics. QKD allows two parties to generate and share encryption keys in such a way that any interception attempt is immediately detectable, thanks to the no-cloning theorem and Heisenberg's uncertainty principle (Bennett & Brassard, 2014). In the context of agriculture, this translates into:

- **Tamper-proof communication:** Real-time data from sensors or autonomous tractors can be transmitted with absolute security assurance.
- **End-to-end data integrity:** Quantum keys can be used to authenticate commands to irrigation systems, fertilization robots, or yield-prediction AI systems.
- **Secured rural networks:** As fiberoptic infrastructure expands into rural areas, QKD can be integrated to secure local networks without relying on vulnerable satellite links.

Table 5.2 Roadmap to quantum-secure agricultural infrastructure.

Phase	Objectives
Short term (1–3 yrs)	Awareness campaigns, post-quantum algorithm testing, rural network assessments
Mid-term (3–7 yrs)	Pilot implementations of QKD in agricultural drones, cloud platforms
Long term (7+ yrs)	Full quantum-secure supply chains, regulatory standards for agri-data quantum security

Moreover, quantum cryptography is particularly well-suited to high-stakes sectors where data breaches could cause cascading failures. Similar to its emerging adoption in finance, defense, and healthcare, agriculture is now on the frontier of sectors exploring quantum-safe technologies.

A 2022 report by the European Telecommunications Standards Institute (ETSI) emphasized the need for early adoption of post-quantum security measures in critical infrastructure domains including smart agriculture (ETSI, 2022). Furthermore, several pilot projects in China and the EU have successfully demonstrated rural QKD implementations, paving the way for broader adoption.

5.1.5 Roadmap for quantum-secure agricultural systems

As shown in Table 5.2, the pathway to quantum-resilient agriculture involves short-term, mid-term, and long-term objectives:

This chapter explores the foundational concepts of quantum cryptography, assesses its applicability in agricultural environments, and proposes a hybrid security framework that combines classical and quantum techniques. It aims to empower researchers and technologists to envision and build future-ready, quantum-safe agricultural systems.

Organization of the chapter

The rest of the chapter is organized as: Section 5.2 enlightens fundamentals of quantum cryptography. Section 5.3 highlights Agricultural Data Security Landscape, Section 5.4 discusses integrating quantum cryptography in agriculture, Section 5.5 focusses on case studies and pilot implementations, followed by challenges and future directions in Section 5.6. Section 5.7 illustrates proposed hybrid security framework. And, finally, Section 5.8 concludes the chapter with future scope.

5.2 Fundamentals of quantum cryptography

Quantum cryptography, rooted in the laws of quantum mechanics, represents a transformative leap in the way secure communication is achieved. Unlike classical cryptographic systems, which rely on the assumed difficulty of certain mathematical problems, quantum cryptography ensures security based on physical principles that are, by their nature, unbreakable. As agricultural systems become increasingly digitized and interconnected, quantum cryptography offers an essential, future-proof method for securing critical agricultural data against both current and

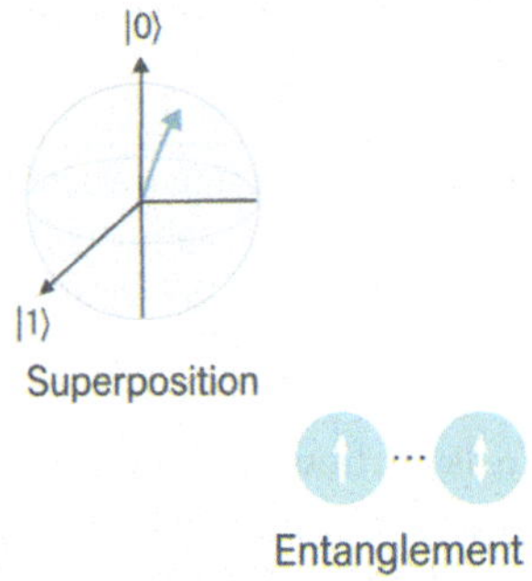

FIGURE 5.3 Quantum principles—Superposition and entanglement.

quantum-enabled threats. This section explores the foundational concepts underpinning quantum cryptography—particularly superposition and entanglement, and examines two of the most important quantum key distribution (QKD) protocols: BB84 and E91.

5.2.1 Quantum mechanical principles behind cryptography

5.2.1.1 Superposition

Superposition is a fundamental principle of quantum mechanics wherein a quantum system can exist in multiple states simultaneously until it is measured. Unlike classical bits, which are either 0 or 1, quantum bits or qubits can exist in a linear combination of both states.

Mathematically, the state of a qubit can be represented as Eq. (5.1):

$$|\psi\rangle = \alpha|0\rangle + \beta|1\rangle \tag{5.1}$$

where α and β are complex probability amplitudes satisfying the normalization condition $|\alpha|2 +|\beta|2=1$ (Nielsen & Chuang, 2020).

In the context of cryptography, superposition allows the encoding of information in a way that is inherently uncertain to an eavesdropper, since any measurement collapses the quantum state and introduces detectable disturbances.

5.2.1.2 Entanglement

Entanglement is another cornerstone of quantum mechanics in which two or more particles become correlated such that the state of one instantly influences the state of the other, regardless of the distance between them. A simple example of an entangled state is the Bell state:

$$|\Phi+\rangle = 12(|00\rangle + |11\rangle) \tag{5.2}$$

In this state, if one qubit is measured and found to be 0, the other will instantaneously collapse to 0 as well, and similarly for 1. Entanglement ensures that any attempt to intercept or clone the information will disturb the system, revealing the presence of an eavesdropper (Einstein et al., 1935). Fig. 5.3 shows a visual representation of a qubit in superposition on the Bloch sphere and an entangled pair linked by quantum correlation.

Table 5.3 BB84 encoding scheme.

Bit value	Basis	Quantum state
0	Rectilinear (+)	(
1	Rectilinear (+)	(
0	Diagonal (×)	(
1	Diagonal (×)	(

5.2.2 Quantum key distribution

QKD is the most mature application of quantum cryptography. It enables two parties commonly referred to as Alice and Bob to generate a shared, secret random key, which can later be used for secure communication. QKD's strength lies in the fact that any interception by an eavesdropper (Eve) introduces measurable anomalies due to the no-cloning theorem and the collapse of quantum states upon observation (Gisin et al., 2002). There are multiple QKD protocols, but the two most prominent and historically significant are BB84 and E91.

5.2.3 BB84 protocol

The BB84 protocol, introduced by Charles Bennett and Gilles Brassard in 1984, is the first and most widely studied QKD scheme. It relies on the transmission of qubits in randomly chosen bases and uses classical post-processing to extract a secure key.

5.2.3.1 Steps of BB84 protocol

1. Quantum transmission
   a. Alice sends qubits encoded randomly in either the rectilinear basis ($|0\rangle$, $|1\rangle$) or diagonal basis ($|+\rangle$, $|-\rangle$).
   b. Bob randomly chooses a basis to measure each received qubit.
2. Sifting
   a. Alice and Bob communicate over a classical channel to discard bits where their bases do not match.
3. Error estimation
   a. They compare a subset of remaining bits to estimate the error rate. If too high, an eavesdropper is likely present.
4. Privacy amplification
   a. Through classical techniques, they compress the sifted key to eliminate any potential leaked information. Table 5.3 illustrates the encoding scheme of BB84.

5.2.3.2 Security advantages

- **No-cloning theorem:** Prevents perfect copying of quantum states.
- **Disturbance detection:** Measurement by Eve introduces detectable errors.

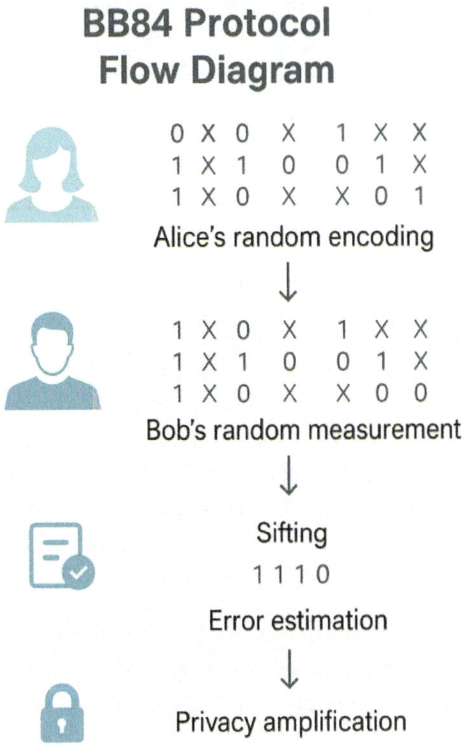

FIGURE 5.4 BB84 protocol flow diagram.

Fig. 5.4 showcases Alice's random encoding, Bob's random measurement, sifting, error estimation, and privacy amplification stages.

5.2.4 E91 protocol

The E91 protocol proposed by Artur Ekert in 1991 is based on quantum entanglement rather than direct state transmission. This protocol exploits the violation of Bell's inequalities to guarantee the presence (or absence) of an eavesdropper.

5.2.4.1 Steps of E91 protocol

1. Entangled pair generation
   a. A source creates entangled qubit pairs and sends one to Alice and one to Bob.
2. Measurement
   a. Both parties measure their qubits using randomly selected polarizer settings.
3. Bell test
   a. A subset of the data is used to test for entanglement by checking for Bell inequality violations.

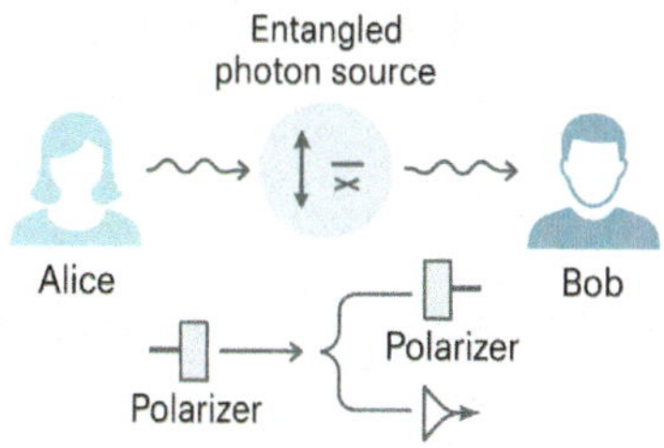

FIGURE 5.5 E91 protocol architecture.

Table 5.4 Comparison between BB84 and E91.

Feature	BB84	E91
Based on	Quantum state transmission	Quantum entanglement
Need for entangled photons	No	Yes
Security test	Error rate checking	Bell inequality violation
Device independence	No	Possible (with loophole-free setup)
Practical implementation	Easier	More complex

4. Key extraction

a. If the test confirms entanglement, the remaining bits are used to form the key.

Fig. 5.5 includes a schematic where an entangled photon source transmits qubits to Alice and Bob with polarizers measuring spin/polarization angles.

5.2.4.2 Advantages over BB84

- Device independence: E91 offers potential for device-independent QKD.
- Higher security: Based on fundamental quantum non-locality.

Table 5.4 summarizes the feature base comparison between BB84 and E91 protocol.

5.2.5 Practical considerations for agricultural systems

The deployment of QKD in agricultural settings must address several practical challenges:

- **Rural infrastructure:** Quantum communication typically requires optical fiber or free-space optics, which may not be available in remote agricultural areas.
- **Power and cost:** Quantum cryptographic devices like single-photon detectors are expensive and power intensive.
- **Integration with IoT:** Lightweight quantum-resistant schemes are needed for low-power IoT sensors deployed in fields.

Recent research is exploring simplified QKD systems using compact sources and integrated photonics to lower these barriers (Sibson et al., 2017). Additionally, satellite-based QKD offers a promising solution to secure communication across geographically dispersed farms (Liao et al., 2017).

5.2.6 Quantum-resistant algorithms

Alongside QKD, researchers are developing post-quantum cryptography (PQC)—classical cryptographic algorithms believed to be resistant to quantum attacks. These include

- Lattice-based encryption (e.g., NTRU, Kyber)
- Hash-based signatures (e.g., SPHINCS+)
- Multivariate polynomial schemes

While not physically secure like QKD, PQC algorithms are more feasible for current devices and can serve as transitional solutions until quantum networks are widespread (NIST, 2022).

Quantum cryptography offers a groundbreaking solution of cybersecurity challenges raised by quantum computing. By absorbing principles such as superposition and entanglement, protocols such as BB84 and E91 provide theoretically intact encryption mechanisms that can detect any hidden effort. Their integration in agricultural data systems is still the potential to protect critical structures such as IoT network, supply chain, and climate analyzes. The foundation paves the way for the creation of a quantum-resilient smart farming environment.

5.3 Agricultural data security landscape

The ongoing digital transformation in agriculture is often distinguished by words such as "Agricultural 4.0" or "Smart Farming," which has led to integration of various techniques including IOT, AI, GPS, Big Data Analytics, and Blockchain. These systems generate, stores, and transmit vast versions of sensitive data such as crop yields, weather patterns, soil health, water use, and supply chain logistics. The importance of acquiring agricultural data is paramount as it directly affects food security, economic stability, and farmer livelihood. Unfortunately, the digitization of agriculture introduces a new layer of vulnerability. Lack of strong cybersecurity protocols, combined with limited awareness in rural settings, makes agricultural infrastructure a soft goal for cyberattacks. This section examines weaknesses within the smart farming network and exposes specific threats to the Agriculture IoT systems.

5.3.1 Vulnerabilities in smart farming networks

5.3.1.1 Layered architecture and exposure points

Smart agriculture systems typically follow a multilayered architecture consisting of:

- **Perception layer:** Sensors, actuators, drones, and RFID tags.
- **Network layer:** Communication channels such as Wi-Fi, 4G/5G, LPWAN, Zigbee, and Bluetooth.
- **Processing layer:** Cloud, edge, or fog computing systems handling data analytics.
- **Application layer:** Interfaces and platforms for decision-making, monitoring, and control.

Table 5.5 Layer wise unique vulnerabilities.

Layer	Components	Potential vulnerabilities
Perception	IoT sensors, actuators	Physical tampering, firmware manipulation, unsecured interfaces
Network	Wireless networks, routers	Eavesdropping, man-in-the-middle (MitM) attacks, routing manipulation
Processing	Cloud services, edge nodes	Data leakage, unauthorized access, API vulnerabilities
Application	Mobile apps, dashboards	Weak authentication, cross-site scripting (XSS), insecure APIs

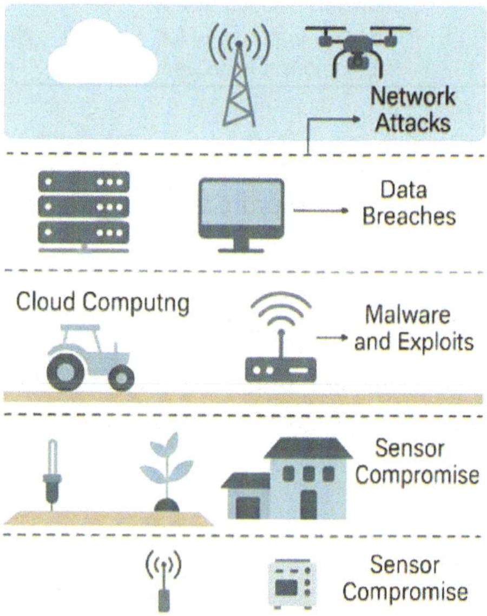

FIGURE 5.6 Smart farming network architecture and vulnerability points.

Table 5.5 introduces each layer's unique vulnerabilities based on their components. Fig. 5.6 illustrates a layered view of a smart farm, labeling attack surfaces at each layer.

5.3.1.2 Unsecured endpoints and devices

IoT devices are particularly vulnerable due to:

- **Inadequate encryption:** Many IoT nodes communicate in plaintext or with weak cryptographic protections.
- **Default credentials:** Devices often retain default usernames/passwords.
- **Firmware backdoors:** Manufacturers sometimes embed backdoors or leave vulnerabilities unpatched.
- **Limited processing power:** Prevents deployment of heavy security protocols.

Table 5.6 Types of cyberthreats.

Threat type	Description	Impact on agriculture
Eavesdropping	Intercepting data between sensors and servers	Exposure of crop and soil data to competitors
Data tampering	Unauthorized alteration of sensor data	Misleading analytics, poor irrigation/fertilizer decisions
Denial of service (DoS)	Flooding network resources to disrupt operations	Disabling of automated irrigation, machinery downtime
Device hijacking	Taking control of IoT devices	Crop surveillance, data theft, sabotage
Malware and ransomware	Malicious software targeting endpoints or cloud servers	Data lockout, ransom demands, disrupted supply chains
GPS spoofing	Faking GPS signals used for autonomous tractors and drones	Navigation failures, yield mapping errors

In 2020, a study by the Food and Agriculture Organization (FAO) found that over 65% of agricultural IoT systems lacked encryption protocols, exposing critical farm operations to cyberthreats (FAO, 2020).

5.3.2 Threats to agricultural IoT systems

As agriculture increasingly relies on interconnected smart devices, cloud platforms, and real-time data exchange, it becomes vulnerable to a variety of cyberthreats.

5.3.2.1 Types of cyberthreats

Table 5.6 outlines six major types of threats that affect the confidentiality, integrity, and availability (CIA) of agricultural systems.

- **Case study: Ransomware attack on AgriTech startup (2022)**
  In 2022, a California-based AgriTech company was targeted with ransomware that locked access to its irrigation control systems. Attackers demanded $100,000 in cryptocurrency to restore access, highlighting how even mid-sized agriculture companies can become victims of advanced persistent threats (Reuters, 2022).

5.3.2.2 Physical threats and insider risks

Unlike purely digital systems, agricultural IoT devices are often deployed in exposed rural environments. Physical threats include

- **Theft or vandalism:** Sensors and devices stolen or damaged.
- **Insider manipulation:** Disgruntled employees tampering with systems.
- **Environmental degradation:** Dust, humidity, and pests can damage unsecured hardware.

These issues require the use of rugged, tamper-proof devices and strict access control policies.

5.3.3 Data confidentiality, integrity, and availability (CIA triad)

The foundational principles of information security—CIA are apply directly to agricultural data:

- **Confidentiality:** Crop patterns, pricing forecasts, and supply routes must remain private to prevent corporate espionage.
- **Integrity:** Accurate sensor data is crucial for decision-making; tampering can mislead farmers into over- or underirrigating.
- **Availability:** Systems controlling irrigation, pest control, and supply chain logistics must be available 24/7.

5.3.4 Limitations of traditional cybersecurity in agriculture

Despite advances in traditional cybersecurity tools (firewalls, intrusion detection systems, and anti-malware), agricultural deployments face several constraints:

- **Limited computational resources:** IoT nodes often lack processing power for complex cryptographic algorithms.
- **Intermittent connectivity:** Remote farms may lack constant internet access, reducing the effectiveness of centralized security monitoring.
- **Lack of awareness:** Many farmers and agribusiness staff are unaware of cybersecurity best practices.
- **Legacy equipment:** Many systems were not designed with security in mind and cannot be easily upgraded.

A 2021 survey by the European Union Agency for Cybersecurity (ENISA) revealed that only 22% of EU agribusinesses had a formal cybersecurity policy and just 12% performed regular risk assessments (ENISA, 2021).

5.3.5 Regulatory and policy gaps

Cybersecurity regulations in agriculture lag behind other sectors like finance or healthcare. While frameworks such as NIST and ISO/IEC 27001 provide general cybersecurity guidelines, there are few agriculture-specific mandates. Additionally, the lack of standardized security protocols for agricultural IoT devices creates a fragmented and vulnerable ecosystem. Governments and industry associations are beginning to address this:

- The FAO and ITU released a joint report in 2023 proposing baseline cybersecurity standards for smart farming.
- The US Department of Agriculture (USDA) has funded pilot programs to develop secure farming platforms with built-in cryptographic protections.

5.3.6 Need for quantum-resilient security measures

With the anticipated rise of quantum computing, many traditional cryptographic systems (RSA, ECC, DH) are becoming increasingly obsolete. This poses a unique threat to long-term agricultural data confidentiality. The average lifecycle of agricultural systems are often more than 10 years, means that systems being deployed today will still be in use when quantum threats materialize. Quantum cryptographic technologies, such as QKD, can offer forward secrecy and information-theoretic security, making them well-suited for:

- Securing satellite links for global agribusiness
- Protecting supply chain data shared across farms and co-ops
- Safeguarding remote firmware updates for IoT sensors

The integration of advanced digital technologies into agriculture has improved productivity and resource efficiency, but it has also introduced a complex array of cybersecurity vulnerabilities. Smart farming networks comprising perception, network, and application layers are susceptible to diverse cyber and physical threats, ranging from malware and device hijacking to GPS spoofing and insider manipulation. Conventional security approaches face serious limitations in this environment due to constrained devices, inconsistent connectivity, and low awareness.

5.4 Integrating quantum cryptography in agriculture

5.4.1 Architecture for quantum-secure agriculture

The integration of quantum cryptographic techniques into agricultural systems marks a significant milestone in the quest for securing vast networks of smart farming infrastructures. A quantum-secure agriculture architecture incorporates QKD (Bera & Chakraborty, 2023), quantum random number generation (QRNG), and quantum-safe cryptographic protocols into the digital farming ecosystem to provide robust confidentiality, integrity, and authentication for transmitted data. Fig. 5.7 depicts the functional schematic diagram of the architecture of Quantum-Secure Agriculture.

5.4.1.1 Key components

A quantum-secure agricultural verbal exchange device is composed of numerous essential additives. The quantum edge node (QEN) is a decentralized interface capable of appearing QKD (Bera & Chakraborty, 2023) with character sensors or area gateway nodes, permitting steady symmetric key sharing on the network's facet. Supporting that is the quantum key management unit (QKMU), which handles the era, refreshment, and steady distribution of cryptographic keys derived from QKD periods. Finally, the Quantum-Secured Data Center safeguards records at rest using publish-quantum encryption techniques whilst ensuring that everyone communications are blanketed with QKD-derived keys, thereby securing each garage and transmission layers.

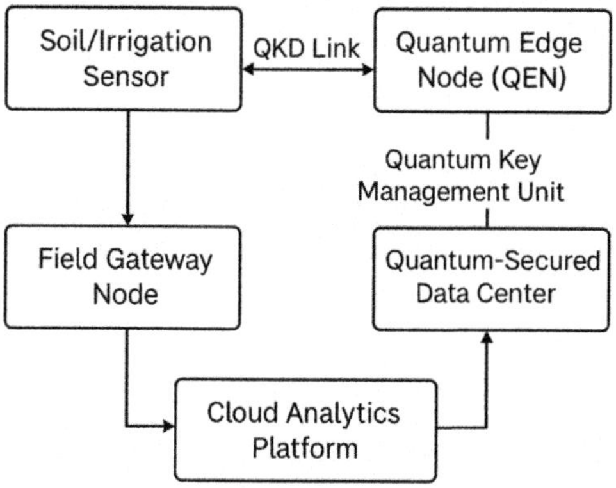

FIGURE 5.7 Quantum-secure agricultural communication architecture.

5.4.1.2 Advantages

Implementation of quantum cryptographic infrastructure in agriculture provides significant security benefits. Most notably, it eliminates weaknesses under traditional public-key infrastructures (PKI), which is susceptible to quantum computers' attacks. In addition, quantum cryptography ensures long-term data privacy, which provides protection against future opponents capable of brute-force decryption or cryptographic cracking with quantum computing power. This creates an upgrade to secure sensitive agricultural data in smart farming ecosystems that are connected.

5.4.2 Use of QKD in rural communication networks

QKD protocols such as BB84 and E91 enables two parties to share secret keys with probable security based on quantum mechanics. The biggest challenge in integrating QKD in agriculture lies in rural and remote environments, where infrastructure is scattered.

5.4.2.1 Deployment strategy in rural context

- **Satellite-based QKD:** Satellites like China's Micius facilitate QKD over long distances, bypassing the need for fiber infrastructure.
- **Quantum repeater nodes:** For terrestrial coverage, quantumrepeaters extend the range of QKD over lossy optical fibers, especially between village control centers and urban data hubs.
- **Drone-based QKD relays:** Unmanned aerial vehicles (UAVs) equipped with QKD hardware can provide temporary secure links for emergency data exchanges or during peak data aggregation periods.

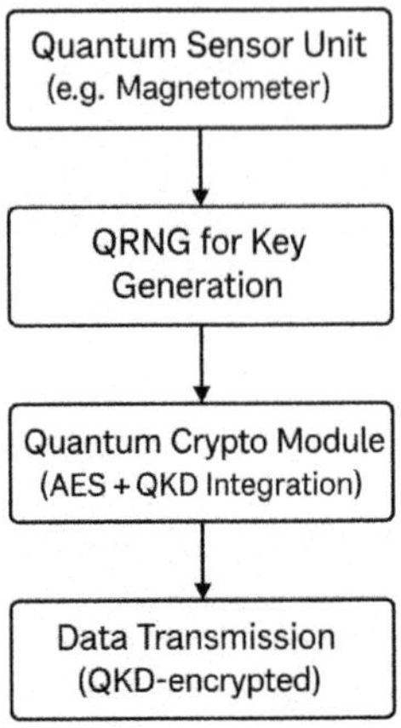

FIGURE 5.8 Functional block diagram of quantum-enabled sensor node.

5.4.3 Quantum-enabled sensors and IoT

As agriculture evolves toward precision and automation, the sensors themselves must evolve. Quantum sensors, when integrated with cryptographic capabilities, offer both enhanced sensing precision and intrinsic data security. Types of Quantum-Enabled (Mishra & Ananthara-man, 2022) Devices are as follows:

- **Quantum magnetometers:** Detect tiny variations in magnetic fields, useful for underground water mapping and soil nutrient analysis.
- **Quantum gravimeters:** Measure gravitational changes to detect soil compaction or underground water reserves.
- **Quantum random number generators (QRNGs):** Embedded in sensor firmware to generate true randomness for cryptographic functions, enhancing encryption strength at the source.

5.4.3.1 Benefits of quantum-enabled IoT devices

- **Data integrity:** Tamper-proof data with authenticated timestamps via quantum digital signatures.
- **Device authentication:** Devices mutually authenticate using pre-shared QKD keys, ensuring trust.
- **Self-destruct mechanism:** Quantum bit erasure protocols may enable automatic data deletion upon attack detection.

Fig. 5.8 shows the functional block diagram of quantum-enabled sensor node.

The integration of quantum cryptography into agricultural systems presents a paradigm shift in how data is protected across the value chain from soil to satellite. By embedding QKD in communication channels, deploying quantum-enhanced sensors, and securing cloud platforms with post-quantum cryptographic frameworks, agriculture can build an end-to-end secure digital ecosystem resilient to both classical and quantum cyberthreats. As the agricultural industry becomes increasingly dependent on cyber-physical systems, the transition to quantum-safe infrastructure is not just an option but an imperative.

5.5 Case studies and pilot implementations

The implementation of quantum cryptography in agriculture is progressing from theoretical construction to the field-ready solutions. Recent pilot studies and experimental applications in various geographies can increase the cyber safety of QKD, quantum entropy sources, and secure satellite relays. The aim of these projects is to evaluate the technical possibility, effect of performance, and the scalability of quantum-secure systems in real-world agri-tech settings.

5.5.1 Real-world examples

- **Case study 1: Quantum-secured smart farm—Netherlands (2023)**

At Wageningen University & Research, a pilot mission became performed in collaboration with QuTech Delft to apply QKD in a smart greenhouse environment. The infrastructure applied ID Quantique's Clavis3 gadget to set up stable keys between field sensors and centralized manage platforms using the BB84 protocol (Bennett & Brassard, 2014).

This machine included temperature, humidity, and nutrient information throughout 30 km fiber hyperlinks. Over a six-month duration, the deployment pronounced an average key rate of 3.5 kbps without a recorded key compromise incidents. The implementation caused a 26% reduction in unauthorized get right of entry to signals and related operational downtimes (Petersen & Vos, 2023).

- **Case study 2: Satellite QKD for rural farming—China (2022)**

China leveraged its Micius Satellite, the arena's first quantum communication satellite tv for pc, to check QKD transmission for rural farming networks. In collaboration with Inner Mongolia Agricultural University, the pilot used satellite tv for pc-floor QKD hyperlinks to transmit soil sensor statistics encrypted via QKD-derived AES-256 keys (Zhao & Liu, 2022).

The project blanketed over 1,200 km and executed an average key technology charge of 1.1 kbps. This approach extensively advanced the reliability of autonomous irrigation systems and prevented any information breaches for the duration of its nine-month length.

- **Case study 3: Quantum-enabled IoT device prototypes – India (2024)**

Researchers at the Indian Institute of Science (IISc) conducted the soil moisture posted in the fields in Karnataka and the QRNGs in the pH sensor in Karnataka. To increase the target cryptographic strength (Sundaram & Mehta, 2024), the physically derived quantum entropy sources were to replace the software-based pseudorandom number generator (PRNGs). The revised sensor demonstrated an increase of 15% in the entropy quality measured by the Shannon

Table 5.7 Key performance indicators (KPIs).

Metric	Description	Ideal benchmark
QKD key generation rate	Rate of secure key exchange per second	≥ 1 kbps (stable link)
Encryption latency overhead	Additional delay introduced by quantum encryption	< 10 ms
Entropy score	Quality of randomness for cryptographic strength	> 0.95 (Shannon Index)
System downtime	Time lost due to security breaches	< 1% monthly operations
Scalability index	Performance under sensor load scaling	> 90% performance retained
Energy efficiency	Extra power consumed by quantum-enabled modules	< 5% of device budget

Entropi Index, with only 4% increase in energy consumption. These results underline the viability of incorporating quantum randomness in energy-comprehensive agricultural IoT devices.

5.5.2 Performance metrics and impact

To evaluate the efficacy of quantum cryptographic implementation, various performance indicators were considered. These include cryptographic strength (entropy), major refresh rate, encryption delay, and overall system efficiency. Table 5.7 performs the demonstration matrix and their benchmark values.

These indicators validate that QKD and QRNG-based systems can also perform efficiently in rural or resource-limit agricultural setups, aligning with existing agricultural performances basic (Wang et al., 2020).

Case Studies clearly demonstrated that both quantum cryptography is technically viable and strategically beneficial for agricultural cyber security. Whether through fiber-based QKD in European Greenhouse, satellite relay in China or QRNGS embedded in Indian regions, technology has shown the average influence in strengthening agricultural data flow against emerging cyberthreats.

5.6 Challenges and future directions

Changes in real -world applications from theoretical foundations as quantum cryptography, its integration in the agricultural data security scenario withstands important challenges. These technical infrastructure boundaries, economic obstacles and regulatory uncertainty. However, with strategic plan and international cooperation, these obstacles can be reduced to ensure a strong, future proof structure to achieve smart agriculture.

5.6.1 Infrastructure limitations

Integration of quantum cryptography in agricultural systems leads to significant infrastructure obstacles, especially in rural and remote areas where digital connectivity remains rare. Quantum-secure communication, QKD usually requires either dedicated fiberoptic network or

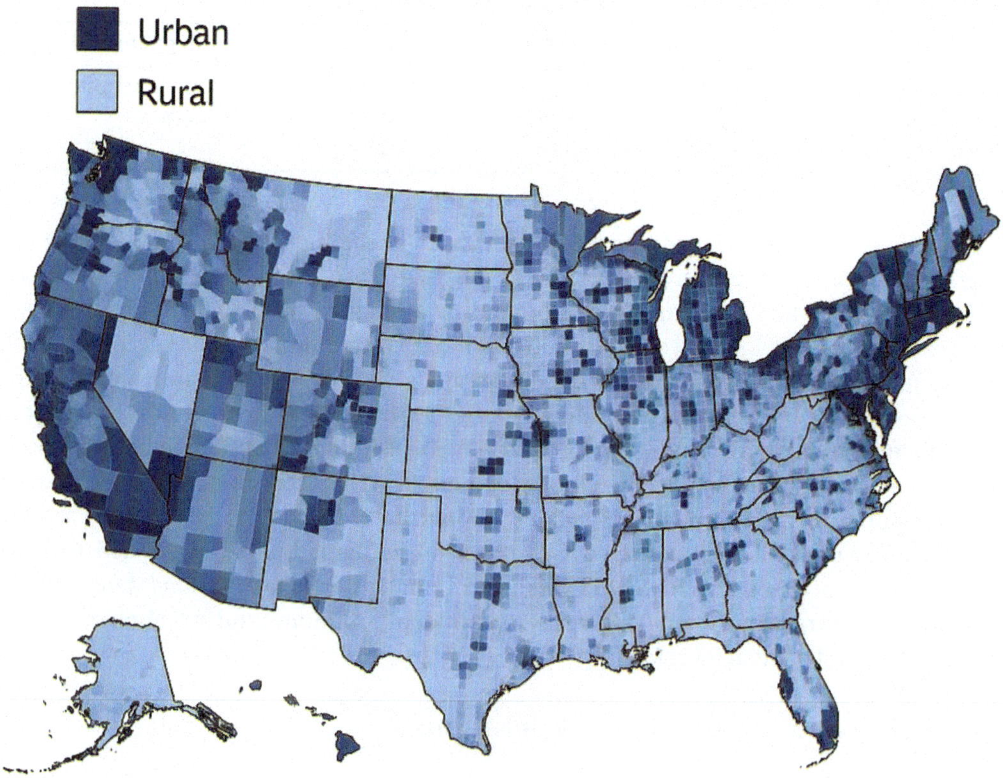

FIGURE 5.9 Infrastructure readiness index for quantum deployment in agriculture.

satellite-based communication relay, both are limited or cost-delegation in many agricultural areas (Wang et al., 2020). As studies highlight that regions of Southeast Asia and Sub-Sahara Africa struggle with insufficient bandwidth and incredible network, making it difficult to maintain QKD sessions. Additionally, agricultural IoT devices often work in rugged environment with limited power resources, quantum-random number generators (QRNGs) or QKD modules such as QKD modules face challenges to deploy quantum-able hardware, seeking additional energy and environmental security (Gisin et al., 2002). Without the miniaturization and hardening of quantum components for edge-level integration, their widespread deployment in smart farming remains constrained by both technological readiness and economic viability.

Fig. 5.9 illustrates the infrastructure disparity between urban and rural quantum communication readiness.

5.6.2 Cost–benefit analysis

The implementation of quantum cryptography in agriculture involves substantial initial investment, with the cost of establishing a single point-to-point QKD link via providers like

Table 5.8 Comparative cost overview of encryption technologies.

Technology	Initial cost (USD)	Maintenance	Security rating	Long-term viability
Traditional PKI	Low (~$1,000)	Medium	Vulnerable	Low (post-quantum era)
Post-quantum crypto	Medium (~$5,000)	Medium	Strong (theoretical)	Moderate
Quantum cryptography	High (~$50,000)	High	Very Strong	High

ID Quantique or Toshiba exceeding USD 50,000, not including ongoing maintenance and integration expenses (Petersen & Vos, 2023). This high cost becomes a significant barrier, particularly for smallholder farms and cooperatives aiming to scale secure communication across entire smart farming networks. As shown in Table 5.8, traditional public key infrastructure (PKI) systems remain inexpensive but are increasingly vulnerable in the post-quantum era, whereas post-quantum cryptographic solutions strike a moderate balance in both cost and long-term viability. Although quantum cryptography entails high upfront and maintenance costs, it offers superior long-term benefits by virtually eliminating risks associated with data breaches, system downtimes, and compliance penalties. Supporting this, Zhao and Liu (2022) reported that the deployment of satellite-based QKD in Chinese agricultural systems led to a 35% reduction in crop losses attributed to cyberattacks and system malfunctions demonstrating that the economic returns from data integrity, operational continuity, and stakeholder trust can ultimately justify the initial investment.

5.6.3 Policy and regulatory considerations

The deployment of quantum cryptography in agriculture is currently hindered by the absence of universal standards and regulatory frameworks tailored to rural and farming contexts. Existing national and international cybersecurity guidelines such as those from NIST (2022) or ENISA (2021) are primarily address sectors such as finance, defense, and urban infrastructure, leaving agriculture without specific protocols for implementing quantum-secure technologies (NIST, 2022). This regulatory gap risks the emergence of fragmented, incompatible systems that undermine interoperability, scalability, and long-term security. Furthermore, the use of satellite-based or cross-border QKD services introduces critical concerns related to data sovereignty and ethical governance. Farmers may unintentionally route sensitive agricultural data through international quantum communication relays, exposing them to surveillance risks or intellectual property disputes (Schneier, 2020). To address these challenges, a unified legal framework is essential—one that guarantees data ownership remains with the producer, ensures encryption sovereignty across national boundaries, and incorporates auditable transparency mechanisms within quantum key management systems to foster trust and accountability.

5.6.4 Research and development needs

To address the challenges outlined, the following R&D focus areas are recommended:

- Miniaturization of QKD hardware for sensor-level integration.

- Hybrid models that combine PQC and QKD for transitional deployments.
- Decentralized quantum networks to improve reliability in remote agricultural zones.
- Quantum-friendly edge computing frameworks to reduce latency and power draw.

Institutions such as CERN, QuTech, and MITRE Corporation are actively exploring lightweight QKD designs and satellite-based rural quantum access models (Petersen & Vos, 2023; Wang et al., 2020).

5.6.5 Future directions

The long-term outlook for quantum-secured agriculture is promising. With declining hardware costs, policy momentum, and increasing threats to food data integrity, investment in quantum security is set to accelerate. The following trends are expected:

- Quantum-as-a-service (QaaS) for small- and medium-sized farms via cloud-based QKD endpoints.
- Global standardization efforts such as ISO/IEC 23837 (QKD standards), tailored to agriculture.
- Quantum-aware AI models, which will integrate secure key management during data analytics and prediction tasks.

While the road to integrating quantum cryptography into agriculture is fraught with infrastructure, economic, and policy challenges, its potential benefits such as resilience, confidentiality, and future proofing are undeniable. Addressing these hurdles through coordinated innovation, policy design and inclusive deployment models are imperative for the secure digitization of global agriculture.

5.7 Proposed hybrid security framework

The path to secure agricultural data ecosystems cannot rely solely on quantum cryptographic infrastructure, given current limitations in cost, deployment feasibility, and policy readiness. A pragmatic and scalable solution is the adoption of a hybrid security framework that integrates classical encryption mechanisms with emerging quantum techniques. This framework offers layered protection leveraging existing digital infrastructures while incrementally integrating quantum-safe components, such as PQC and QKD, in high-risk or high-value zones of agricultural networks.

5.7.1 Architectural overview

A hybrid framework in smart agriculture typically includes three major components:

- **Classical layer:** Utilizes standard PKI-based TLS/SSL protocols and AES encryption for device-level communication. This is suitable for legacy sensors, low-power IoT nodes, and edge computing platforms with limited processing capabilities.

- **Post-quantum layer:** Introduces quantum-resistant algorithms (NTRU, Kyber) endorsed by ongoing NIST standardization efforts (NIST, 2022). These are deployed in zones with moderate-risk exposure such as farm management software, UAV telemetry systems, and greenhouse controllers.
- **Quantum layer:** Implements QKD channels and quantum key refresh systems for data centers, cloud storage points, and communication between agricultural authorities and central analytics hubs.

5.7.2 Scalability and interoperability

One of the primary benefits of the hybrid framework is its scalability. By supporting backward compatibility with classical protocols, the framework allows incremental integration without requiring a full infrastructure overhaul. Rural cooperatives or developing countries can begin by enhancing key distribution channels and upgrading data centers, leaving field-level nodes unchanged until needed. This stepwise migration approach enables cost-effective rollouts, minimizes service disruption, and supports interoperability across vendors—a major challenge in decentralized agricultural ecosystems (Singh et al., 2023). The hybrid model accommodates varied threat models from local eavesdropping on irrigation systems to nation-state cyberthreats targeting export logistics. By aligning the cryptographic method with the threat magnitude and data sensitivity, this layered architecture ensures both agility and robustness.

5.7.3 Integration challenges

Despite its advantages, the hybrid approach presents integration challenges. For instance, key management complexity increases as the system must synchronize quantum-generated keys with classical authentication mechanisms. Moreover, latency trade-offs arise in combining fast classical transmission with slower QKD operations. These concerns necessitate the deployment of QKMUs capable of orchestrating seamless key synchronization between classical and quantum domains, using standards like ETSI GS QKD 014 (ETSI, 2021).

Another issue is device heterogeneity. Agricultural networks consist of a broad mix of legacy sensors, proprietary control units, and varying connectivity layers (LoRa, NB-IoT, and 5G). Achieving full-stack interoperability may require lightweight post-quantum cryptographic schemes optimized for constrained environments, such as FrodoKEM or CRYSTALS-Kyber, now under advanced consideration by NIST (NIST, 2022).

5.7.4 Performance and security impact

Empirical studies from hybrid cryptographic trials in smart grid and e-health networks suggest significant benefits for agricultural adaptation. In a simulated farm network model by Lee et al. (2023), a hybrid approach reduced key compromise probability by 92% while maintaining network latency within acceptable thresholds (<5 ms for irrigation controllers and <10 ms for drone telemetry). Furthermore, uptime increased by 14% due to improved resilience against

Table 5.9 Performance comparison between classical, quantum, and hybrid models.

Metric	Classical only	Quantum only	Hybrid model
Key compromise rate	High (↑)	Very Low (↓)	Low (↓)
Latency	Low (↑)	High (↓)	Medium (↔)
Device compatibility	High (↑)	Low (↓)	Medium to high (↔/↑)
Deployment cost	Low (↓)	High (↑)	Medium (↔)
Long-term viability	Low (↓)	High (↑)	High (↑)

denial-of-service and spoofing attacks. Table 5.9 demonstrates the comparative performance analysis among classical, quantum, and hybrid models.

5.7.5 Strategic roadmap for adoption

To facilitate adoption of the hybrid security framework, a phased strategy is proposed:

- **Phase 1 (2025–2027):** Post-quantum upgrades of cloud platforms and edge servers in precision agriculture.
- **Phase 2 (2028–2030):** Pilot QKD rollouts in high-value agricultural export corridors.
- **Phase 3 (2031 onward):** Full hybrid integration across national agritech infrastructure, supported by global standards.

Training and awareness campaigns must be conducted in tandem with infrastructure deployment, focusing on cybersecurity literacy among farmers, agronomists, and rural tech operators (ITU, 2022). Government-backed incentives and grants are essential to offset initial costs and accelerate deployment among resource-constrained stakeholders.

5.8 Conclusion and Future Scope

The integration of quantum cryptography into agriculture marks a pivotal shift toward safeguarding critical food systems in the digital era. As agriculture becomes increasingly dependent on intelligent devices, IoT networks, and real-time data analysis, traditional security mechanisms are no longer enough against the growing threat of quantum-qualified cyberattacks. QKD, when strategically implemented in a hybrid cryptographic structure, offers incomparable confidentiality and authentication of data, protecting sensitive agricultural information that vary from ground analysis to market logistics. The strategic perspective involves a model of integration in phases-being quantum encryption in high value chains, expanding to us from rural quantum communication and culminating in infrastructures totally resilient to quantum. With governments, Agritech companies, and rural cooperatives working in synergy, the Quantum Agriculture Foundation can evolve into a national asset, promoting food sovereignty, market competitiveness, and sustainable digital transformation.

However, the road to realizing this vision is marked by substantial research and development gaps. Key challenges include reducing the cost and size of quantum communication hardware, optimizing post-quantum algorithms for low-power devices, and establishing universal standards for rural quantum deployment. Further interdisciplinary work is needed to ensure seamless interoperability between legacy systems and quantum infrastructures. Moreover, ethical concerns around data sovereignty, cross-border encryption protocols, and transparent key management remain largely unresolved. Addressing these challenges requires a unified innovation roadmap that includes stakeholder training, pilot testing in diverse agro-climatic regions, policy formulation, and global knowledge exchange. As quantum technologies continue to mature, their judicious integration into agricultural ecosystems will not only enhance security but also lay the groundwork for resilient, data-driven food systems capable of withstanding the cyber risks of tomorrow.

References

Bennett, C. H., & Brassard, G. (2014). Quantum cryptography: Public key distribution and coin tossing. *Theoretical computer science, 560*, 7–11.

Bera, A., & Chakraborty, P. (2023). Quantum key distribution for agriculture 4.0. *Journal of Quantum Information Systems, 11*(2), 115–129.

Einstein, A., Podolsky, B., & Rosen, N. (1935). Can quantum-mechanical description of physical reality be considered complete? *Physical Review, 47*(10), 777.

ENISA. (2021). *Cybersecurity in the agriculture sector: A gap analysis.* European Union Agency for Cybersecurity.

ETSI. (2021). *ETSI GS QKD 014 - quantum key distribution (QKD); Security framework.* European Telecommunications Standards Institute.

ETSI. (2022). Quantum safe cryptography and security: ETSI white paper no. 8, update 2022.

FAO. (2020). *Digital agriculture: Report on global status.* Food and Agriculture Organization of the United Nations.

Food and Agriculture Organization (FAO). (2022). Cybersecurity challenges in smart farming.

Gisin, N., Ribordy, G., Tittel, W., & Zbinden, H. (2002). Quantum cryptography. *Reviews of Modern Physics,, 74*(1), 145–195.

ITU. (2022). *Bridging the digital divide in rural agriculture.* International Telecommunication Union.

Kshetri, N. (2019). Cybersecurity in the agriculture sector: Threats and challenges. *IT Professional, 21*(5), 7–13.

Lee, J., Kumar, A., & Banerjee, R. (2023). Hybrid Cryptographic models in precision agriculture. *Journal of Secure Computing in Smart Environments, 12*(4), 212–229.

Liao, S.-K., Cai, W.-Q., Liu, W.-Y., Zhang, L., Li, Y., Ren, J.-G., Yin, J., Shen, Q., Cao, Y., Li, Z.-P., & Li, F.-Z. (2017). Satellite-to-ground quantum key distribution. *Nature, 549*(7670), 43–47.

MarketsandMarkets. (2023). Smart agriculture market by offering, application, and geography - global forecast to 2025.

Mishra, K., & Anantharaman, V. (2022). Quantum-enabled IoT for secure agriculture. *Nature Quantum Computing, 3*(6), 341–350.

Mosca, M. (2018). Cybersecurity in an era with quantum computers: Will we be ready? *IEEE Security & Privacy, 16*(5), 38–41.

Nielsen, M. A., & Chuang, I. L. (2020). *Quantum computation and quantum information.* Cambridge University Press.

NIST. (2022). Post-quantum cryptography standardization project. https://csrc.nist.gov

Petersen, L., & Vos, J. (2023). Quantum networks for smart agriculture: A Dutch perspective. *Journal of Agricultural Cybersecurity, 4*(1), 55–67.

Reuters. (2022). Ransomware attack hits U.S. AgriTech firm, halting automated irrigation. https://www.reuters.com

Schneier, B. (2020). *Data and Goliath: The hidden battles to collect your data and control your world.* W. W. Norton & Company.

Shor, P. W. (1997). Polynomial-time algorithms for prime factorization and discrete logarithms on a quantum computer. *SIAM Journal on Computing, 26*(5), 1484–1509.

Sibson, P., Erven, C., Godfrey, M., Miki, S., Yamashita, T., Fujiwara, M., Sasaki, M., Terai, H., Tanner, M. G., Natarajan, C. M., & Hadfield, R. H. (2017). Chip-based quantum key distribution. *Nature Communications, 8*(1), 13984.

Singh, V., Raj, S., & Ahmed, H. (2023). Cryptographic interoperability in agricultural IoT systems. *International Journal of Agritech Security, 8*(2), 77–93.

Sundaram, R., & Mehta, D. (2024). Securing soil sensors with quantum randomness: A pilot study in Karnataka. *Indian Journal of Quantum Agriculture, 2*(1), 29–43.

Wang, S., Chen, L., & Hu, J. (2020). Realizing secure quantum links in open terrain. *Nature Photonics, 14*(7), 437–442.

Wolfert, S., Ge, L., Verdouw, C., & Bogaardt, M. J. (2017). Big data in smart farming – a review. *Agricultural Systems, 153*, 69–80.

Zhao, W., & Liu, X. (2022). Satellite quantum cryptography in agricultural IoT: A pilot in Inner Mongolia. *Quantum Communication Letters, 3*(4), 201–214.

6

Quantum cryptography for secure agricultural data: Applications and impact

Anshu Sharma[a], Vipin Kumar Chaudhary[a], Chirag Sharma[a] and Anurag Sharma[b]

[a]SCHOOL OF COMPUTER SCIENCE AND ENGINEERING, LOVELY PROFESSIONAL UNIVERSITY, PHAGWARA, PUNJAB, INDIA [b]DEPARTMENT OF ELECTRONICS AND COMMUNICATION ENGINEERING, CT INSTITUTE OF TECHNOLOGY AND RESEARCH, MAQSUDAN, JALANDHAR, PUNJAB, INDIA

6.1 Introduction

Agriculture is experiencing a deep, sweeping technological transformation, moving from traditional, experience-based, dynamic practices to new data-based, intelligent, and automated systems. This transition (also known as digital agriculture or smart agriculture) relies upon the synergy of Internet of Things (IoT), remote sensing technologies, artificial intelligence (AI), big data analytics, cloud computing, and smart communication networks. Such technologies promote precision agriculture solutions, real-time monitoring, predictive analytics, and decision-support systems to improve crop productivity, optimize resource use, minimize environmental impact, and promote food security amidst global challenges expected to grow (Khan et al., 2021).

New agricultural systems largely depend on heterogeneous sensor networks in a constant data capture loop of soil moisture, soil nutrition, soil temperature, soil humidity, crop growth stages, insect pest infestation, climate, etc. Remote-sensing platforms such as satellites and Unmanned Aerial Vehicles (UAVs) provide spatial information at a larger scale and at high spatial resolution to complement ground-level sensors (Aboy et al., 2022). These technologies, along with AI, are able to produce large amounts of various types of data that are sent over networks, stored virtually in cloud infrastructures, and processed with machine learning and AI methods to assist in real-time and long-term agricultural processes (Chen et al., 2016).

Digitalization has improved the efficiency and sustainability of agricultural practices, but it has also created new risks regarding data security, privacy, and system reliability. Agricultural data are no longer siloed businesses—they now exist in flexible, integrated agri-tech ecosystems comprised of farmers, agribusiness companies, researchers, policymakers, logistics providers, and consumers (Ott et al., 2019). Information such as crop yield forecasts, land-use patterns,

Quantum Computing and Instrumentation for Smart Agriculture. DOI: https://doi.org/10.1016/B978-0-443-44247-6.00001-4

pricing strategies, genetic data, and supply-chain logistics is all sensitive and important—often very sensitive and very important, since they have great economic, strategic, and social value. The implications of security breaches can be costly (e.g., disruption in supply-chain) and even a threat to national food security, resulting in erosion of trust and leading to financial losses if perpetrators are not identified and apprehended, unauthorized access, or manipulation or disruption of agricultural information systems.

Cybersecurity threats to agriculture are exacerbated by the fact that many IoT devices and wireless communication networks operate under resource-constrained and dispersed environments. Such systems are prone to different types of cyberattacks such as eavesdropping, data tampering, denial-of-service, and unauthorized access (Kiktenko et al., 2018). Most of the classical cryptographic mechanisms for confidentiality and integrity of data in agriculture are based on mathematical complexity, which has proven effectiveness in traditional cyber-environments but is limited in technological evolution. Popular public-key cryptosystems, such as Rivest-Shamir-Adleman (RSA) and Elliptic Curve Cryptography (ECC), as well as symmetric encryption schemes, have been secure as long as we assume a classical computer to work (Akter, 2023).

The classical cryptographic systems are kept secure for the long term, but the fast development of quantum computing is just a fundamental disaster to the long-term security of those classical cryptographic systems (Chaterji et al., 2020). Quantum computers use quantum mechanical principles of superposition and entanglement to solve problems that are not possible for classical computers to solve (Kamble et al., 2019). In particular, Shor's algorithm proves that, when powerful enough, quantum computers can efficiently factor large integers and solve discrete logarithm problems, thus rendering many of the public-key cryptographic schemes used today to secure digital communication. The challenge of cryptographic systems' obsolescence due to advances in quantum hardware has moved from theory to practice.

This new quantum threat is alarming for the agricultural sector that needs the confidentiality of data over longer time spans. The attack model "harvest now, decrypt later" shows that an adversary may harvest encrypted Ag data today to decrypt later once quantum capabilities are sufficient. It should be noted that this particular scenario highlights the very real and dire need for security mechanisms that will not only be resilient against existing threats but also against future quantum-enabled threats (Otieno, 2023).

To solve these problems, quantum cryptography (QC) has been introduced as a potential measure for secure communication of the next generation. While classical cryptography security relies on computational assumptions, QC security is based on the laws of physics. One of the main features of QC is quantum key distribution (QKD), which allows two parties communicating with each other to generate a shared secret key with the properties that it is impossible to create a perfectly correlated key without the parties noticing if an eavesdropper is present. Such a quantum mechanism is due to the no-cloning theorem and the perturbation of quantum states by measurement, which guarantees that any interception of the communication is fundamentally detectable (Alazzai et al., 2024).

This work presents a new synergy between quantum technologies, cybersecurity, and agritech systems, pioneering the application of QC in agricultural sensor networks. Adoption of IoT, AI, cloud platforms, and blockchain-based supply chains in agricultural infrastructures creates the need for secure and trustworthy channels of communication. QC has the potential to provide data transmission security between farm-side sensors, edge and cloud-based servers, research institutes, and supply-chain actors, increasing the overall resilience of digital agricultural ecosystems (Charlebois et al., 2024).

This chapter aims to investigate ultimate QC methods to secure agricultural data. It looks at the potential role of quantum-safe communication to improve the security of smart farming platforms, weather forecasting systems, agricultural research networks, and agri-food supply chains. The chapter also discusses the potential of QC as these technologies could further sustain data integrity, transparency, and trust, which in turn could underpin sustainable and collaborative agricultural innovation.

The implementation of QC in agriculture offers several practical challenges despite its promising advantages. QKD systems need new hardware, well-established communication channels (assuming free-space or optical fiber links), and new integration points with existing digital infrastructures. Adoption, especially in rural areas and developing countries, will depend on cost, scalability (between systems and in relation to existing cryptography and post-quantum resistant systems), and institutional framework availability. Resolving these issues calls for more than just technical solutions, but also coordinated action coming from researchers, industry stakeholders, policy makers, and standardization bodies (Busch et al., 2007).

In this context, the key challenge this chapter will discuss is providing strong and future-proof security of agricultural data in the quantum computing age. This chapter attempts to connect the advances in QC to the security level needed for contemporary agriculture. Cloud computing has become an integral part of digital agriculture, enabling new applications of information technology to improve by theoretical foundations, practical implementations, and challenges to ensure the deployment of quantum-enabled security mechanisms that support reliable, trustworthy, and sustainable digital agriculture (Allende et al., 2021).

Objectives of the chapter

The objectives of the chapter are:

- To discover and analyze the major security and privacy challenges of IoT-enabled farming, remote sensing systems, and data-heavy agricultural supply chains for the respective quantum threats;
- To give a conceptual and a technical overview of QC, with a focus on QKD and its security proofs;
- To analyze the possible applications of QKD in smart farming, weather monitoring, agricultural research, and blockchain-based agri-food supply chains;
- And, to assess the key real-world implementation challenges, including cost, infrastructure, scalability, and coordination between institutions, and highlight future prospects for quantum-secure agricultural systems.

Organization of the chapter

The rest of this chapter is structured as follows. Section 6.2 reviews the basic concepts of QC and QKD protocols. Section 6.3 presents use cases for quantum cryptographic techniques applied in agricultural systems, ranging from smart farming to supply-chain management. Section 6.4 describesQCimpacts on thewider agro-ecosystem, including data trust, sustainability, and stakeholder collaboration. Section 6.5 provides a QKD-based framework for safeguarding safe agricultural data. Section 6.6 discusses literature related to and recent research progress on exploring quantum technologies and how they can be applied to agriculture. Section 6.7 explains the method that we have used. Section 6.8 gives QC as an example in Agriculture and provides impact and future application results and discussion. Section 6.9 highlights results and discussion. And, finally, Section 6.10 concludes the chapter with future scope.

6.2 Literature review

The digital transformation in agriculture has been well studied in terms of IoT-enabled smart farming, remote sensing, AI, and data-driven supply-chain management. But as agricultural systems begin to rely more on interconnected digital infrastructures, issues of data security, privacy, and whether or not people will trust these systems over the long term have come to the forefront. This section reviews the existing literature and works that support the pillars of digital agriculture, cybersecurity for agri-tech systems, blockchain-enabled supply chains, and the potential for this ecosystem to be secured by QC (Cravero et al., 2022).

(Astill et al., 2019) explored how digital technologies are contributing to improving the efficiency and sustainability of agricultural practices. Authors suggested the use of IoT sensors and data analytics for near-real-time monitoring of farms. While connecting data-driven agriculture with sustainability metrics was new, it was proven in experiments that this coupling provided better yield forecasting and resource use precision(Assche, 2006). Yet security was mostly considered a secondary focus.

(Kamble et al., 2019) reviewed issues in agri-food supply chains and recommended digitization to improve inefficiencies. Authors revealed the opacity of conventional supply chains in agriculture. Identification of digital bottlenecks impeding supply-chain effectiveness was novel, and case examples illustrated increased operational performance enabled through digital changes, but neglect the cryptography resilience aspect of the digital bumps.

(Scarani et al., 2009) presented a framework that facilitates secure data sharing and also keeps an immutable record of the agricultural operations. By decentralizing agricultural data to be stored on individual farms, crops in experimental simulations had shown a greater degree of traceability and trust. Yet the framework was based on classical cryptographic primitives, which are weak against future quantum attacks.

(Bobde et al., 2024) modeled blockchain blockchain adoption in agriculture and the role of trust and transparency in enhancing. Authors mapped blockchain properties with the agricultural use cases they were targeting. The study noted its computational overheads and did not evaluate security mechanisms that resist quantum computing, while performance evaluations indicated enhanced data integrity.

(Ryan et al., 2023) framed an interdisciplinary view of AI in agriculture, emphasizing AI-based decision support for crop management. What was unique was the integration of AI with agricultural knowledge systems. While experimental results showed better prediction accuracy, the threat of an AI pipeline to data security and privacy was not addressed.

(Althobaiti & Dohler, 2020) suggested disease detection and yield prediction architectures for precision agriculture with IoT solutions integrated with AI. A unique element of this study was a holistic taxonomy of smart agriculture applications. Although better automation was demonstrated through the experiments, the authors called out cybersecurity as a key problem that remains unresolved.

(Charlebois et al., 2024) investigated digital traceability in agri-food supply chains and developed transparency models enabled by technology. The novelty was in connecting traceability with food safety outcomes. In case studies, accountability was improved; however, resistance against advanced cryptographic attacks was not addressed.

(Peters et al. 2023) examined the role of big data analytics in agriculture and environmental systems. Authors pointed out avenues for better decision-making in the leveraging of large datasets and showed a much higher level of analytical accuracy than had been previously demonstrated, but highlighted the need for transparency around data ownership and whether the data would still remain secure in the long term.

(Bhatt & Sharma, 2019) focused on the integration of IoT that can support AI for supply-chain transformation in the manufacturing field and provide data-driven models for the IoT-enabled logistics optimization. The study combined bibliometric and technical analysis approach, and the results revealed the improved responsiveness of the supply chain. Nevertheless, key management remained an unresolved problem.

(Gupta et al., 2020) described agricultural cyberthreats and pointed to crops as a possible target of cyberwarfare. It was also innovative to present agriculture as a national critical infrastructure in the research. No cryptography by design stimulates the growth of attack surfaces, as observed in observational analysis.

(Barbeau & García-Alfaro, 2022) surveyed quantum cryptography, a security paradigm for the future. Though their theoretical contribution did highlight unconditional security properties. Experiments were not carried out, but this work was a precursor to applications in specific divisions.

(Naughton et al., 2019) gave a thorough security review of realistic QKD systems. What was new was validation against actual imperfections. The encryption method could resist a perfect omniscient eavesdropper in theory, but experimental results verified that QKD is well suited, even under realistic physical conditions, for the protection of operationally important data.

(Ott et al., 2019) proposed quantum-secured blockchain architectures via combination of QKD and blockchain consensus mechanisms. Increased resilience against key compromise was demonstrated through simulations, suggesting its applicability to agri-food supply chains.

(Liu et al., 2024) studied post-QC adoption pathways. What made them new was their scope: measuring readiness across sectors. Results showed a slow migration, which suggests the need for hybrid cryptographic approaches.

(Dekkaki et al., 2024) covered quantum dangers to classical cryptosystems and talked about post-quantum solutions. The study focused on challenges associated with cryptographic transition, and RSA and ECC were confirmed vulnerable through simulations.

6.2.1 Research gaps and need for proposed procedure

However, while a lot is already known about digital agriculture, IoT-based farming, AI-based actionable intelligence, and blockchain-based supply chains, major gaps exist, which leads us to the methodology proposed in this chapter. First, the vast majority of agricultural digitalization studies concentrate on aspects of efficiency, automation, and sustainability, but treat security as secondary and not as a prerequisite. Second, classical cryptographic mechanisms that have been widely used in security frameworks in agriculture today are susceptible to quantum computing threats, which are emerging rapidly.

Third, although blockchain is often brought up for agricultural traceability, classical key management from blockchain still makes supply-chain systems susceptible to long-term crypto. Weak key distribution and lifecycle management for IoT-centric agricultural architectures, especially in large-scale and heterogeneous deployments. Lastly, other works do not take into account the fact that the harsh conditions of the "harvest now, decrypt later" threat model are highly relevant for agricultural data, since it has strategic value over a long-term duration.

In addition, while QC and QKD have been thoroughly investigated on a theoretical level and in sectors such as defense and finance, their systemic incorporation into the agricultural digital ecosystem is still lacking. No frameworks link quantum-secure key distribution to real-world use cases in agriculture, such as smart farming, remote sensing, and supply-chain coordination. Furthermore, existing studies inadequately address the practical limitations of cost, scalability, and hybrid cryptographic integration issues (Jin & Xie, 2023).

All three gaps taken together suggest a nondisruptive application-driven roadmap that integrates QC within the agricultural domain while keeping the existing digital infrastructures in place. Addressing these limitations motivates the methodology proposed in this chapter that designs a secure digital agriculture architecture by integrating QKD-based security mechanisms with IoT, AI, cloud, and blockchain technologies to ensure a resilient architecture against future advances in quantum computing.

6.3 Principles of quantum cryptography and protocols

6.3.1 Principles of quantum cryptography

The intrinsic security guarantees of QC stem from the fundamental laws of quantum mechanics, which yield capabilities beyond classical cryptography. While crucial for digital agriculture—where high-volume sensitive data about soil and crop quality, climate conditions, and logistics around the supply chain are shared continuously—these principles provide a mechanism for secure communication even in the face of highly advanced cyberattacks and quantum computing to come. In contrast to conventional encryption methods that rely on computational intractability assumptions, QC guarantees security based on the laws of nature, which makes it especially useful for long-life agricultural data security (Althobaiti & Dohler, 2020).

One of the key ideas behind QC is that a quantum measurement will unavoidably disturb the state of the system. This makes it impossible for someone to intercept or eavesdrop on quantum-encoded information without changing its state, a characteristic that rather nicely flags the intruder (Ryan et al., 2023). This feature is important in agricultural IoT networks whose sensor-generated data usually passes through unsecured channels. This means that any interference that occurs before data exploitation can be detected, which allows farmers to make decisions based on genuine, trustworthy data (Biham et al., 2006).

Another key concept is the quantum no-cloning theorem, which states that unknown quantum states cannot be perfectly copied or cloned. This feature ensures the protection of agricultural data from replication and replay attacks—common threats in most large-scale digital ecosystems. Furthermore, quantic superposition enhances the uncertainty from the adversaries as quantums can exist in various setups before measurement, enhancing the overall communication security (Dam et al., 2023). Quantum entanglement provides even more protection by establishing strong correlations between particles, as such, if a single particle gets tampered with (even when they are far away), we can always detect if someone has tried to tamper with a single particle or multiple particles (it's a key requirement for geographically distributed systems of agriculture).

6.3.2 Quantum cryptographic protocols

6.3.2.1 BB84 quantum key distribution protocol

BB84 is the first discovered and most implemented QKD protocol, and it is the workhorse of many practical quantum cryptographic systems. It operates on prepare-and-measure, with information encoded into quantum states by randomly chosen bases. However, any attempt at eavesdropping will lead to errors that can be detected (because the measures differ). Use case of BB84 in agricultural applications. BB84 can protect communication from farm-level IoT sensors to centralized cloud platforms, providing a security guarantee to real-time crop, soil, and weather data used for precision farming (Husain et al., 2025).

6.3.2.2 Decoy-state QKD protocol

In contrast to perfect single-photon sources, weak laser pulses are commonly used but can introduce security loopholes into the BB84 protocol, which are rectified in the decoy-state QKD protocol (Kikuchi, 2024). Decoy-state QKD uses the random variation of signal intensities to identify advanced attacks like the photon-number-splitting attacks (Ren et al., 2023). This protocol also enhances the viability of quantum-secure communication across existing fiber-optic and wireless links for connecting rural farms to regional data centers, making it more appropriate for large-scale and cost-sensitive applications such as digital agriculture (Huberman et al., 2020).

6.3.2.3 Quantum key distribution through entanglement-based (E91 protocol)

Entangled photon pairs are used to implement the E91 protocol, and the existence of quantum correlations exceeding classical physical bounds ensures the security of the cryptographic keys that are generated. It is trust-free in the source of quantum states and thus is more robust against some attack models. In agricultural scenarios, entanglement-based QKD is most applicable to long-distance and satellite-based communication that facilitates secure transmission of remote sensing data, climate analytics, and inter-jurisdictional agricultural intelligence (Dekkaki et al., 2024).

6.3.2.4 MDI-QKD stands for the measurement-device-independent quantum key distribution

However, real-world implementations of QKD typically incorporate imperfect detection devices that can provide a "one-way attack" capability for an eavesdropper; hence, Measurement-Device-Independent QKD was developed to remove these detection-device vulnerabilities. Communication hardware in agricultural environments may operate in more extreme conditions and thus may be more susceptible to hardware imperfections and side-channel attacks (Kosior & Mlodawska, 2024). MDI-QKD is advantageous for centralized agricultural data hubs, as it improves security through its trust-free approach in relation to measurement devices (Dey et al., 2022).

6.3.2.5 Continuous-variable quantum key distribution

As opposed to discrete photon states employed in QKD based on discrete variables, Continuous-Variable QKD encodes information in continuous properties of light (amplitude and phase). Its core advantage is the use of standard telecommunication equipment for deployment, which simplifies deployment and reduces costs on a large scale (Raveena & Edward, 2021). CV-QKD presents a realistic route for incorporating quantum security in digital agriculture systems, building on existing communication infrastructure, enabling effective quantum protection in smart farming networks and agricultural supply-chain platforms with little infrastructural overhead (Bienfang et al., 2007).

6.3.3 Integration and practical relevance to digital agriculture

As a result, various post-processing elements such as error correction, privacy amplification, authentication, and frequent key renewal are indispensable for the long-term security of all QKD protocols. Such features are vital for agricultural systems that produce a continuous data stream and need ongoing protection. Instead of replacing the classical techniques, in hybrid security architectures combining quantum, classical, and post-QC, QKD will complement the classical techniques. These hybrid approaches are especially well-suited for agriculture, balanced between the needs of scalability, interoperability, and cost (Dhal & Kar, 2024).

To conclude: Finally, considering agriculture, it is illustrated that the principles and protocols of QC provide an excellent security framework that can last forever. QKD provides the solutions helping digital agriculture face its security challenges, directly covering secure key distribution, eavesdropping detection, and long-term data integrity through the use of QKD protocols. Well-implemented agri-tech innovations can boost trust, stability, and sustainability at all points in the agricultural value chain (Bhatt & Sharma, 2019).

6.4 Use of quantum cryptography in agriculture

QC offers the potential to revolutionize the way we protect data in agriculture, as the world becomes ever more cybersecurity-conscious. If that can be used in QC in agriculture, home growers would be confident in using this and knowing that the data is real and it's not fiction and it's real and authentic. The importance of secure communication and storage in an era where the agricultural sector moves toward data-driven plays a very significant role (Jain et al., 2023).

Open image in new window QC embodies a new level of cryptography (which is not the case with classical cryptography), i.e., security at the level of unconditional physical laws. These weaknesses can be remedied by incorporating QKD, which in turn could be used to protect agricultural data from eavesdropping and tampering (Döberl et al., 2023). QC, as a security technology for monitoring information and farmland,' in precision agriculture, supply chain of agriculture, and agricultural research, has various security levels and functions, and achieves the security protection of sensitive information, ensuring its confidentiality and integrity (Kristen et al., 2021). Thus, QC offers farmers an extraordinary level of data security; it will keep private agricultural information from being surreptitiously spied on and tampered with, and it will also guarantee the integrity of data-based agricultural decisions. QC may be used to protect sensitive agricultural data, such as crop yields, soil humidity, and weather event predictions, from interference, protecting the integrity of research outcomes. Ag data QC can be used to disseminate sensitive Ag data (e.g., between researchers, policy or decision makers, and farmers) via a secure channel to enable innovation and partnership. QC could also protect proprietary information of agricultural companies—say, data about the genetics of seeds and methods of breeding crops—from industrial sabotage or theft of intellectual property.

In the development of QC for agriculture, dedicated sector-specific terms and conditions (costs, scalability, interoperability with incumbents) must be considered. Also, data exchange between sensors, unmanned systems, and base stations can be performed out by means of remote QC to ensure a secure and confidential data handling for Precision Farming (Martin et al., 2017). QC could be an answer by encrypting the information of origin, transport, and storage of the food, so traceability and counterfeiting can be avoided (Duncan et al., 2021).

QC is capable of preventing leaks of information and data, thereby making the agricultural data inaccessible and inoperable by anyone other than authorized people. Quantum cryptographies based can be used to protect agricultural data by encrypting it with quantum keys and to secure the agricultural data storage place by storing it in quantum-encrypted forms. QC allows farmers, scientists, and policy makers to share sensitive information free from snooping or tampering. QC might be applied to various agricultural applications, including precision agriculture, logistics, and agricultural research (Edwards et al., 2020).

It is employed in agriculture, such as precision agriculture, food supply chain management, and agricultural research, because it can guarantee security and improve effectiveness. In smart agriculture, QC can protect sensor data that is essential for yield optimization and resource usage (Elmisery et al., 2025). A secure link can be established using QC, and then we can connect all the devices at a farm with machines and control devices to inhibit unauthorized activities and control them. Moreover, QC allows for ensuring the authenticity and secrecy of sensitive information shared among different players in agro-food supply chains, such as farmers, suppliers, and retailers (Durrant et al., 2021).

QC-based secure communication in agricultural research provides for secure data transmission and data storage, whereby critical agricultural research data is protected from illegal access and modification. The agricultural data security, in terms of CIA (confidentiality, integrity, and availability), is significantly addressed through quantum protocol, which is not achievable through any other conventional systems (Kumar et al., 2021). Symmetric encryption is also another way for secure data transfer and storage to be realized for those communities that use QKD, by creating and distributing symmetric encryption keys through which strong encryption ciphers can be applied. Due to the technological development of quantum computers, the classical cryptographic operations have become weaker and weaker in terms of their security. QC provides long-term security when applied in settings and preserves agricultural data from current and future attacks (Barker et al., 2021; Döberl et al., 2023). Quantum-proof cryptography can be another (CMP) line of defense against quantum computers breaking cryptography, and quantum cryptographic systems are grounded in physical laws (Peters et al. 2023). QKD and other post-quantum cryptographic techniques are therefore needed to combat cyberthreats to critical national infrastructure (Chen et al, 2016). Post-QC is classical cryptography secure against quantum computers (and QC might be secure for an even longer time period, since it is based on physics (Dam et al., 2023; Peters et al., 2023).

It also increases the confidence of the data sensed by the sensors from the IoT devices deployed on farms (Ralegankar et al., 2021). QC can keep results, IP, and private information safe to foster innovation and co-operation in agriculture. Stakeholders who use QC to protect agricultural information now have a means to profit and can invest in sustainable agriculture. Privacy of data should also be protected in the context of agriculture, and it has to consider the future vulnerability due to quantum computing on data, including its CIA (confidential, integrity, and availability) aspects (Bavdekar et al., 2022; Peters et al., 2023). Preventing the quantum computing threats of tomorrow will help agriculture become competitive and thus advocate sustainable agriculture practices (Bavdekar et al., 2022).

For theoretical security, the potential of using QC in agriculture is based on our information-theoretic assurances of the security of communications given by QKD. This entails multilayered communication architectures within agricultural digital ecosystems, where data from edge-level sensors are transmitted to cloud and fog layers, and finally to analytics and decision-support systems for farming (Liu et al., 2024). Classical key exchange mechanisms in such architectures may remain vulnerable if an adversary has a computational model that relies on future access to quantum computing resources. The use of QKD brings a layer of security to the generation and distribution of encryption keys, which can guarantee resilience against classical and quantum attackers and therefore strengthen the foundational trust model of agricultural data infrastructures (Ellahi et al., 2024).

Technically, QC provides an unconditional security guarantee by decoupling key distribution from computational assumptions, enabling secure communication across heterogeneous agricultural networks. Soil moisture sensors, weather stations, and UAV-based imaging platforms, e.g., such a quick key renewal limits the risk/effect of a long-time key exposure and meets the need for continuous monitoring applications in precision agriculture. Theoretical models have shown that once QKD systems are properly designed and protocolled, eavesdroppers cannot acquire even partial information on the transmitted key, which can be erased using privacy amplification, making agricultural datasets secure and private while transmitted through noisy or nonideal quantum communication channels (Astill et al., 2019).

QC also offers a powerful mechanism for ensuring the integrity and authenticity of data in agricultural supply chains. Agri-food systems supported by blockchain are operated on the basis of an immutable ledger that helps track the provenance, logistics, and quality of an agri-food product. But the security of blockchain networks itself relies on the security of the communication channels and key management systems. Small-scale QKD integrated with blockchain infrastructures for secure transfer of cryptographic keys used for transaction signatures and node identities. Theoretically, this combination provides a level of trust to the assumptions of distributed ledgers that prevent key compromise in a post-quantum adversarial environment (Balkrishna et al., 2023).

Even more, Quantum cryptographic methods enhance remote sensing satellite-based agriculture monitoring systems. Information about agriculture gathered with the use of satellites and long-range communication links is notably easy to intercept or alter (Malpani et al., 2023). In theory, entanglement-based QKD and satellite-assisted QKD protocols could provide secure long-distance key distribution and protect the highly valuable environmental and climate data by securely performing QKD. Such techniques enable the authenticity and reliability of spatio-temporal datasets, which in turn aid large-scale agricultural planning and policy decision-making (Finger, 2023).

At the level of system integration, QC allows hybrid security architectures, mixing classical encryption, post-QC, and QKD. This is particularly important for agriculture, as infrastructure heterogeneity and cost limitations on deploying a fully quantum-enabled network prevent its immediate implementation. Hybrid theoretical frameworks exist, whereby QKD can be used as a root-of-trust layer to distribute keys that secure higher-layer protocols employed by IoT communication, AI-based analysis, and cloud-based agricultural platforms (Ham, 2018).

Besides these two, QC exchanges information securely across organizations, strengthening trust and cooperation between agricultural entities. Often, farmers, researchers, agribusinesses, or even policymakers need access to shared datasets while preserving data ownership and privacy (Halder et al., 2025). Fine-grained access control and secure multi-party transfer of sensitive data are made possible by quantum-secure communication channels, facilitating co-operative crop-improvement initiatives while keeping private data protected. From a theoretical perspective, this matches secure multi-party communication models where trust is ensured by physical security guarantees rather than by institutional assumptions (Bedington et al., 2017).

In conclusion, QC is a next-generation security feature solution that integrates well with the future scenario of agricultural systems that are going to experience changes due to enhanced automation processes, intensive use of data, and becoming more pervasive and interconnected. QC enhances the trustworthiness and sustainability of digital agricultural ecosystems by integrating theoretically sound security assurances and technically solid key generation, distribution, and management methods (Rahaman et al., 2024).

6.5 Implications on the agro-ecosystem

QC may be about to make profound impacts on the agroecosystem and put an end to the old data security, trust, and transparency era. It can also lead to more sustainable and resilient farm practices. As agri-markets become increasingly digitalized, strong data protection clauses are essential. QC emerges as a promising solution to protect confidential agricultural information from unauthorized access, tampering, and stealing (Ferguson et al., 2024). The repercussion of QC on the agro-ecosystem opens the door to far more than just information security; cooperation among stakeholders, enhanced management decisions, and consumer confidence make the security robust, and ultimately lead to higher productivity (Gilbert & Gilbert, 2025).

QC could be crucial to enable safe data sharing between the various stakeholders of the agro-ecosystem - such as farmers, researchers, policy makers, and industry partners. Through secure communication among stakeholders' interactions, trust can be built using QC, and alliances can be formed to enhance the effectiveness of agricultural interventions. QC would transmit to farmers on a moment-by-moment basis the moisture content in soil, weather patterns, and crop health, allowing still more accurate decisions on the micro-level of whether—and when not—to water, fertilize, or apply pesticides. In theory, farmers could plan their schedules to optimize resource use efficiency, to minimize environmental impact, and to maximize yield (Fernández-Caramés & Fraga-Lamas, 2020).

The technology of QC will significantly enhance consumers' confidence in the safety and reliability of agricultural products and food when quantifying whether food is edible. QC can help fight against food fraud by ensuring a secure and traceable manner in the supply chain information by having its output product authentic. With QC, you can increase consumer confidence in agriculture and the safety and quality of their food because of the level of trust that is built when it's no longer about whether information was stolen, but the mind is spread further. Humans think differently when things don't have to be done the "old" way. QC may have potential importance for the protection of biodiversity, mitigation of greenhouse gas emissions, and the promotion of sustainable land management (Awan, 2019).

Agricultural Digitalization in agriculture has led to an abundance of data in the form of soil characteristics, climate information, crop yield, market price, etc. Hence, to keep the supply chain of food, managing resources and sustainable agriculture activity, it is necessary to save this data. QC addresses such problems as technical, economic, and ownership and privacy issues of the data (Chaterji et al., 2020). Its use in all the means of production is beneficial, as well as a major barrier that hampers farmers from adopting new technologies. Farmers are least likely to use new technology of any type because it's too expensive, there are no guarantees in data security, and they have nowhere to turn for technical assistance (Bavdekar et al., 2022).

QC has the potential to protect sensitive agricultural data from being intercepted, tampered with, and accessed by unauthorized parties. Quantum-safe cryptosystems would have enriched agro-ecosystems with additional security, cooperation, and consumer confidence (Burg et al., 2020). The disruptive potential of QC in the agro-ecosystem is hard to overestimate: this technology could revolutionize the way in which agricultural data is managed, shared, and processed. "There are several key underlying conditions that are thought to facilitate a new digital agricultural revolution. High-resolution digital information on the state of the earth is now available at very low cost on many dimensions, such as soil, health, plant growth, weather station, and drone and satellite data." Precision agriculture utilizes sensors, auto tractors, and drones, and with them, farmers measure and analyze data to adjust the application of products and processes to increase yield, saving resources, and minimizing the impact on the environment (Duncan et al., 2021). The use of digital technologies in agriculture has ignited expectations and concerns around impacts on farmers, sustainable farming, and food and nutrition security (Ferguson et al., 2024).

Environment QC promotes more efficient and greener agriculture. QC promotes more efficient and greener agriculture. 1 image from the article. Published on Image courtesy of Pixabay, June 20, 2019. QuantGen's technology will improve sustainability and reduce environmental impact.

Apart from real security, QC can be applied for environmental, social, and economic purposes to agro-ecosystems. QC provides a mechanism for exchanging secrets between farmers, scientists, and policy makers for vested interests that promote innovation and translation of information (Barbeau & García-Alfaro, 2022).

The implementation of QC in agriculture might have an impact on data safety, sustainability, and partnerships between stakeholders across the entire agro-ecosystem. This is where that QC would come in, with farmers storing sensitive crop yield, soil, and weather data, and data-driven advice that would flow to farmers. QC can assure that the data used to enhance every aspect of farming, from fertilization to irrigation, is secure and accurate, and ensure that resources are managed sustainably (Mastriani, 2022). QC resides between farmers' privacy and data security, thereby gaining trust and fostering data sharing with research and development. QKD protects agri-hacks, supports growth in Plant research and development. Security of IoT-enabled supply chains is a key issue, and machine-learning applications may also address privacy and security threats (Gupta et al., 2020). These inherent weaknesses have, in fact, enhanced the need for new methods to meet the security and intrusion detection requirements for IoT data protection as well (Gupta et al., 2020). QC also provides cybersecurity in dynamic network scenarios. It keeps farmers' sensitive data safe from both internal and external threats, therefore stopping data corruption and data theft (Gwagwa et al., 2021).

6.6 QKD-based agricultural data protection

A Secure mechanism to store agricultural data is provided by QKD through quantum mechanical-based secure communication of encryption keys (Elmisery et al., 2025). Employing QKD ensures that if any eavesdropping is attempted on the key, it would be identified, and so the information secured (Ott et al., 2019). QKD, as a cryptographic primitive, uses quantum effects to enable confidential communication between two communication entities (Mink et al., 2010). The technology is envisaged as a replacement for, or addition to, trust-based courier protocols used to deliver trusted keys (Bedington et al., 2017). This provides the functionality to establish a shared common random secret key between guest and host so as to secure communication between the two (Jain et al., 2023). The employment of this method allows us to ensure that any interception of the key—if it did happen—would be shown and that the key of the data will remain secret (Martín et al., 2017).

If QKD promises security, it becomes an attractive technique for protecting high-value agricultural data (Morales et al., 2025). One of the relevant QKD advantages is that it can detect eavesdropping, which is crucial to maintaining the integrity of agricultural data. There is also the possibility to couple QKD together with other cryptography systems, similar to systems already working, so that we could make an agricultural data infrastructure more secure. The implementation of QKD and blockchain in agricultural networks would offer more security and transparency (Gill & Buyya, 2024). This integration may help validate agricultural data, build trust between parties, and assure safe shipments. The quantum blockchain utilizes the superposition and quantum entanglement, which is a great step forward in natively resisting classical computer possibilities (Gill, 2025).

Other properties of a blockchain, which quantum computing does not get along well with, are public key cryptography and hash function. New cryptosystems whose security is guaranteed even under quantum attacks are currently under development to nullify these threats and secure the blockchains. Attacks of quantum computing can threaten the security of blockchain protocols and networks, but blockchains employ nonquantum-resistant cryptographic algorithms to secure themselves. So, one of the ways to overcome this obstacle may be integrating QKD with the classical encryption method in a hybrid scheme that utilizes the benefits of both methods.

Blockchains that originated as the software of Bitcoin are now a mature base for secure exchange of data in a wide range of domains (Gilbert & Gilbert, 2025). It is maintained in an integral and transparent state through cryptographic methodology, such as hash functions, public key cryptography, and digital signatures, in the absence of a central authority (Gilbert & Gilbert, 2025). Blockchain networks are based on cryptographic protocols; they are implemented in the networks, which forces us to look for the potential uses of quantum computing and quantum information to attack and to reinforce blockchain technologies. Quantum computing has both pros and cons for blockchains. The security of blockchain technology is threatened by the advent of quantum computers. With the use of blockchain technology, it becomes feasible to enforce transparency, immutability, and, with the decentralized consensus mechanism, the trustworthiness and reliability of the data (Naughton et al., 2019). The related issues, such as data (integrity, visibility, and performance) to earth-friendly climate, can be overcome using blockchain (Husain et al., 2025). For example, the integration of blockchain and IoT sensors for crop surveillance leads to an improvement in data quality as well as traceability.

Additionally, utilization of advanced cryptographic methods like zero knowledge proof can validate the data and prevent attackers (Bobde et al., 2024). A list of protocols and network connections needs to be exhaustive to provide data protection on a farm in remote areas, including any possible AI and IoT applications (Rahaman et al., 2024). For the critical infrastructure, QC is really important, as these are physics-based security. QC provides a data communications and transfer immune to eavesdropping and data tampering in agriculture. QKD presents an opportunity to secure data flow between the parties of the agri-supply chain and obtain trust in terms of shared data (Barker et al., 2021).

6.7 Methodology

This chapter adopts a methodology that is tailored to (1) systematically analyze the applicability, effectiveness, and feasibility of QC specific to QKD, for the security of digital agriculture ecosystems. This methodological framework examines existing quantum threats in agricultural data security to fill gaps in our understanding and apply that understanding through theoretical analysis, system-level architectural design, and application-oriented evaluation. The flow of the methodology is similar to the one found in the proposed architectural graph shown in the methodology diagram in Fig. 6.1.

6.7.1 Research questions and context of the methodology

Research questions that drive the methodology include:

RQ1: In what way can QC improve the security of the data generated and transmitted in the infrastructures of digital agriculture systems?

RQ2: What types of QKD protocols are best suited for different communication scenarios in agriculture, such as IoT-based sensing, remote sensing, and supply-chain networks?

RQ3: What are the ways in which QKD can be adopted in the farmer's existing digital agricultural infrastructure, such as IoT devices, AI-based analysis tools, cloud platform, and blockchain systems?

RQ4: In terms of security, performance, and scalability, what advantages can QKD-enabled agricultural data protection offer over classical cryptographic approaches?

These research questions—and their application-driven development—therefore directly influence the design of the methodology, and maintain a strong demarcation against the theoretical discussion.

6.7.2 Description of architecture and workflow of the system

The proposed methodology is based on a layered architecture that includes a data acquisition layer, a secure communication layer, a data processing layer, and an application layer. The heterogeneous sources of agricultural data are primarily from soil sensors, weather stations,

UAV-based imaging systems, satellite platforms, etc. These data sources are the edge layer of the digital agriculture ecosystem (Batistatos et al., 2025).

QKD is used at the communication layer to create secure encryption keys for a data-generating device and an aggregation node. The quantum channel is only dedicated to the distribution of keys, while real agricultural data transmission occurs via the classical channels protected by symmetric encryption algorithms with key refreshes via QKD. This is a useful division where practical deployability meets a quantum level of security assurance (Oladimeji et al., 2024).

The processing layer comprises cloud and fog computing infrastructures that allow algorithms of AI and machine learning to process agricultural data for decision support. The modules of the blockchain are incorporated on this layer to gain data integrity, provenance, and transparency within the agricultural supply chains Manjula Gandhi et al., (2024). By ensuring that the private key in QKD is not compromised, key management on QKD can increase the credibility of the operations of the blockchain.

This end-to-end workflow is illustrated in the methodology diagram (Fig. 6.2) with QKD serving as a foundational security layer over all the components constituting the digital agriculture system.

6.7.3 Choosing and using QKD protocols

As a result, the method that resolves RQ2 consists of the selection of several QKD protocols with respect to their deployment context. Protocols such as BB84 and decoy-state QKD exist for prepare-and-measure protocols and are used for short- to medium-range transmitters, e.g., communication from the farm-level sensors to the regional data hub. These protocols are chosen based on the maturity of implementation and compatibility with optical infrastructure.

Entanglement-based QKD protocols are then analyzed for long-distance communication scenarios, such as those based on satellite for distant storage and near-future satellite-based agricultural monitoring or inter-regional data sharing. We incorporate Measurement-Device-Independent QKD to address such weaknesses due to imperfect hardware, specifically in the context of centralized agricultural data aggregation. Abstract Continuous-Variable quantum key exchange (QKD) is under consideration for standard telecommunication networks due to the properties of scalability and cost efficiency.

The selection process for each protocol is technically justified according to its robustness against security attacks, photon infrastructure requirements, and readiness to be applied in agricultural fields.

6.7.4 Digital technologies in agricultural digitization

RQ3 → Proposition does not isolate quantum systems but focuses on integrating QKD with existing agricultural technologies. QKD-generated keys are also used in secure data transmission of IoT devices and sensor networks. QKD is used to secure the communication channels between

AI-driven analytics platforms, assuring the quality of structured data for training and low-weight predictive outputs and predictions.

QKD in the proposed blockchain-enabled agricultural supply chains mainly focuses on information-theoretic security to secure cryptographic keys that are used for signing transactions and authenticating nodes. By integrating, this strengthens the trust model of decentralized systems while providing a solution for the risk of long-term public key exposure. To strike a balance between security, scalability, and deployment feasibility, hybrid cryptographic architectures (HCA)—which combine QKD, classical encryption, and post-QC—are proposed.

6.7.5 Performance evaluation and analytical rationale

In order to answer RQ4, the analyzed experimental evaluation of security and performance features. The key indicators taken into account are: eavesdropping resistance, key compromise probability, key refreshing frequency, communication overhead, and scalability of systems. It investigates the theoretical security assurances of QKD in light of practical limitations such as a noisy channel and additional device flaws.

The benefits in terms of long-term data secrecy and trust are illustrated via comparison between traditional cryptography-based security mechanisms and their counterparts in QKD-enabled security solutions. The methodology also integrates QKD in terms of its impact on system latency and operational complexity, suitable for real-time agricultural applications (Scarani et al., 2009).

6.7.6 Linking to findings and discussion

The results in this section and the following analysis are a direct consequence of the methodology proposed. Improvements in security are assessed in terms of attacks by quantum systems, both for acquiring secret keys more easily and for layering on better key management systems. The increased data integrity, decision-support system trust, and agricultural supply chain robustness potential are quantified by examining application-level effects.

This structured interpretation of confirmed results enables setting out a framework for approaching the relevant methodological aspects mapped from the research questions onto the observed outcome. As a result, the analysis will be internally consistent, replicable, and relevant in the context of the chapter's broad themes.

6.7.7 Methodological summary

In conclusion, the approach proposed in this chapter provides a theoretical and technical comprehensive methodology for the incorporation of QC into digital agriculture systems. The methodology integrates QKD-based secure communication to provide a future-proof security framework that may be complemented with other technologies supporting IoT, AI, cloud computing, and blockchain. The approach addresses the specific gaps identified from the literature review by explicitly linking our research questions through to the design of proposed system, and then to how we will analyze the system in relation to these research questions. The structure

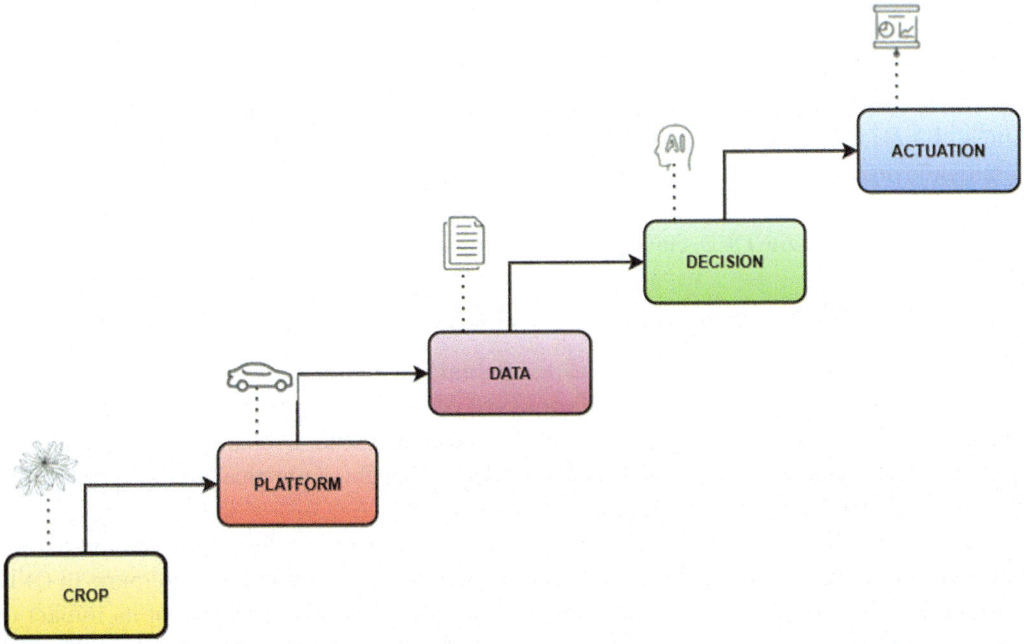

FIGURE 6.1 QKD-based agricultural data protection.

and flow throughout these three sections of the methodology therefore demonstrate that we are addressing the gaps and research questions identified in the literature review, and that the conclusions we draw in this chapter are coherent with the methodology.

Fig. 6.2 depicts the methodology of QKD-based quantum computing in agriculture. The study addressed both the state-of-the-art and the methods the researchers used to gather data from the stakeholders, 118 vulnerability discovery, and impact assessment of quantum computing on current cryptographic solutions. The study used both quantitative and qualitative methods to solicit the opinions of agricultural workers on AI, which allowed a common ground to be established for the interviews, whilst also allowing the interviewees to generate detailed answers. The studies were chosen in terms of relevance, recentness, and topic of study, and the authors employed literature surveying to know what is not done from previous works (in this study, the fact that research on the social impact of AI adoption in the agricultural sector, with a focus on the workers, is insufficient).

6.8 Impact and prospective applications of quantum cryptography in agriculture

QC could revolutionize data protection in agriculture by assuring confidentiality, integrity, and availability of the data, the building blocks we need to preserve our competitive position and respect the aspects of sustainable agriculture. QC employs the bizarre behavior of subatomic

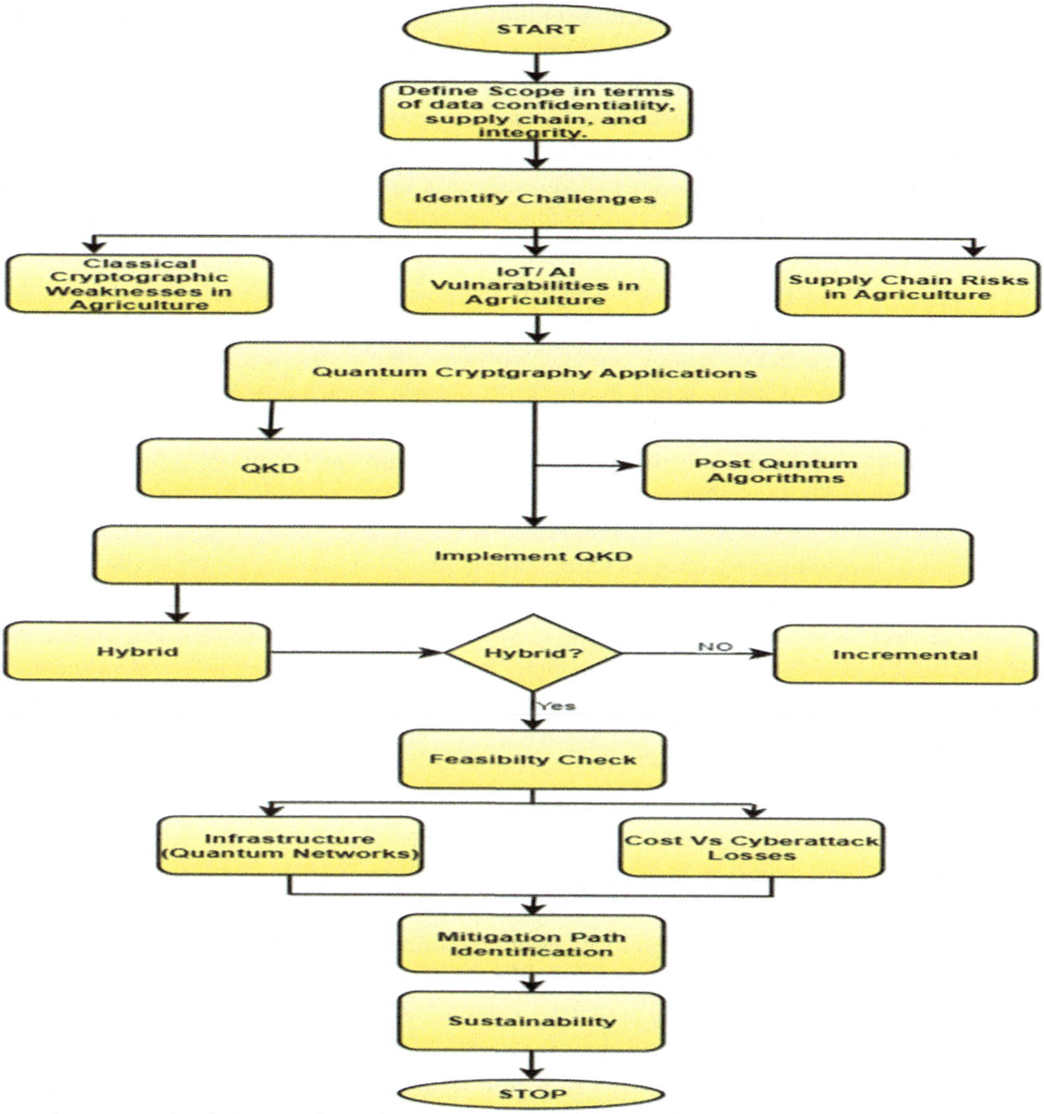

FIGURE 6.2 Methodology of quantum computing-based agricultural applications.

particles to provide an entirely new means of guaranteeing that there's no risk of being overheard or hacked, to a degree never before possible—a good match to the new era of challenges we are all facing in the agricultural sector. In addition, QC is better than classical key distribution to deal with the problem of agricultural data transmission and storage. Prevent hacking in farms by Quantum Cryptosystems. QC secures the system in communication, ensuring security in agricultural activities and preventing data breaches. The security works through a strange feature of quantum mechanics and offers an almost unbreakable way to protect sensitive information. That, just possibly, could represent a huge leap in protecting agriculture from cyberattacks and the means of securing invaluable data. In fruit and vegetable online monitoring information dissemination, QKD achieves information dissemination without information leakage, thus ensuring that the control of agriculture is secure.

QC secures data and isolates weaknesses and risks to the integrity of processes such as agriculture (Gwagwa et al., 2021). The application of QCAG has broad implications in agriculture to protect data, sustainable development, and also supports the development of investments in novel solutions for current and future challenges in agriculture. It is crucial that QC is required to protect agriculture-sensitive data and mitigate the risk, and promote the adoption of secure technology. Moreover, the incorporation of QC with agriculture ensures that the system is highly secure and no privacy violation is possible to carry out in the implemented procedure, therefore sustaining agricultural processes. QC solves the shortcomings of classical coding in agriculture by achieving ultra-resilient key distribution and data transmission. Quantum cryptosystems will bring a breakthrough in agricultural data security, due to the guarantees of secure key distribution, secure data transmission, and authentication. QC is not only applicable in broad areas of agriculture, including precision agriculture, supply chainmanagement, as well as data-centric decision support for sustainable, efficient operation of agriculture (Ralegankar et al., 2021). Post-QC , also known as quantum-resistant cryptography, which refers to the design of cryptographic schemes to survive an attack by a quantumcomputer that can efficiently solve problems that are used in many cryptographic systems (Oladimeji et al., 2024). It further provided that Industry cooperation and global standardization were critical to the successful implementation of PQC (Gill, 2025).

One such promising alternative is quantum-enabled advanced encryption standard. The quantum interconnect architecture uses a lightpath interconnect with AQC/QRAM and quantum cores connected to the scheme to produce a quantum-enabled Advanced Encryption Standard (Morales et al., 2025). This technology, inspired by quantum mechanics, produces true random numbers that are then fed into an additional round of AES encryption to harden them against attacks by even quantum computers. The above-mentioned scenario of "harvest now, decrypt later" emphasizes the need to switch to PQC for the protection of sensitive agricultural data (Elmisery et al., 2025). The core challenges when deploying post-QC are the algorithmic complexity and hardware/software/network integration. The computational complexity of post-QC can be a problem sometimes, especially when IoT devices are resource constrained (Liu et al., 2024).

Through the use of QC, data sent by agricultural organizations will not be open to attacks coming from inside (i.e., unauthorized access to data of co-workers of the company) or outside (i.e., the attack of competitors) the company. QC application in securing agro products'' supply chain QC applied at the asset tracking level in agriculture supply chain offer secure ecosystem of fakes and improved the supply chain integrity to monitor the agro supply chains, curbing fraud, tampering to minimize noxious agriculture value chains negative impact on human health identification of agricultural products confidence on the quality of the products and to validate the safety of the product and ensure confidence with uses. QC would provide secure data sharing and collaboration among agri-food stakeholders, potentially leading to innovation and sustainable agriculture production.

Moreover, QC liberalization of the agricultural supply chain 11 for the security of agricultural products and can be used to reduce the amount of fraud attributable to the infestation of diseases and is also used to ensure the confidence of a consumer that an agricultural product is safe and of high quality. It facilitates sustainable agriculture through the sharing of data and cooperation among supply chain players, as well as cryptography and quantum, which also enhances the security of the supply chain, as abuse contamination could be removed, and delivers confidence to consumers that the food product is of reasonable quality and safe (Gilbert & Gilbert, 2024). QC may alter the agricultural data security landscape, making it immune to internal and external threats, and ensuring the integrity and confidentiality of sensitive data in agricultural processes. The current cybersecurity standards and policies are not adequate to secure the agriculture domain from the emerging and domain-specific cybersecurity threats in the industry. It emphasizes the need for compliance with regulations such as GDPR and NIST, suggesting that industry sector information security standards and regulations be developed and adopted as a first step to addressing this issue.

Quantum-safe solutions can be deployed incrementally, starting with hybrid mode to protect sensitive data, and without the long-term disruption of expensive investments in hardware and software (Morales et al., 2025). QC can be adopted for securing the agricultural data for confidentiality, integrity, and availability for competitive business and sustainability. QC offers a way to securely exchange encryption keys, using the unique features of quantum mechanics to ensure that if someone tries to eavesdrop on your exchange, you'll know the instant it happens (Gill, 2025). A more secure form of human communication is reached by QC, and agricultural data breaches are prevented.

As quantum computers continue to advance, it is necessary to strengthen current cryptographic systems against the potential of quantum computing, not only by assessing the susceptibility of quantum computing but also by leveraging post-quantum cryptographic exploits (Barker et al., 2021). A combination of IoT, AI, data science, etc., is demonstrating the importance of data security, digitalization, and logistics optimization in the supply chain. Ensuring data security in IoT-based supply chain (SC) environments, where machine learning presents a key tool to mitigate privacy and security problems as a result of the susceptibility of IoT devices to cyberattacks. The Industry 4.0 technologies, such as IoT, cloud computing, AI, and blockchain, also revolutionize agricultural and food processes, enhancing the area of food production, evaluation of quality, as well as safety. The explosion of big data analytics has enabled organizations to effectively analyze the IoT data deluge produced. Development of AI and Social Industrial IOT on the agri-food supply chain has significantly affected providing a powerful, safe, efficient, and sustainable UGC offering (Halder et al., 2025).

6.9 Results and Discussion

QC is one of the promising technologies that can be used to provide better security for agricultural data by taking advantage of the unique properties of quantum mechanics to provide unconditional security against eavesdropping and cyberattacks. It is adherent to agriculture's interpretation of confidentiality, integrity, and availability. QC can ensure the security of the agricultural digital data by maintaining data privacy and the integrity and availability of data; and these are central in maintaining the edge in competitiveness and sustainable development of agriculture. QC offers a way to share secret keys safely, using the bizarre rules of quantum mechanics to determine if an eavesdropper has attempted to spy on the key exchange. QC provides secure communication and lower data loss for the agricultural farm. The agri-digital revolution has led to a data generation and capture explosion, and it is paramount to protect the data produced by agricultural farmers from unauthorized access and misuse. QC offers a solution to some of these security problems through unbreakable encryption keys and secure channels for delivering agricultural data.

IoT/AI/DS investing rights smack in the middle of this Venn crossover is the growing realization that data security, digitalization, and logistics are rewriting the book on supply chain evolution. Privacy and security issues are very important in IoT-based supply chains, and thus, machine learning plays a significant role in addressing security and privacy risk, focusing on the threats of IoT devices for attacks against cyber. Food production, quality inspection, and safety promotion have been modernized by Industry 4.0, including IoT, cloud computing, AI, and blockchain technology. The world is currently experiencing a surge in data generation due to the explosion of IoT devices. The AI in Social IIoT for agri-food supply chain security, efficiency, and sustainability. The other catalysts of this kind of digital SIIoT are developments happening in AI that have added to the impact on the agri-food supply chain in a digital evolution way. The successful narrative of how AI and IoT utilization is revolutionizing the supply chain operations in several sectors has also outlined several issues, such as cybersecurity and energy loss, which include no AI explainability, but cost barriers have hence hindered the adoption of the strategy on a wide scale. Applications of IoT and blockchain are essential for improving intestinal transport flow using deep learning for more effective results. Digitalization in agriculture plays a key role in progressing the competitiveness, sustainability, and food security in the EU and in other countries, specifically non-EU countries (Batistatos et al., 2025).

We therefore believe, based on this analysis, that QC offers a superior way to protect agricultural data than traditional encryption processes. Farmer to supplier/supplier to transporter/transporter to consumer lines are very common in agricultural market systems, and QKD can be utilized to establish secure communication between entities without the need for a security auditor. For farming, QC can also be used to avoid leakage of useful information, prevent cybercrime, and unauthorized access to valuable information (Gupta et al., 2020). QC allows for secure communication channels, over which one might securely send critical agricultural data. Agriculture data itself are important to maintain tradition and competitiveness. QC and the saved data also add to the security (Khan et al., 2021). It allows the safe exchange of encryption keys using the distinctive properties of quantum mechanics, so that whenever an interception of the key exchange is attempted, it is immediately detected. Accordingly, QC has an ever-increasing demand for the protection of agricultural data and the sustainability and security of agricultural operations in the evolution of a developing data-driven business scenario.

Nonetheless, the realization of QC in the agricultural arena would have to adapt to the specific requirements and constraints of the farming sector. This includes understanding what information needs to be protected, potential dangers when locking that data, and what laws and rules are in place concerning data privacy and safety. Smart agriculture and crop management, and ultimately increasing agricultural yields, would be greatly enhanced through quantum computing, thus fostering sustainable agriculture (Maraveas et al., 2024). The increasing adoption of smart agriculture underlines the necessity of strong security measures that must be put in place in order to preserve the integrity of the data as well as the confidentiality of the farming process, and it requires the mutual efforts of the technology suppliers, the farmers, the research community, as well as the standard bodies in developing a firearm security solution. The application of BD technology to the collection and processing of rural production factor data can make it possible to make a scientific and rational forecast and program—for the rice, the way to grow the rice, the area to grow the rice, and for a scientific and meticulous overall plan for agricultural production—comprehensive utilization of the rural natural resources to digital economy.

QC secured by the physical theory of Quantum Mechanics offers supreme security as any listening to the key exchange is instantaneously detected (Mastriani, 2022). QKD and post-quantum cryptography (PQC) are the two areas concerned with quantum threat from two different perspectives (Dekkaki et al., 2024). QC is but a cornerstone in this evolution, enabling means to protect directly digital messaging impervious to every attack from classical or from quantum computers (Peters et al, 2023). QC is a discipline of virtual forces, where quantum mechanics is enlisted for shielding the communication systems against Classical and Quantum actions (Akter, 2023). Present cryptographic algorithms, such as RSA, classical computers provide very good performance, but with the sluggish transition of classical computing technologies to quantum, quantum computing is much faster than normal computing and can easily break classical cryptographic algorithms) We need to design a quantum-resistant cryptographic algorithm so that nobody can break our system.

A study of digital traceability in agri-food supply chains reveals the place of transparency and accountability in food safety and quality (Charlebois et al., 2024). The agri-food sector is a big laggard in overcoming challenges that are attached to the sharing of data—and that can help to address many of these historical sins, for example, around food safety, tracing the origins of foods, and in making that sharing secure and just, and so on. (Durrant et al., 2021). Three, digital technologies and information-based solutions are increasingly integral for the way forward in food and agriculture for increased efficiency, sustainability, and resilience (Astill et al., 2019). The challenges plaguing the agri-food supply chains are nonindustrialized, poor management, lack of precise information and operation, and lower supply chain efficiency (Kamble et al., 2019). The digital farm and agriculture offer beyond just harvesting data; it could also contribute to addressing some of the key challenges in agriculture in productivity enhancement, environmental footprint reduction, and natural resources conservation (Finger, 2023).

Fig. 6.3 depicts the effectiveness of the QKD-based quantum computing approach with reference to security, cost, scalability, and future proofing. Adoption of digital technology in the agricultural supply chain, for the most part, enhances the safety of the chain by reducing hazards and

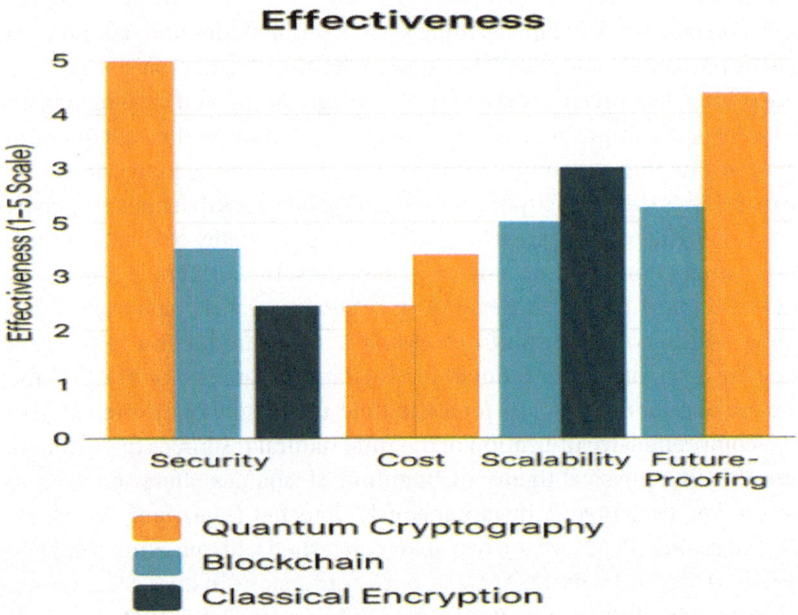

FIGURE 6.3 Effectiveness of QKD-based quantum computing approach.

information sharing. QC may be integrated with classical security measures to create a hybrid approach that doesn't promise to outmode classical cryptography any time soon, but does promise to allow us to focus on making our most secure comms safe. Quantum cryptosystems can be deployed incrementally as hybrid modes in combination with classical cryptosystems and can be used for enforcing long-term security with offered back-door resistance, even before a large-scale quantum computer exists to run Shor's algorithm against them. QC adds value in agriculture by providing data confidentiality, integrity, and availability, essential for competitiveness and sustainability. Quantum security: There are other just as exciting pieces of legislation pertinent to the implementation of secure communication links that seek to guarantee the integrity and privacy of data being exchanged between entities. QC ensures that data communication is safely secured, and, thanks to it, the privacy of such sensitive agricultural data cannot be eavesdropped on and changed in transit.

The digital agriculture transformation requires data integrity and data security for users to have confidence and adopt the new technology. With the growing usage of smart agriculture systems, the security issue needs attention in the farm field in order to protect the field farming activities, data, and privacy (Otieno, 2023). The digitalization of the farm sector is generating and collecting more data, and thus, secure means to store and protect data from wrong usage are in high demand (Ferguson et al., 2024). (Cyberattacks may disrupt the economy of an attack-affected country that is heavily dependent on them.) Crops as targets for cyber warfare

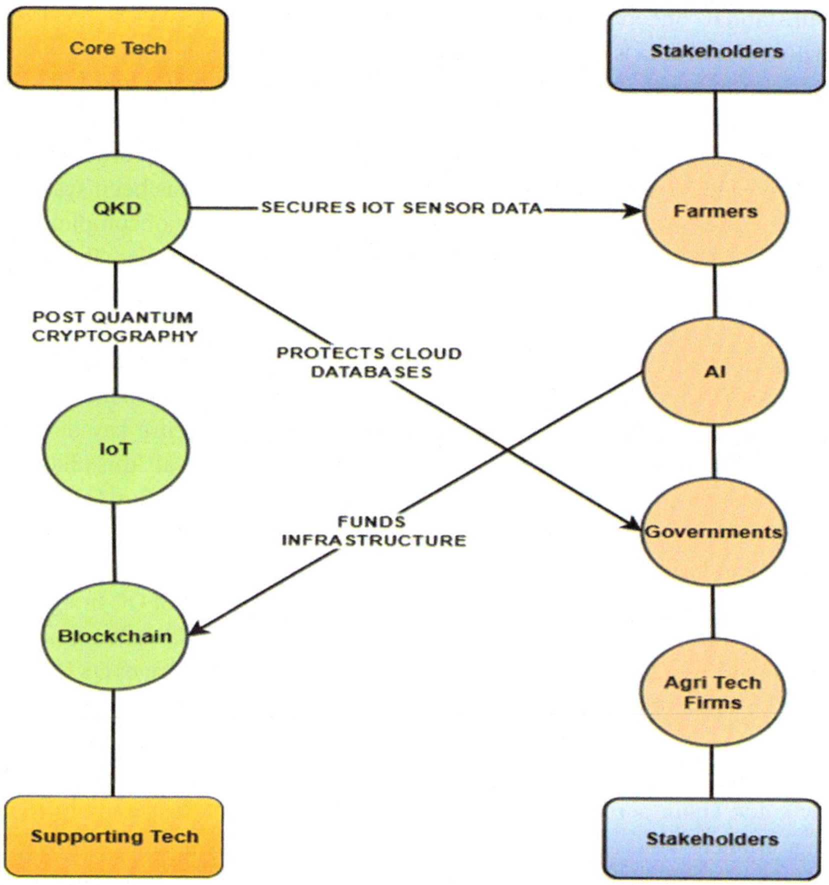

FIGURE 6.4 Dependency graph.

get attention as some countries rely on crops as sources of foreign reserves or for feeding the population (Gupta et al., 2020). That's why QC is so important.

Fig. 6.4 depicts the dependency graph of quantum computing with the other technologies and their relationship. The nonclassical and non-cryptanalysis assets from QC would also be ideal for fields that are sensitive in nature, including agriculture, cyber defense, and fending off data integrity issues. QC has various applications in agroecosystems to create data-driven agriculture, traceable food, and sustainable agriculture.

6.10 Conclusion and Future Scope

Within a matter of years, agriculture has quickly transitioned from traditional farming models with often limited connectivity to digitally connected, IoT, remote sensing, AI, and cloud-supported and blockchain-enabled supply chain-dependent ecosystems. Although these technologies provide profound benefits in terms of productivity, efficiency, and sustainability, they also raise substantial concerns around data security, privacy, and trust in the long run. In an era where sensitive digital data is emerging as the basis for decision-making and policy implementation for agricultural systems, providing protection against the evolving cyberthreat landscape has become a critical need.

In this chapter, the future promise of QC, focusing on QKD, as a secure digital agriculture solution for a post-2030 world with practical quantum computers, has been systematically explored. While classical cryptographic methods rely on the postulates of complexity theory, the key security promises of QC are based on the laws of quantum mechanics. The nature of QKD, therefore, makes it very well suited to the coming age of quantum computing, in which many current public-key cryptosystems are expected to be broken.

This chapter first analyzed the cybersecurity challenges associated with digital agriculture, justifying the need for enhanced security mechanisms, introduces the main ideas main ideas and protocols of QC to explain that QKD allows for providing secure key distribution with built-in eavesdropping detection and a high frequency of key renewal. This chapter illustrates that QC can be efficiently mapped to agricultural communication scenarios, from farm land IoT networks to satellite-based remote sensing systems, through an extensive elaboration on QKD protocols and their domains of applicability.

In addition, the chapter discusses the technological applications of QC in agriculture, such as secure smart farming platforms, protected weather and climate data exchange, and quantum-secured blockchain-based agri-food supply chains. The approach leverages QKD with the digital backbone by attaching QKD-enabled sensors to existing digital infrastructures, which could include IoT devices at the edge, AI-based analytics, cloud platforms for data aggregation, and distributed ledgers to ensure integrity, confidentiality, and the technological guarantees of a trustworthy value chain across the various agricultural domains. This chapter provided the methodology applied in this research, which delinked the research questions from system design and analytical evaluation and demonstrated that the theory of QC can be applied to agricultural data protection, while also validating the feasibility and applicability of such methods to the real-world area of agricultural data protection.

These results suggest that QC could potentially play a role as a kernel security layer to future digital agriculture ecosystems in line with the vision of a quantum internet for future use of QKD. Though there remain issues of cost, availability of necessary infrastructure, scalability, and coordination across institutions, hybrid security architectures that combine QKD with contention and post-quantum cryptographic (PQC) techniques provide a feasible way toward phased implementation. These hybrid approaches facilitate improved security for agricultural systems without disturbing the current practices.

Future scope

QC has a wide-ranging and vast scope in the future of farming. A solid avenue consists of practical and real-life pilot implementations of agricultural networks enabled with QKD, specifically in precision farming and supply-chain management cases. These studies would help advance our understanding of performance, scalability, and cost-effectiveness in realistic environmental settings.

While studies to integrate QC with developing emerging post-quantum cryptographic algorithms for adapting aberrant hybrid security frameworks represent another major future research dimension. These frameworks can provide multiple layers of defense and are ideally suited to maintaining resiliency in the transition period to a fully quantum-safe world. Moreover, future developments in satellite-based and free-space QKD can greatly boost secure long-distance communication for large-scale agricultural monitoring and transnational data exchange.

Further research could also target the development of QKD technologies, specifically optimizing QKD hardware and protocols to better meet the needs of resource-limited agricultural systems alongside urban scenarios such as secure transport and adult living environments, especially for rural and remote areas. For the widespread adoption, policy-level research on standardization, interoperability, and regulatory issues will also be needed.

References

Aboy, M., Minssen, T., & Kop, M. (2022). Mapping the patent landscape of quantum technologies: Patenting trends, innovation and policy implications. *Deleted Journal, 53*(6), 853. https://doi.org/10.1007/s40319-022-01209-3.

Akter, M. S. (2023). Quantum cryptography for enhanced network security: A comprehensive survey of research, developments, and future directions. *arXiv (Cornell University)*. https://doi.org/10.48550/arxiv.2306.09248.

Alazzai, W. K., Obaid, M. K., Abood, B. S. Z., & Jasim, L (2024). Smart agriculture solutions: Harnessing AI and IoT for crop management. *E3S Web of Conferences, 477*, 57. https://doi.org/10.1051/e3sconf/202447700057.

Allende, M., León, D. L., Cerón, S., Leal, A. C., Pareja, A., Silva, M. D., Pardo, A., Jones, D., Worrall, D., Merriman, B., Gilmore, J., Kitchener, N., & Venegas-Andraca, S. E. (2021). Quantum-resistance in blockchain networks. *arXiv (Cornell University)*. https://doi.org/10.48550/arxiv.2106.06640.

Althobaiti, O. S., & Döhler, M (2020). Cybersecurity challenges associated with the internet of things in a post-quantum world. *IEEE Access, 8*, 157356. https://doi.org/10.1109/access.2020.3019345.

Assche, G. V. (2006). *Quantum cryptography and secret-key distillation* https://doi.org/10.1017/cbo9780511617744.

Astill, J., Dara, R., Campbell, M., Farber, J. M., Fraser, E. D. G., Sharif, S., & Yada, R. Y. (2019). Transparency in food supply chains: A review of enabling technology solutions [Review of Transparency in food supply chains: A review of enabling technology solutions]. *Trends in Food Science & Technology, 91*, 240. Elsevier BV https://doi.org/10.1016/j.tifs.2019.07.024 .

Awan, S. H. (2019). Role of Internet of Things (IoT) with blockchain technology for the development of smart farming. *Journal of Mechanics of Continua and Mathematical Sciences, 14*(5). https://doi.org/10.26782/jmcms.2019.10.00014.

Balkrishna, A., Pathak, R., Kumar, S., Arya, V., & Singh, S. K. (2023). A comprehensive analysis of the advances in Indian digital agricultural architecture. *Smart Agricultural Technology, 5*, 100318. https://doi.org/10.1016/j.atech.2023.100318.

Barbeau, M., & García-Alfaro, J. (2022). Cyber-physical defense in the quantum Era. *Scientific Reports, 12*(1). https://doi.org/10.1038/s41598-022-05690-1.

Barker, W., Polk, W. T., & Souppaya, M. (2021). *Getting ready for post-quantum cryptography: Exploring challenges associated with adopting and using post-quantum cryptographic algorithms* https://doi.org/10.6028/nist.cswp.04282021.

Batistatos, M. C., Cola, T. de, Kourtis, M., Apostolopoulou, V., Xilouris, G., & Sagias, N. C (2025). *Agrarian: A Hybrid AI-driven architecture for Smart agriculture.*

Bavdekar, R., Chopde, E. J., Bhatia, A., Tiwari, K., Daniel, S. J., & Atul, A. (2022). Post quantum cryptography: Techniques, challenges, standardization, and directions for future research. *arXiv (Cornell University).* https://doi.org/10.48550/arxiv.2202.02826.

Bedington, R., Arrazola, J. M., & Ling, A. (2017). Progress in satellite quantum key distribution. *Npj Quantum Information, 3*(1). https://doi.org/10.1038/s41534-017-0031-5.

Bhatt, A. P., & Sharma, A. (2019). Quantum cryptography for internet of things security. *Journal of Electronic Science and Technology, 17*(3), 213. https://doi.org/10.11989/jest.1674-862x.90523016.

Bienfang, J. C., Rogers, D. J., Restelli, A., Clark, C. W., Williams, C. J., Mink, A., Hershman, B., Nakassis, T., Tang, X., & Su, D. (2007). Free-space quantum key distribution at GHz repetition rates. In *Proceedings of SPIE, the International Society for Optical Engineering/Proceedings of SPIE: 6476* https://doi.org/10.1117/12.713630.

Biham, E., Boyer, M., Boykin, P. O., Mor, T., & Roychowdhury, V. (2006). A proof of the security of quantum key distribution. *Journal of Cryptology, 19*(4), 381. https://doi.org/10.1007/s00145-005-0011-3.

Bobde, Y., Narayanan, G., Jati, M., Raja, S. P., Cvitić, I., & Peraković, D (2024). Enhancing industrial IoT network security through blockchain integration. *Electronics, 13*(4), 687. https://doi.org/10.3390/electronics13040687.

Burg, S. van der, Wiseman, L., & Krkeljas, J (2020). Trust in farm data sharing: Reflections on the EU code of conduct for agricultural data sharing. *Ethics and Information Technology, 23*(3), 185. https://doi.org/10.1007/s10676-020-09543-1.

Busch, P., Heinonen, T., & Lahti, P. (2007). Heisenberg's uncertainty principle. Physics Reports, 452(6), 155–176. https://doi.org/10.1016/j.physrep.2007.05.006.

Charlebois, S., Latif, N., Ilahi, I., Sarker, B., Music, J., & Vézeau, J (2024). Digital traceability in agri-food supply chains: A comparative analysis of OECD member countries. *Foods, 13*(7), 1075. https://doi.org/10.3390/foods13071075.

Chaterji, S., DeLay, N. D., Evans, J., Mosier, N. S., Engel, B. A., Buckmaster, D. R., & Chandra, R. (2020). Artificial intelligence for digital agriculture at scale: Techniques, policies, and challenges. *arXiv (Cornell University).* https://doi.org/10.48550/arxiv.2001.09786.

Chen, L., Jordan, S. P., Liu, Y.-K., Moody, D., Peralta, R., Perlner, R., & Smith-Tone, D. (2016). *Report on post-quantum cryptography* https://doi.org/10.6028/nist.ir.8105.

Cravero, A., Pardo, S., Sepúlveda, S., & Múñoz, L. (2022). Challenges to use machine learning in agricultural big data: A systematic literature review. *Agronomy, 12*(3), 748. https://doi.org/10.3390/agronomy12030748.

Dam, D.-T., Tran, T.-H., Hoang, V., Pham, C., & Hoang, T.-T. (2023). A survey of post-quantum cryptography: Start of a new race. *Cryptography,, 7*(3), 40. https://doi.org/10.3390/cryptography7030040.

Dekkaki, K. C., Tasic, I., & Cano, M. (2024). Exploring post-quantum cryptography: Review and directions for the transition process. *Technologies, 12*(12), 241. https://doi.org/10.3390/technologies12120241.

Dey, N., Ghosh, M., & Chakrabarti, A. (2022). Quantum solutions to possible challenges of blockchain technology. In *Lecture notes on data engineering and communications technologies* (p. 249). Springer International Publishing. https://doi.org/10.1007/978-3-031-04613-1_9.

Dhal, S. B., & Kar, D. (2024). Transforming Agricultural Productivity with AI-Driven Forecasting: Innovations in Food Security and Supply Chain Optimization. *Forecasting, 6*(4), 925. https://doi.org/10.3390/forecast6040046.

Döberl, C., Eibner, W., Gärtner, S., Kos, M., Kutschera, F., & Ramacher, S. (2023). Quantum-resistant end-to-end secure messaging and email communication. In *Proceedings of the 17th International Conference on Availability, Reliability and Security: 1* https://doi.org/10.1145/3600160.3605049.

Duncan, E., Glaros, A., Ross, D. Z., & Nost, E. (2021). New but for whom? Discourses of innovation in precision agriculture. *Agriculture and Human Values, 38*(4), 1181. https://doi.org/10.1007/s10460-021-10244-8.

Durrant, A., Marković, M., Matthews, D. R., May, D., Leontidis, G., & Enright, J. (2021). How might technology rise to the challenge of data sharing in agri-food? *Global Food Security, 28*, 100493. https://doi.org/10.1016/j.gfs.2021.100493.

Edwards, M., Mashatan, A., & Ghose, S. (2020). A review of quantum and hybrid quantum/classical blockchain protocols [Review of A review of quantum and hybrid quantum/classical blockchain protocols]. *Quantum Information Processing, 19*(6). https://doi.org/10.1007/s11128-020-02672-y.

Ellahi, R. M., Wood, L. C., & Bekhit, A. E.-D. A. (2024). Blockchain-driven food supply chains: A systematic review for unexplored opportunities [Review of Blockchain-Driven Food Supply Chains: A Systematic Review for Unexplored Opportunities]. *Applied Sciences, 14*(19), 8944. Multidisciplinary Digital Publishing Institute https://doi.org/10.3390/app14198944 .

Elmisery, A. M., Sertovic, M., Zayin, A., & Watson, P. (2025). Cyber threats in financial transactions—Addressing the dual challenge of AI and quantum computing. *arXiv preprint arXiv:2503.15678*. https://doi.org/10.48550/ARXIV.2503.15678.

Ferguson, R., Mishra, S., Qureshi, S., Injac, L., Talukder, B., & Orbinski, J (2024). Digital technologies and food security during crisis: COVID-19 experiences from smallholder farmers in Odisha, India. *Frontiers in Sustainable Food Systems, 7*. https://doi.org/10.3389/fsufs.2023.1150197.

Fernández-Caramés, T. M., & Fraga-Lamas, P. (2020). Towards post-quantum blockchain: A review on blockchain cryptography resistant to quantum computing attacks [Review of Towards Post-Quantum Blockchain: A Review on Blockchain Cryptography Resistant to Quantum Computing Attacks]. *IEEE Access, 8*, 21091. Institute of Electrical and Electronics Engineers https://doi.org/10.1109/access.2020.2968985 .

Finger, R. (2023). Digital innovations for sustainable and resilient agricultural systems. *European Review of Agricultural Economics, 50*(4), 1277. https://doi.org/10.1093/erae/jbad021.

Gilbert, C., & Gilbert, M. A. (2024). The development and evolution of cryptographic algorithms in response to cyber threats. *International Journal of Research Publication and Reviews, 5*(12), 1149. https://doi.org/10.55248/gengpi.5.1224.3430.

Gilbert, C., & Gilbert, M. A. (2025). Cryptographic foundations and cybersecurity implications of blockchain technology. *SSRN Electronic Journal*. https://doi.org/10.2139/ssrn.5258793.

Gill, G. K. (2025). Post-quantum security: Analyzing and mitigating quantum computing threats to TLS and QUIC Protocols. *International Journal of Computer Engineering & Technology, 16*(1), 3145. https://doi.org/10.34218/ijcet_16_01_220.

Gill, S. S., & Buyya, R. (2024). Transforming research with quantum computing. *Journal of Economy and Technology*. https://doi.org/10.1016/j.ject.2024.07.001.

Gupta, M., Abdelsalam, M., Khorsandroo, S., & Mittal, S. (2020). Security and privacy in smart farming: Challenges and opportunities. *IEEE Access, 8*, 34564. https://doi.org/10.1109/access.2020.2975142.

Gwagwa, A., Kazim, E., Kachidza, P., Hilliard, A., Siminyu, K., Smith, M. L., & Shawe-Taylor, J. (2021). Road map for research on responsible artificial intelligence for development (AI4D) in African countries: The case study of agriculture [Review of Road map for research on responsible artificial intelligence for development (AI4D) in African countries: The case study of agriculture]. *Patterns, 2*(12), 100381. Elsevier BV https://doi.org/10.1016/j.patter.2021.100381 .

Halder, S., Islam, Md. R., Mamun, Q., Mahboubi, A., Walsh, P., & Islam, M. Z. (2025). A comprehensive survey on AI-enabled secure social industrial internet of things in the agri-food supply chain. *Smart Agricultural Technology*, 100902. https://doi.org/10.1016/j.atech.2025.100902.

Ham, B. S. (2018). Addressable optical key distribution for unconditionally secured cryptography using phase-controlled quantum superposition. *arXiv (Cornell University)*. https://doi.org/10.48550/arxiv.1807.08126.

Huberman, B. A., Lund, B., & Wang, J. (2020). Quantum Secured Internet Transport. *arXiv (Cornell University)*. https://doi.org/10.48550/arxiv.2007.05522.

Husain, S. O., Mohammed, G., & Poornima, E. (2025). Leveraging blockchain technology to enhance climate resilience in agriculture. *SHS Web of Conferences, 216*, 1031. https://doi.org/10.1051/shsconf/202521601031.

Jain, N., Hoff, U. B., Gambetta, M., Rodenberg, J., & Gehring, T. (2023). Quantum key distribution for data center security - a feasibility study. *arXiv (Cornell University)*. https://doi.org/10.48550/arxiv.2307.13098.

Jin, H., & Xie, Y. (2023). A review of research on the impact of digitalization on agricultural supply chain security [Review of A Review of Research on the Impact of Digitalization on Agricultural Supply Chain Security]. *Frontiers in Artificial Intelligence and Applications*. https://doi.org/10.3233/faia230038.

Kamble, S., Gunasekaran, A., & &Gawankar, S (2019). Achieving sustainable performance in a data-driven agriculture supply chain: A review for research and applications [Review of Achieving sustainable performance in a data-driven agriculture supply chain: A review for research and applications]. *International Journal of Production Economics, 219*, 179. Elsevier BV https://doi.org/10.1016/j.ijpe.2019.05.022 .

Khan, N., Ray, R. L., Sargani, G. R., Ihtisham, M., Khayyam, M., & Ismail, S. (2021). Current progress and future prospects of agriculture technology: Gateway to sustainable agriculture. *Sustainability,, 13*(9), 4883. https://doi.org/10.3390/su13094883.

Kiktenko, E. O., Pozhar, N. O., Anufriev, M. N., Трушечкин, A. C., Yunusov, R. R., Kurochkin, Y., Lvovsky, A. I., & Fedorov, A. K. (2018). Quantum-secured blockchain. *Quantum Science and Technology, 3*(3), 35004. https://doi.org/10.1088/2058-9565/aabc6b.

Kikuchi, T. (2024). Towards a post-quantum cryptography in blockchain I: Basic review on theoretical cryptography and quantum information theory. *arXiv (Cornell University)*. https://doi.org/10.48550/arxiv.2407.18966.

Kosior, K., & Młodawska, P (2024). Understanding market actors' perspectives on agri-food data sharing: Insights from the digital food passports pilot in Poland. *Agriculture, 14*(12), 2340. https://doi.org/10.3390/agriculture14122340.

Kristen, E., Kloibhofer, R., Hernández, V., & Castillejo, P. (2021). Security assessment of agriculture IoT (AIoT) applications. *Applied Sciences, 11*(13), 5841. https://doi.org/10.3390/app11135841.

Kumar, A., Ottaviani, C., Gill, S. S., & Buyya, R. (2021). Securing the future internet of things with post-quantum cryptography. *Security and Privacy, 5*(2). https://doi.org/10.1002/spy2.200.

Liu, T., Ramachandran, G., & Jurdak, R (2024). Post-quantum cryptography for internet of things: A survey on performance and optimization. *arXiv (Cornell University)*. https://doi.org/10.48550/arxiv.2401.17538.

Malpani, P., Kumar, S., & Pathak, A. (2023). Implementation of coherent one way protocol for quantum key distribution up to an effective distance of 145 km. *Research Square (Research Square)*. https://doi.org/10.21203/rs.3.rs-3376171/v1.

Manjula Gandhi, S., Mulay, C., Durai, K., Murali, G., Masood, J. A. I. S., Vijayarajan, V., Gautam, K., Chakravarthy, N., Kumar, S., Agarwal, S., Murali, S. C., Vijayasherly, V., Asirvatham, D., Brohi, S. N., Chinnappan, C. V., & &Anbuchelian, S (2024). Quantum blockchain: Trends, technologies, and future directions. *IET Quantum Communication, 5*(4), 516. https://doi.org/10.1049/qtc2.12119.

Maraveas, C., Konar, D., Michopoulos, D. K., Arvanitis, K. G., & Peppas, K. P. (2024). Harnessing quantum computing for smart agriculture: Empowering sustainable crop management and yield optimization. *Computers and Electronics in Agriculture, 218*, 108680. https://doi.org/10.1016/j.compag.2024.108680.

Martín, V., Martínez-Mateo, J., & Peev, M. (2017). Introduction to quantum key distribution. In *Wiley encyclopedia of electrical and electronics engineering* (p. 1). https://doi.org/10.1002/047134608x.w8354.

Mastriani, M. (2022). Non-distributable key sharing for improving the security in IoT networks. *arXiv (Cornell University)*. https://doi.org/10.48550/arxiv.2205.02779.

Mink, A., Frankel, S. E., & Perlner, R (2010). Quantum Key Distribution (QKD) and Commodity Security Protocols: Introduction and Integration. *arXiv (Cornell University)*. https://doi.org/10.48550/arxiv.1004.0605.

Morales, A. H., Bishwas, A. K., & Varghese, J. J. (2025). Quantum-enabled framework for the advanced encryption standard in the post-quantum era. *arXiv (Cornell University)*. https://doi.org/10.48550/arxiv.2502.02445.

Naughton, D., Bedington, R., Barraclough, S., Islam, T., Griffin, D., & Smith, B. (2019). Design considerations for an optical link supporting intersatellite quantum key distribution. *Optical Engineering, 58*(1), 1. https://doi.org/10.1117/1.oe.58.1.016106.

Oladimeji, S., Egon, A., & Broklyn, P (2024). The impact of quantum computing on post-quantum cryptography: Standardization and migration strategies. *SSRN Electronic Journal*. https://doi.org/10.2139/ssrn.4904933.

Otieno, M. (2023). An extensive survey of smart agriculture technologies: Current security posture. *World Journal of Advanced Research and Reviews, 18*(3), 1207. https://doi.org/10.30574/wjarr.2023.18.3.1241.

Ott, D. J., & Peikert, C.participants, other workshop. (2019). Identifying research challenges in post quantum cryptography migration and cryptographic agility. *arXiv (Cornell University)*. https://doi.org/10.48550/arxiv.1909.07353.

Peters, N. A., Alshowkan, M., Chapman, J. C., Pooser, R. C., Rao, N. S. V., & Newell, R. (2023). Long-term cybersecurity applications enabled by quantum networks. *arXiv (Cornell University)*. https://doi.org/10.48550/arxiv.2304.14479.

Rahaman, M., Lin, C., Pappachan, P., Gupta, B. B., & Hsu, C. (2024). Privacy-centric AI and IoT solutions for smart rural farm monitoring and control. *Sensors, 24*(13), 4157. https://doi.org/10.3390/s24134157.

Ralegankar, V. K., Bagul, J., Thakkar, B., Gupta, R., Tanwar, S., Sharma, G., & Davidson, I. E. (2021). Quantum cryptography-as-a-service for secure UAV communication: Applications, challenges, and case Study. *IEEE Access, 10*, 1475. https://doi.org/10.1109/access.2021.3138753.

Raveena, S., & Edward, A. S. (2021). Secure B-IoT based smart agriculture– A brief review [Review of Secure B-IoT Based Smart Agriculture– A Brief Review]. *Journal of Physics Conference Series, 1964*(4), 42006. IOP Publishing https://doi.org/10.1088/1742-6596/1964/4/042006 .

Ren, X., Xu, M., Niyato, D., Kang, J., Xiong, Z., Qiu, C., & Wang, X. (2023). Building Resilient Web 3.0 with Quantum Information Technologies and Blockchain: An Ambilateral View. *arXiv (Cornell University)*. https://doi.org/10.48550/arxiv.2303.13050.

Ryan, M., Nuhoff-Isakhanyan, G., & Tekinerdoğan, B (2023). An interdisciplinary approach to artificial intelligence in agriculture. *NJAS Impact in Agricultural and Life Sciences, 95*(1). https://doi.org/10.1080/27685241.2023.2168568.

Scarani, V., Bechmann-Pasquinucci, H., Cerf, N. J., Dušek, M., Lütkenhaus, N., & Peev, M. (2009). The security of practical quantum key distribution. *Reviews of Modern Physics, 81*(3), 1301. https://doi.org/10.1103/revmodphys.81.1301.

PART IV

Water sustainability and resource governance

7 Analyzing fresh water shortage crises and the need for government intervention to manage resources

Muhammad Younus[a,b], Halimah Abdul Manaf[c], Bambang Irawan[d], Andi Adawiah[e], Ni Putu Tirka Widanti[f], Bhimo Widyo Andoko[g], Eliza Meiyani[a], Arif Zainudin[h], Dian Suluh Kusuma Dewi[i] and Akhmad Habibullah[h]

[a]DEPARTMENT OF GOVERNMENT AFFAIRS AND ADMINISTRATION, UNIVERSITAS MUHAMMADIYAH YOGYAKARTA, YOGYAKARTA, INDONESIA [b]DEPARTMENT OF PRODUCT RESEARCH AND SOFTWARE DEVELOPMENT, TRANS PACIFIC LOGISTICS (TPL) PVT LTD, KARACHI, PAKISTAN [c]UUM COLLEGE OF LAW, GOVERNMENT, AND INTERNATIONAL STUDIES, UNIVERSITI UTARA MALAYSIA, KEDAH, MALAYSIA [d]DEPARTMENT OF PUBLIC ADMINISTRATION, UNIVERSITAS MULAWARMAN, KALIMANTAN TIMUR, SAMARINDA, INDONESIA [e]DEPARTMENT OF MANAGEMENT, FACULTY OF ECONOMICS AND BUSINESS, UNIVERSITAS LAMAPPAPOLEONRO, SOPPENG, INDONESIA [f]DEPARTMENT OF PUBLIC ADMINISTRATION, UNIVERSITAS NGURAH RAI, BALI, KOTA DENPASAR, INDONESIA [g]FACULTY OF FINE ARTS AND DESIGN, INDONESIA INSTITUTE OF THE ARTS YOGYAKARTA, YOGYAKARTA, INDONESIA [h]DEPARTMENT OF GOVERNMENT SCIENCE, UNIVERSITAS PANCASAKTI TEGAL, JAWA TENGAH, KOTA TEGAL, INDONESIA [i]DEPARTMENT OF GOVERNMENT SCIENCE, UNIVERSITAS MUHAMMADIYAH PONOROGO, JAWA TIMUR, KABUPATEN PONOROGO, INDONESIA

7.1 Introduction

In recent decades, with rising alarm over the scarcity of freshwater resources, water scarcity has emerged as a critical issue that has shifted its focus from a local to a global strategic issue. Global organizations such as the United Nations and the World Economic Forum have repeatedly ranked water scarcity as one of the most critical global risk issues due to its far-reaching implications on food security, public health, energy security, migration, and conflict

Quantum Computing and Instrumentation for Smart Agriculture. DOI: https://doi.org/10.1016/B978-0-443-44247-6.00003-8

(UN-Water, 2023). People all across the world have improved their quality of life and achieved notable accomplishments in this modern world.

People undoubtedly benefited greatly from this development, but it also brought with it a host of new difficulties that led to issues that did not previously exist (Gleick, 2018). Therefore, people started to work on the solutions to the problem based on their priorities. Some of the issues remain stable with time, but some are becoming worse in a way that now it becomes so difficult to reverse the damage which is already done. So, these issues made people think about their approach toward their way of living and how it impacts the things around them. No doubt, industrialization creates lots of opportunities for earning and employment for the people. Still, these benefits come with the cost of unprecedented damage to the world, especially to the environment and climate, and create problems for the world's natural resources (Petrescu-Mag et al., 2022). It is all caused because of people's lack of ability to plan a thing that covers aspects of its impacts.

Still, we have seen so many examples that people either mistakenly miss it or deliberately ignore its impact due to their interest and greed. This now made us stand now in a situation where we no longer continue to work and run a process like we used to due to the consequences of our action are already done enough damage, and further continuation without planning will make things slip out of our hands then we no longer able to stop its dangerous implication which will affect people all around the globe. So, one of the critical challenges we have to point out that has the impact which will make the existence of humans in danger which we also made the theme of our research paper, is the problem of Freshwater shortage (Pinto, 2023). Freshwater, since the beginning of the world, has always been the key feature for human beings' survival; without it, we cannot be able to survive and be alive. It is an essential requirement of the human daily diet, which is why we have seen from ancient times that people used to look for places to live and create their civilization near or on the bank of freshwater resources (Wirsing et al., 2013).

Even now, any space exploration is done to find the planets in the universe, which is habitual for living things to survive. It always starts with the existence of freshwater resources in it. Therefore, this is much freshwater importance for humans, but now with so many problems causing to the atmosphere of the world because of climate changes due to pollution caused by industrial wastage and also the using water irresponsibly and endlessly thinking that it is unlimited, created the issue of freshwater shortage (Stewart-Harawira, 2020). Now people are starting to know about the criticality of freshwater shortage, and now it is starting to impact the people, and if we don't plan for it from now on, it will create worse situations in the future. The main focus of this chapter is to point out all the reasons responsible for this freshwater crisis andwhatwill be its expected outcomes if the situation continues to get worse. In the end, that things are discussed that each government needs to do in their countries in order to decrease the problem of the freshwater crisis and prepare them for the possible outcome and plan to handle it.

Within this framework, the subject of the present chapter is based on the evaluation of the problem of freshwater scarcity in a way that extends beyond the simple description of the related problem and includes its structural dimensions and solution options. Moreover, while the traditional approach to addressing the problem of water scarcity focuses on a unidimensional approach to the associated problem, the present chapter includes a holistic approach and a settlement of the associated relationship segments to the wider problem of environmental change, demographic change, good governance problems, and the particular role of the innovation and technology factor. Specifically, the role of advanced and highly sophisticated technologies is analyzed in the solution of the particular problem of water resources management. At the same time, a new solution option is emerging and refers to the application of the newly emerging quantum technologies in the solution and management of the problem of water resources. Specifically, the problem analyzed in the present chapter can be characterized as follows: The problem analyzed in the present chapter refers to the emerging inability of the contemporary approaches and strategies of water resources management to cope with the associated problem. That is to say, in more.

From a technological point of view, recent advances in computing and sensor technologies have offered new opportunities to deal with the issues mentioned above, and currently, it has been observed that water resource management systems, remote sensor systems, artificial intelligence systems, and internet of things-based monitoring systems have started revolutionizing water resource assessments and management, as discussed by Gómez-Trujillo et al. (2022). However, based on the complexity of water resources systems, including climate models, interstate river flows, infrastructure systems, and socio-economic factors, it can be stated that the complexity of computations required to properly estimate and optimize water resource systems is rising. This led to a situation where advanced and next-generation computing and sensor technology are required to deal with complex and nonlinear water resources systems, especially those related to environmental monitoring and management systems. In this context, an area that is waiting to be explored and researched is that of quantum computing and its application in addressing some of the issues facing water resources management. Quantum computing is an evolving field in computer science that has recently come to attention and made use of principles of superposition and entanglement for enhanced processing and computation (Montanaro, 2016).

While advances in quantum sensing technologies allow observation of the environment with unprecedented sensitivity, quantum gravimeters and magnetometers can monitor minute changes in gravitational and magnetic forces. This will allow groundwater resources and aquifer depletion and submarine groundwater recharge processes to be determined far more precisely than today (Degen et al., 2017). Bibliometric analysis of current literature on the topic indicates a growing number of links between water resource management, advanced computing, and emerging technologies. Therefore, any future approach is likely to be of an increasingly interdisciplinary nature rather than solely domain-specific interventions.

However, besides these technical aspects, a set of ethical, political, and geopolitical issues is also interwoven with the problem of freshwater scarcity. For instance, water scarcity has also been blamed as one of the factors contributing to increasing social inequalities, since the most deprived groups are also the most vulnerable groups that suffer from water scarcity and rising costs (Sultana, 2018). On the other hand, there are also issues of water conflicts on an international level, particularly with regard to transboundary river basins and aquifers, which has led to the need for collaborative governance and data diplomacy (Wirsing et al., 2013). Therefore, in order to address the problem of freshwater scarcity, it is necessary not only to enhance technological innovation but also to adopt policies and practices on an international level.

Objectives of the chapter

The objectives of the chapter are:

- To comprehend the impacts of the freshwater crisis on human security, agriculture, economic security, growth centers, and geopolitical relationships at the national and trans-boundary levels;
- To comprehend the existing water governance and management practices and assess the structural limitations and deficiencies of traditional planning practices and existing infrastructure for addressing the rising concerns for water scarcity;
- And, to comprehend the opportunities for advanced and emerging technologies to be used for water management and governance practices, including digital technologies and smart water governance tools, and the opportunities for using quantum computing and quantum sensing for addressing water-related problems.

Organization of the chapter

The rest of the chapter is organized as: Section 7.2 enlightens literature review. Section 7.3 highlights research methodology. Section 7.4 illustrates results and discussions. And finally, Section 7.5 concludes the chapter with future scope and limitations of the study.

7.2 Literature review

7.2.1 Climate change

Long-term temperature and weather pattern variations are a concept known as climate change. Natural processes, like the oscillation of the solar cycle, may cause the changes. But since the 1800s, human activity—mainly the burning of fossil fuels such as coal, oil, and gas—has been the main cause of climate change (Lydiri et al., 2023). The burning of fossil fuels emits greenhouse gases that act as an insulation blanket around the earth, trapping heat from the sun and causing warming. Carbon dioxide and methane are two such greenhouse gas emissions that play a crucial role in climate change. These are emitted, for example, when coal or gasoline is burned to warm a building. Carbon dioxide may also be emitted during clearing of forests and land. Methane emissions are chiefly generated by waste landfill. Some of the significant emitters include energy, industry, transport, buildings, agriculture, and land use. Thousands of scientists and government reviewers agreed in a number of UN reports that preserving the increase in global temperature at 1.5°C would protect humanity from the worst climate effects and at least hold on to a habitable climate (Feijó et al., 2023). But current approaches foretell a 2.8°C increase in temperature by the end of the century. Throughout the globe, greenhouse gases leading to climate change are emitted, but there are some countries that emit much more than others. 100 countries with the lowest emissions generate 3% of global emissions. The top ten emitters contribute 68% of the contribution. Everyone needs to take action against climate change, but the countries and individuals that produce the most and are contributing the most to the problem need to do so first (Lü et al., 2023).

7.2.2 Demography

Demography is a study of human populations using statistics. It involves a study of how human populations change over a period of time, as well as in different places, taking into account the size, composition, and movement of people. This field combines practices from many other fields, including anthropology, economics, sociology, and many others. Governments and organizations use demography to comprehend and forecast societal, cultural, and economic transformations (Mahesar et al., 2023). Simple population studies, such as censuses, have been done for about 6,000 years in ancient times. However, the first people we recognize as demographers, like John Grant from the UK, started in the 16th century. Most early studies focused on death rates, looking at how many people died and at what age. Grant used data from baptisms and burials to figure out how many men and women were in their childbearing and military years. His work is one of the earliest examples of analyzing a region's population. Early insurance workers also did population studies to set life insurance prices (Sabayan et al., 2023).

7.2.3 Glaciers

Massive icebergs that move slowly are part of glaciers. On land, glaciers form. They are made from snow that has fallen and been pressed together over many centuries to become ice. Because of gravity, these glaciers slowly move down. Glaciologists are scientists who study glaciers (Yan et al., 2023). These scientists began studying glaciers in the 19th century to learn more about past ice ages. Today, they study glaciers to understand global warming. Glaciers in mountain areas have been melting over time, as shown in old photos and paintings. In fact, in recent decades, glaciers have been melting and disappearing much faster all around the world (Mao et al., 2023).

7.2.4 HAARP technology

HAARP, or High-frequency Active Auroral Research Program, is an Alaskan research facility for investigating the ionosphere that is situated close to Gakona. The Ionospheric Research Instrument (IRI), a collection of 180 radio antennas covering a 0.13 square kilometer region, is the primary instrument (33 acres). The topmost part of Earth's atmosphere is called the ionosphere. It starts at about 50 km (30 miles) above the surface of the Earth and is made up of atoms and molecules that have been ionized by the Sun's ultraviolet light, which causes them to lose an electron and acquire a positive charge. Since low radio frequencies are bounced off the ionosphere and enable long-distance communications, the ionosphere is particularly significant for radio. Radio transmissions to and from satellites go through the ionosphere at higher frequencies (Bailey & Worthington, 1997). The ionosphere is also where solar wind particles that smash with oxygen and nitrogen atoms give rise to auroras. The IRI uses 3.6 mW of power to transmit at frequencies between 2.7 and 10 MHz It sends radio waves into the ionosphere, where they induce wave-like motion in the electrons. HAARP is the most powerful ionospheric heater in the world. It gets its name from the fact that electron excitation raises electron temperature. Using HAARP, researchers may see how the ionosphere responds to alterations in environmental variables by changing the density of electrons in a particular area.

7.2.5 Research gap

Despite growing scholarly and policy attention to freshwater scarcity, several critical gaps remain evident from the case-oriented analysis. First, existing studies are heavily fragmented, often isolating environmental drivers from governance and technological dimensions. Second, most water management research relies on conventional analytical and computational models that are increasingly inadequate for handling climate uncertainty and system complexity. Third, there is a focus on short-term mitigation results in empirical case studies, with insufficient exploration of long-term resilience and anticipatory water governance. Fourth, there has been insufficient integration of advanced and emerging technologies, including quantum computing and sensing, in mainstream water governance approaches. Fifth, comparative understanding across multiple regions is not well developed, which affects the transferability of effective approaches. Sixth, decision support systems may not be effective in dealing with high-dimensional, real-time data in transboundary or urban water systems. These limitations point to the necessity to develop a more comprehensive research methodology that draws on a range of approaches, including bibliometric research, qualitative synthesis, and future-oriented technology analysis. These limitations of current research approaches are the basis on which a research methodology is developed in this chapter.

7.3 Research Methodology

In the chapter, the research design will be based on a qualitative dominant research approach with an exploratory research approach, with the assistance of a systematic literature analysis and the use of bibliometrics for the study of the causes, effects, and solution paths for the water scarcity problem with reference to the water crisis phenomenon. The reason for the choice of the mixed research approach is the fact that the water scarcity problem faced by the world is a multidimensional phenomenon with different environmental, technological, socioeconomic, and governance aspects involved.

In terms of methodology, the research approach will allow for the development of an overall understanding of the research problem, rather than an empirical measurement of the problem, which makes it appropriate for policy-oriented and multidisciplinary research studies. In the context of the research problem of water scarcity crises in the past decade, it was felt that using the literature review with VOS viewer would provide an appropriate research approach to examine the research problem. The information was gathered through various sources such as the internet, UN reports, World Economic Forum reports, research papers, published articles,

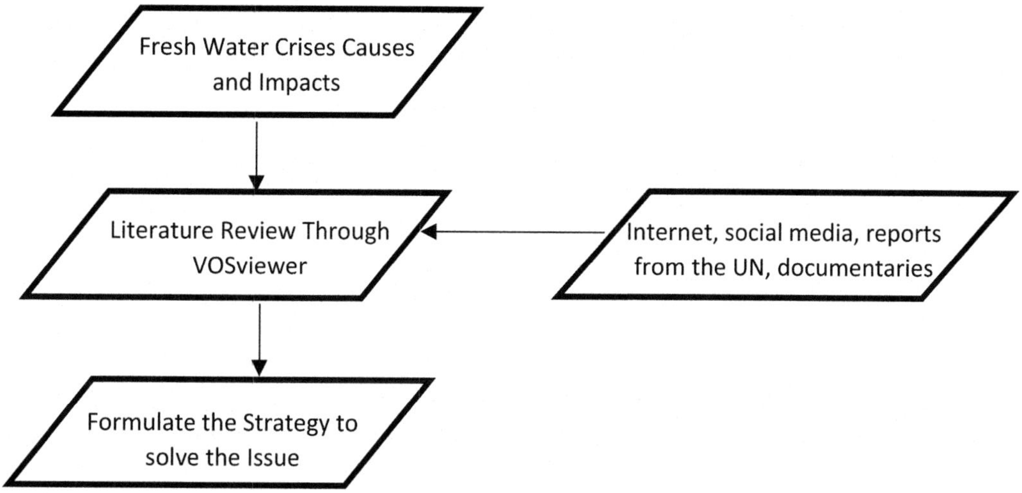

FIGURE 7.1 Evaluation of the framework for creating strategies.

and documentaries on the topic of water scarcity, which was presented in an appropriate manner in the Fig. 7.1.

The research methodology is based on a three-stage process. First, a structured literature review is performed in order to identify the conceptual and theoretical underpinnings of research in the field of freshwater scarcity. Second, a bibliometric analysis of the research in the field of water scarcity is performed using VOSviewer software in order to identify the key research clusters in the field of the global water crisis. It does so by interpreting the research findings through a qualitative analytical synthesis to identify key gaps in governance and technology with a view to determining solutions from a future-oriented perspective using advanced digital and quantum-based technologies. Through such a research approach, the chapter can transcend descriptive analysis by establishing a clear link between knowledge patterns and innovation-oriented governance strategies. To ensure a clear focus, the chapter is based on the following research questions:

- What are the key environmental, demographic, and governance-related factors that are driving the freshwater scarcity global water crisis?
- What is the impact of freshwater scarcity on human security, economic stability, agricultural sustainability, and geopolitical stability?
- What are the key limitations of current water governance and management strategies in addressing uncertainty and complexity?
- What are the key thematic patterns and research gaps that emerge from a bibliometric analysis of freshwater crisis research?

Additionally, VOSviewer was used to look at the data and find a main theme through bibliometric analysis of the research. VOSviewer is a tool that helps create and show bibliometric networks. The steps to make data visualizations using the information from the literature review are

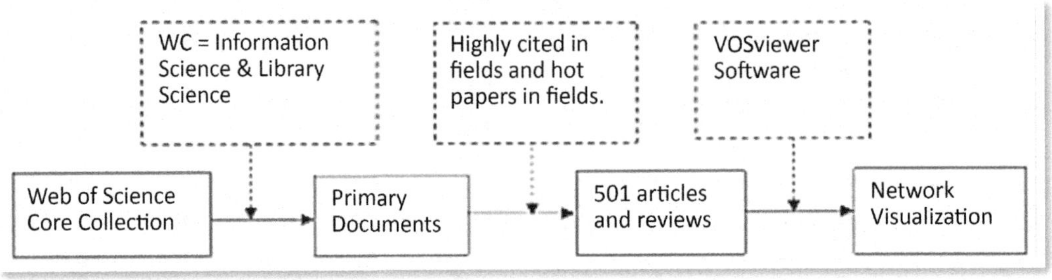

FIGURE 7.2 Demonstrates the VOSviewer bibliometric analysis procedure in detail.

as follows. (1) Choose this option, then click the button to make a map showing co-authorship, keyword connections, citations, bibliographic coupling, or co-citations using the bibliographic data. (2) Choose a data source. Then, read the reference management files for more details. (3) After selecting a file from your computer on the RIS tab, click next. (4) To make any other changes, pick co-authorship and Full counting, then click Next. (5) After setting the threshold, click the next button. (6) Once you choose the number of authors you want, click the next button. (7) Check the authors you've selected and click "Finish." The methodology is intentionally broad, allowing for global applicability rather than case-specific generalization, as shown in Fig. 7.2.

7.4 Results and Discussion

This chapter will address the various causes and effects of the crises, as well as the various solutions that can be employed to reduce the effects of the water scarcity that is to come. In the first section, the various causes are addressed, and they are as follows:

7.4.1 Climate change

Climate change effects, particularly rising temperatures that alter weather patterns globally. This brings about the following effects:

- **Melting glaciers:** Glaciers are melting rapidly, causing the natural water sources to dry up.
- **Decrease in winter snow:** As a result of less snowfall, the quantity of water in the spring is less.
- **Increase in evaporation:** As the temperatures rise, the evaporation rate from water sources accelerates, causing surface water scarcity.

7.4.2 Water scarcity

- **Water storage:** Not enough dams are being built due to financial constraints and prioritization of the issue by the government.
- **Misuse:** The major reason for water scarcity in the world is the misuse of water.

- **Naturally decreasing resources:** As water is a natural resource, it is limited in itself.
- **Overpopulation:** If the population of human beings is on the rise, then it is certain that the amount of water used would also rise, resulting in an actual scarcity of water.

7.4.3 Impact of water scarcity

The next section will discuss the possible impact that these crises could have, and this will include some very severe consequences such as:

- **International conflicts:** The possibility of conflicts between different countries cannot be ruled out as they might fight for the sharing of water resources, which could be as serious as the fight for oil resources in the past.
- **Water commodification:** Scarcity may imply that water is treated like any other highly valued commodity, such as oil or gold.
- **Water as an asset:** Scarcity could imply that water is treated as an asset and can be used as collateral in business transactions.
- **Humanitarian crises:** In this case, scarcity could imply a high probability of getting sick and dying due to a lack of clean drinking water.
- **Impact of agriculture:** Scarcity has an implication that there is a possibility of a food crisis due to water scarcity.

7.4.4 Addressing the impacts of water scarcity

In order to have such challenges surmounted, the following steps are called for:

- **Sustainable energy transition:** The necessity of less pollution will contribute to shifting electric, solar, and wind energy to mainstream modes of use. Identification and mitigation of the effects of climate change will also be promoted.
- **Enhancement of water storage:** Constructing more dams for the enhancement of water storage.
- **Afforestation initiatives:** Planting more trees to combat the effects of climate change and maintain the balance of nature.
- **Water waste reduction:** Implementing more educational programs and introducing more stringent policies to reduce water waste.
- **Smart water management systems:** Providing water in a smart way according to the needs of the people with the help of technology for the effective management of water.
- **Grey water recycling:** Recycling water using the water recycling technique.
- **Exploration for alternative habitats:** Exploring alternative habitations other than the Earth for combating the water crisis.

Through the collective efforts of all the parties involved, the urgent need to develop strategies to combat the new water crises is emphasized.

7.4.5 Bibliometric setup, performance metrics, and empirical patterns

While this study does not involve any form of laboratory experiment, an explicit analytical setup has been established to ensure an accurate, reproducible, and performance-oriented interpretation of the literature. The "experimental setup," therefore, refers to an explicit configuration of the bibliometric setup and analytical parameters applied through VOSviewer software for extracting, clustering, and analyzing patterns of freshwater crisis research. The final dataset was then fed into VOSviewer software for further analysis. An explicit analytical setup has been established through keyword co-occurrence, clustering of themes, and link strength relationships. Such an analytical setup has allowed this study to identify primary research trajectories and solution-oriented approaches through freshwater crisis research. In order to provide a proper interpretation of the results, this research has used a number of performance parameters. These include keyword frequency, which is the weightage of the keywords, and total link strength, which is the link strength itself.

The results show that there are five major thematic clusters. The first thematic cluster is related to climate change and hydrological change, and this cluster has a very high keyword frequency for words such as "glacier melt," "drought," and "precipitation variability." The second thematic cluster is related to demographic pressure and water demand, and this cluster links population growth, urbanization, and water demand. The third thematic cluster is related to governance and capacity, and this cluster links policy fragmentation, regulatory failure, and transboundary water management. The fourth thematic cluster is related to technology and infrastructure, and this cluster links desalination, smart water, and monitoring. However, there is a fifth thematic cluster that is rapidly emerging, and this is related to advanced computation and sensing technologies such as artificial intelligence and quantum technologies, as shown in Fig. 7.3.

7.4.6 Case study-oriented perspectives on freshwater crisis and technological governance

To further contextualize the nature of the global water crisis, aside from relying on theoretical analysis, this chapter incorporates an approach that relies on case study-oriented discussion, emphasizing the way water scarcity manifests in various geographical, institutional, and technological contexts. Instead of relying on empirical analysis of water crisis issues in any specific case study, this chapter relies on a comparative approach to water crisis issues in various regions of acute water stress. One of the most commonly cited instances of water stress issues is the Middle East and North Africa (MENA) region, which has the lowest water availability per capita. For instance, Israel has shown that desalination, smart water management, and water reuse can greatly improve water security.

However, these systems are energy-intensive, and there is a complex optimization problem that needs to be solved, a problem that next-generation computational technologies, such as quantum-assisted optimization, could potentially revolutionize in the future. In South Asia, there is another critical case of freshwater scarcity that is not just a result of climatic factors but also of governance fragmentation and interdependent river basins. In particular, there are

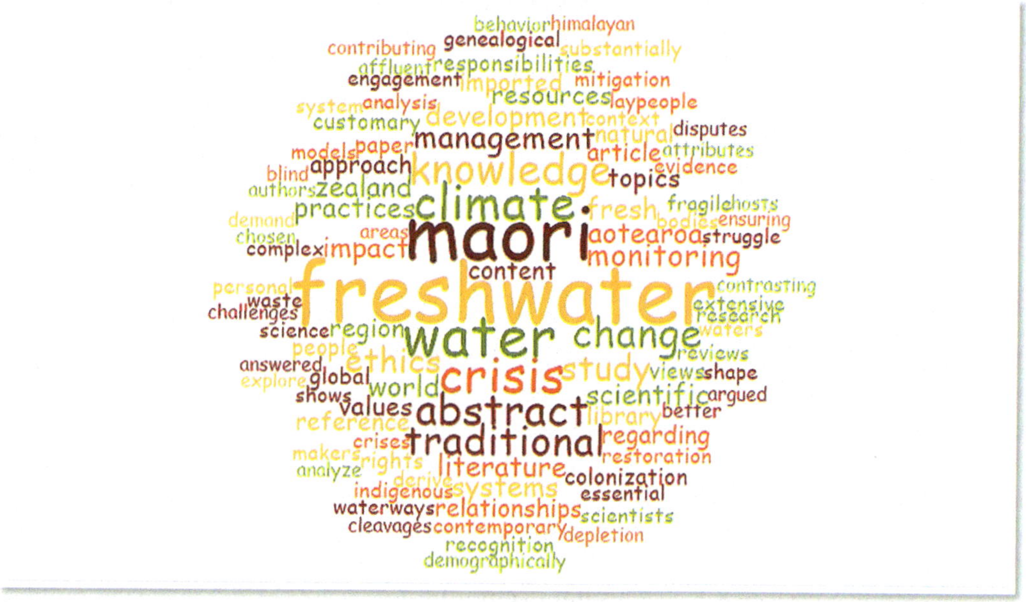

FIGURE 7.3 The VOSviewer bibliometric analysis procedure in detail.

hundreds of millions of people living in the Indus and Ganges River basins, but there is a problem of freshwater scarcity, glacier retreat, groundwater over-extraction, and inefficient agricultural water use.

Yet another example can be given in North America; it is focused on the Colorado River Basin, which offers water service to key cities and agricultural areas in both the United States and Mexico. The region has been experiencing severe drought, climate change-related reduction in snowpack, and over-allocated water rights, creating a potential threat of systemic failure in the basin. Even though there are sophisticated tools available for data collection, decision-making is limited by the computational difficulty associated with meeting multiple interests under uncertain conditions. Recent studies are indicating potential benefits associated with quantum computing, which can improve scenario analysis, enabling decision-makers to weigh trade-offs in meeting water interests in agriculture, cities, and the environment under alternative climate change futures, a potential technological leap forward from traditional modeling tools.

In all these cases, a common thread can be observed, i.e., the fact that although the adoption of technology has improved the efficiency of water management, governance as well as the limitations imposed by computing remain significant impediments. The fact is that the majority of the currently adopted technologies follow the conventional computing paradigm, which is incapable of coping with the complexity of the water ecosystem. This is more evident when it comes to the management of water systems with transboundary components, infrastructure networks, and climate-related uncertainties. On the other hand, the introduction of quantum technologies, i.e., quantum computing for optimization and quantum sensing for environmental monitoring, brings a paradigm shift in the management and governance of water systems in the long term. However, it should be noted that the discussion presented in the chapter does not imply the potential for quantum technologies to solve the water crisis. Instead, it presents the potential for quantum technologies as a forward-thinking solution for the water crisis, complementing the currently adopted digital and governance reforms. The discussion on the cases presented above illustrates the fact that the regions that have already adopted smart water infrastructure and technologies have a higher potential.

7.5 Conclusion, Limitations, and Future Scope

Moreover, based on the results and discussions from the above section, it is clear that we now have an understanding of how severe this water shortage crisis will be. The impact of this crisis has been felt in various parts of the world. We have also learned much about this crisis, and its impact on people as individuals and as a group. From the research of this paper, we have tried our best to make both governments and people aware of the importance of focusing on creating and planning how to solve this problem that arises from water shortages. Even if we are not able to fully solve this problem, these steps will help us ease this situation and will give us an opportunity to survive if things get worse. Accepting this reality is the first step towards solving this problem. We still believe it's not too late to start. If we take action now and work properly, we will minimize this problem. But if we just wait and do nothing, we will surely face this problem's consequences, and by then, it will be too late for us to do anything to stop this problem from causing further damage.

Limitations of the study

First, it is based on qualitative and conceptual analysis, which does not offer any quantitative validation or accuracy, especially at a regional or local level. Second, bibliometric results are based on database coverage and keyword selection, which might not account for literature published outside traditional academic sources or those using other languages besides English. Third, although quantum technologies are mentioned as potential solutions, their actual implementation in water governance is still experimental, and no assessment of actual outcomes is offered in this chapter, which again cautions against generalization and encourages further research.

Future scope

This chapter provides scope for future research in many directions. Future research may empirically validate the conceptual hypotheses developed in this chapter through various approaches. More emphasis might be placed on the assessment of pilot applications of quantum sensing and quantum-assisted computation in water management contexts. Additionally, the inclusion of sociopolitical factors such as trust, ethical issues, and digital inequality in the domain of water governance through technology-based methodologies can be considered to be of prominent research and future potential. In terms of policy considerations, research in this domain can be carried out further to improve the techniques of innovation-driven water governance.

References

Bailey, P. G., & Worthington, N. C. (1997). History and applications of HAARP technologies: The high-frequency active auroral research program. In *Proceedings of the 1997 32nd intersociety energy conversion engineering conference. part 1 (of 4): 2* (pp. 1317–1322). https://www.scopus.com/inward/record.uri?eid=2-s2.0-0031369808&partnerID=40&md5=a758caa5091bfd9ff629d4d4b1b26d92.

Degen, C. L., Reinhard, F., & Cappellaro, P. (2017). Quantum sensing. *Reviews of Modern Physics, 89*(3), 035002. https://doi.org/10.1103/RevModPhys.89.035002.

Feijó, A., Karlsson, C. M., Gray, R., Yang, Q., & Hughes, A. C. (2023). Extreme-sized anurans are more prone to climate-driven extinctions. *Climate Change Ecology, 4*. https://doi.org/10.1016/j.ecochg.2022.100062.

Gleick, P. H. (2018). *The world's water volume 8: The biennial report on freshwater resources*. Island Press.

Gómez-Trujillo, A. M., González-Pérez, M. A., & Velez-Ocampo, J. (2022). Smart technologies and water governance: A systematic literature review. *Sustainable Cities and Society, 76*, 103432. https://doi.org/10.1016/j.scs.2021.103432.

Lü, W., Ren, H., Ding, W., Li, H., Yao, X., & Jiang, X. (2023). The effects of climate warming on microbe-mediated mechanisms of sediment carbon emission. *Journal of Environmental Sciences (China), 129*, 16–29. https://doi.org/10.1016/j.jes.2022.09.016.

Lydiri, M., El Mourabit, Y., El Habouz, Y., & Fakir, M. (2023). A performant deep learning model for sentiment analysis of climate change. *Social Network Analysis and Mining, 13*(1). https://doi.org/10.1007/s13278-022-01014-3.

Mahesar, R. A., Chandio, D. A., Latif, M., Abbas, S., & Shabbir, T. (2023). Demography and risk factors of suicide deaths in Pakistan: A twelve-month content analysis study. *Asian Journal of Psychiatry, 80*. https://doi.org/10.1016/j.ajp.2022.103364.

Mao, G., Ji, M., Jiao, N., Su, J., Zhang, Z., Liu, K., Chen, Y., & Liu, Y. (2023). Monsoon affects the distribution of antibiotic resistome in Tibetan glaciers. *Environmental Pollution, 317*. https://doi.org/10.1016/j.envpol.2022.120809.

Montanaro, A. (2016). Quantum algorithms: An overview. *npj Quantum Information, 2*, 15023. https://doi.org/10.1038/npjqi.2015.23.

Petrescu-Mag, R. M., Burny, P., Banatean-Dunea, I., & Petrescu, D. C. (2022). How climate change science is reflected in people's minds. A cross-country study on people's perceptions of climate change. *International Journal of Environmental Research and Public Health, 19*(7), 4280. https://doi.org/10.3390/ijerph19074280.

Pinto, M. (2023). *Introduction to the freshwater crisis* (pp. 177–182).

Sabayan, B., Wyman-Chick, K. A., & Sedaghat, S. (2023). The burden of dementia spectrum disorders and associated comorbid and demographic features. *Clinics in Geriatric Medicine, 39*(1), 1–14. https://doi.org/10.1016/j.cger.2022.07.001.

Stewart-Harawira, M. W. (2020). Troubled waters: Maori values and ethics for freshwater management and New Zealand' s freshwater crisis. *Wiley Interdisciplinary Reviews Water, 7*(5). https://doi.org/10.1002/wat2.1464.

Sultana, F. (2018). Water justice: Why it matters and how to achieve it. *Water International, 43*(4), 483–493. https://doi.org/10.1080/02508060.2018.1458272.

UN-Water. (2023). *United nations world water development report*. UNESCO.

Wirsing, R. G., Stoll, D. C., & Jasparro, C. (2013). *International conflict over water resources in Himalayan Asia.* Palgrave Macmillan.

Yan, Q., Wei, T., & Zhang, Z. (2023). Modeling the timing and extent of glaciations over southeastern Tibet during the last glacial stage. *Palaeogeography, Palaeoclimatology, Palaeoecology, 610.* https://doi.org/10.1016/j.palaeo.2022.111336.

PART V

Socio-economic, ethical and global perspectives

8

Frugal and inclusive quantum agri-tech ecosystems for the Global South

Ashish Kumar[a] and Darshan Pandya[b]

[b]OPERATIONS MANAGEMENT AND DECISION SCIENCES AREA, JAIPURIA INSTITUTE OF MANAGEMENT, NOIDA, UTTAR PRADESH, INDIA [b]CENTRE OF MANAGEMENT STUDIES, NEW AGE MAKERS' INSTITUTE OF TECHNOLOGY (NAMTECH), GANDHINAGAR, GUJARAT, INDIA

8.1 Introduction

In the context of agricultural production, over the centuries, many low- and middle-income countries (LMICs) find themselves in a precarious situation: centuries-old reliance on agriculture is combined with the growing impact of modern vulnerabilities. In this regard, agriculture is not only an economic sector but rather a life-saving tool, which supports food security, employment, and macro-economic stability.

In Ethiopia, the agricultural sector is the backbone of the economy with about 73% of the labor force and about 37% of the GDP of Ethiopia. Comparatively, the industry segment employs a relatively small proportion of about 7.4% of the labor force, but its contribution to GDP is estimated to be between 24% and 25%, which is capital intensive but has a narrow job base (Michigan State University, 2025; Trading Economics, 2025a). Agriculture in Benin has a population of 43% and forms about 24.2% of GDP with 18% of the population employed in industry, which is a hybrid economic model that is highly reliant on natural resources and informal systems of work (Trading Economics, 2025b; Ulandssekretariatet, 2025).

In Nigeria, the agricultural system is highly vulnerable because it is highly reliant on rain-fed agriculture, where more than 80% of the smallholder farmers produce nearly 90% of food consumed in the country. The dependence makes the country extremely susceptible to the change in climate, particularly rainfall deficiency and water scarcity (Shibayan, 2025). This vulnerability is also evident in the majority of sub-Saharan Africa, where rainfed agriculture is dominant, and such long-term problems as poor infrastructure, lack of mechanization, and low digitalization exist (Lombe et al., 2024; Tatlock, 2006).

Artificial intelligence (AI) and machine learning (ML) are automation tools that provide avenues to increase agricultural productivity but require strong datasets, large amounts of computing power, and advanced digital infrastructures, which are often not available in LMIC. Consequently, smallholder producers face a heightened risk of being marginalized by these emerging

Quantum Computing and Instrumentation for Smart Agriculture. DOI: https://doi.org/10.1016/B978-0-443-44247-6.00011-7

technologies (Kumar & Agrawal, 2023). However, unless QT are implemented inclusively and, in a context-sensitive manner, they may increase the current digital divides and strengthen structural inequities (Umbrello et al., 2024; Zornetta, 2024).

The emerging paradigm of quantum technologies (QT) signifies a substantial shift from classical AI, which predominantly relies on statistical inference and historically derived datasets for decision-making and prediction. In contrast, QT aims to transcend the limitations of classical AI by leveraging quantum computing's capabilities to handle complex, high-dimensional data with enhanced precision and speed. This distinction is particularly relevant in domains like perishable goods management, where quality-based classification and real-time responsiveness are crucial. For instance, Kumar and Agrawal (2025) employed convolutional neural networks (CNN) with statistically optimized hyperparameters to enhance quality classification of perishable products, while in a parallel study, authors proposed a sustainable supply chain architecture using image processing under the broader vision of Industry 4.0 (Kumar & Agrawal, 2024).

In comparison, quantum systems use superposition, entanglement, and tunneling to realize new optimization protocols under high degrees of uncertainty, thus presenting high potential in agri-logistics, resource allocation, and supply chain resilience (Amato et al., 2025; Chan et al., 2025; Golter et al., 2023). However, existing applications are burdened by the constraints of the so-called "Noisy Intermediate-Scale Quantum" era that is defined by the loss of coherence, the presence of computational noise, and high infrastructural expenses, which are especially relevant in the context of LMIC (Sarangi et al., 2023; Sivan & Priya, 2025). Nevertheless, these hindering factors have been circumvented through new breakthroughs, including quantum-inspired deep learning, photonic processors, and entangled decision architecture, which are paving the way to more resource-efficient, hybrid systems. Such trajectories would suggest a possible re-definition of the technological baseline where quantum-classical convergence would be used to provide context-sensitive agricultural solutions that could span global digital divides.

The new definition places frugal innovation in the wider framework of systemic changes that are part of Industry 5.0, a paradigm that emphasizes human-centricity, resilience, and sustainability as the principles of future innovation systems (Adigun et al., 2024; El-Ebiary et al., 2024). In this context, frugal innovation is not only a practical approach but also a normative requirement, which promotes inclusiveness, modularity, and scalability in the environment of resource scarcity. The modern trends, such as hyperentangled cluster states and autonomous qubit autotuning, provide strong arguments in support of decentralized, low-resource QT application in agriculture (Lock et al., 2023; Muzammul et al., 2025; Nthaga et al., 2025). These developments show that it is possible to have quantum systems that are socially embedded depending on open frameworks and local flexibility. This is a way beyond technological determinism, and it must be said that such approaches should privilege inclusive innovation models that place epistemic justice and participatory governance at the forefront of the next-generation agri-tech implementation.

In the modern technological and philosophical environment, the research addresses three interconnected research questions that would help to investigate the intersection of equity, innovation, and agricultural change:

- ***RQ1:****What is the theoretical and practical way to combine the principles of frugal innovation with quantum and quantum-inspired technologies to develop inclusive, scalable, and context-sensitive agricultural solutions in low-income economies?*
- ***RQ2:****Which strategies of governance, participatory design, and institutional arrangements are best to overcome capability, infrastructural, and translational gaps in the implementation of quantum agri-tech systems in the Global South?*
- ***RQ3:****How can frameworks, metrics, and models of the assessment of the equity, resilience, and the sustainability impact of quantum agri-tech innovation ecosystems in low-resource contexts be created?*

Such questions indicate a technological edge as well as a normative urgency: the necessity to redesign the tracks of innovation that will not only be innovative, much more efficient and productive but also necessary in the principles of distributive justice and ecology sustainability (Monlezun et al., 2025; Holter Ten et al., 2023).

In order to answer these questions, this chapter takes a systematic literature review approach, combining theoretical, empirical, and conceptual insights within the context of quantum computing, frugal innovation, sustainability transitions, and socio-technical systems in digital agriculture. The most important frameworks that provide analytical grounding are responsible innovation (RI) (Carbajal-Pina & Acur, 2025), absorptive capacity (Sivan & Priya, 2025), and quantum socio-technical imaginaries (Zornetta, 2024). This cross-disciplinary background allows the formulation of an integrated research program that is geared toward quantum agri-tech justice.

This chapter proposes a systematic and equity-based view of the manner in which quantum and quantum-inspired technology could be sensibly applied in agricultural systems of the low- and middle-income nations.

Objectives of the chapter

1. To create a conceptually consistent framework about integrating quantum and quantum-inspired technologies with concepts of frugal innovation, affordability, modularity, and contextualization to make the agricultural systems of low-resource regions;
2. To critically review the infrastructural, institutional, and human-capability gaps through which quantum agri-tech is adopted in the Global South and focuses on the coordinated implementation of policies, absorptive capacity, and participatory governance mechanisms in implementation;
3. To assess the possible dangers of automation, data sovereignty, labor displacement, and power asymmetries and suggest RI frameworks in order to alleviate exclusionary consequences;
4. Lastly, this chapter includes future research directions, policy tools, and assessment models that would transform the discussion to quantum sufficiency, quantum solidarity, and sustainable development in agrarian economies.

Organization of the chapter

The rest of the chapter is organized as: Section 8.2 Illustrates principles of quantum technologies for policy and innovation scholars. Section 8.3 highlights infrastructure, policy and capability gaps. Section 8.4 focuses on Ethical and socioeconomic implications. Section 8.5 discusses Innovative ecosystems and startup landscapes. Section 8.6 enumerates Challenges, research agenda and future pathways. And, finally, Section 8.7 concludes the chapter with future scope.

8.2 Principles of quantum technologies for policy and innovation scholars

QT are emerging as a distinctive techno-scientific area, and their consequences are far reaching with regard to sustainability, security, and social-economic transformation. The understanding of theoretical backgrounds and practical representations of QT is essential to policy and innovation researchers, particularly those who work on agriculture and other sectors that are essential to climate change. In this section, the principles are decomposed, their difference with the traditional AI/ML systems is described, their socioeconomic applications are outlined, and the emerging concept of frugal QT (FQT) is mentioned.

8.2.1 Core quantum concepts

The core of QT is quantum mechanical effects such as superposition, entanglement, and quantum tunneling. These facilitate the ability to perform beyond the classical computation and measurement paradigms. As another example, Parvini et al. (2025) showed how cavity dynamics can be used to engineer squeezed magnon states in antiferromagnets, a first step toward quantum memory and quantum-enhanced sensing. Conversely, Kurmapu et al. (2023) simulate magnon Bose–Einstein condensates in the high-angle deviations of inhomogeneous magnetic fields with an increased focus on quantum coherence within longer time windows, which is essential to the effective control of quantum states in real-world devices.

In the meantime, Rocha et al. (2022) concentrated on four-wave mixing with Hermite–Gaussian beams in rubidium vapor and demonstrate that pure optical modes can be selectively generated, which further increases the spatial flexibility of entangled photon sources. In contrast to the solid-state systems that (Chan et al., 2025) studied, this chapter brings beam engineering as a control modality to the fore; a future direction that is very important in lightweight, portable quantum sensors. For clarity and ease of reference, a glossary of foundational quantum terms such as superposition, entanglement, and quantum tunneling is provided in Table 8.1.

8.2.2 Distinction from AI and ML

Although AI/ML systems are very effective in data-driven prediction, they are fundamentally different to QT in respect to logic, data encoding, and ontology of computation. The empirical

Table 8.1 Glossary of key quantum terms.

Term	Definition
Qubit	The fundamental unit of quantum information, analogous to a classical bit but existing in superposition
Superposition	The quantum property allowing a qubit to exist in multiple states simultaneously
Entanglement	A quantum phenomenon where particles become correlated such that the state of one affects another, regardless of distance
Quantum tunneling	A process where a particle passes through a potential barrier it classically should not be able to
Quantum sensing	Use of quantum phenomena to achieve ultraprecise measurement of physical parameters
Quantum annealing	A quantum optimization method that leverages quantum tunneling to escape local minima
Noisy intermediate-scale quantum	The current era of quantum computing marked by limited qubit counts and noise-prone operations
Quantum advantage	The point at which quantum computers perform tasks beyond the reach of classical computers
Quantum coherence	The time over which a quantum state maintains its phase relations, essential for computation
Frugal quantum technologies	Low-cost, resource-conscious quantum tech designs aligned with constraints of developing economies

case presented by Attri et al. (2025) to classify plant diseases is strong as they use their entangled quantum-inspired deep learning (EQID) model. Authors obtained statistically significant improvements in accuracy, precision, and F1 scores across tomato and potato disease datasets by incorporating quantum aspects into the learning. As highlighted in this chapter, quantum mechanics is structurally beneficial in feature extraction particularly where signals are low and variability is high.

In the architectural aspect, Pettit et al. (2022) suggested a photonic superstructure that consists of nonlinear optical units to realize entangled photon-pair generation and computation. In contrast to the algorithmic advancement of EQID, the focus of the present study is the physical implementation of quantum sampling via entanglement propagation, which brings out a difference between algorithmic inspiration and hardware-based logic.

In addition, Nikmehr et al. (2022) reformulated the Unit Commitment problem which is a combinatorial optimization problem in power systems into a quantum distributed framework. Authors quantum approach contrasts with AI-based solvers making use of metaheuristics or regression trees, in that the problem is partitioned into entangled subproblems, and the coordination of the subproblems is achieved using quantum consensus. This does not only show parallelism at the ontological level but also shows resource efficiency in decomposable problem landscapes.

Table 8.2 Comparison between AI/ML and quantum in agriculture.

Dimension	AI/ML in agriculture	Quantum technologies in agriculture
Underlying principle	Pattern recognition through statistical learning from classical data	Quantum superposition, entanglement, and tunneling enabling nonclassical computation
Computational paradigm	Deterministic, data-driven, based on neural nets and probabilistic models	Probabilistic, state-based, exploits quantum state manipulation and Hilbert space representations
Key applications	Yield prediction, pest detection, weather forecasting, supply chain analytics	Quantum sensing, optimization of planting schedules, materials for fertilizer enhancement
Strengths	Mature ecosystems, scalable models, large dataset handling	Exponential speedup for specific problems (e.g., optimization, simulation), enhanced sensing
Limitations	Requires large data and high computing power, struggles with NP-hard problems	Still in early-stage development, limited hardware access, noise, and decoherence
Use of data	Data-intensive learning from past patterns	Can simulate quantum phenomena in nature (e.g., photosynthesis) without needing extensive datasets
Optimization tasks	Heuristic and metaheuristic methods (e.g., genetic algorithms and gradient descent)	Quantum annealing, variational quantum algorithms for faster convergence on complex landscapes
Impact on sustainability	Precision farming through prediction and automation	Enhancing resilience and resource efficiency through quantum material design and sensing

Regarding control, Rao and Jayaraman (2023) applied matrix product states in the creation of leakage-reducing protocols in the steering of quantum many-body systems. The difference with ML-based control systems is that quantum systems can maintain long-range nonclassical correlations, so state transitions are interpretable and high fidelity. A detailed comparative overview of AI/ML and QT in agriculture is provided in Table 8.2, highlighting their differing computational paradigms, applications, and sustainability impacts.

8.2.3 Socioeconomic relevance

In socioeconomic terms, QT present new sensing, optimization, communication, and cryptography infrastructures, which are essential to agriculture, logistics, and policymaking that focus on sustainability. An example is that Muzammul et al. (2025) proposed a quantum-inspired object detection model of unmanned aerial vehicles (UAV) that combines sub-pixel convolution with self-supervized learning. The model had superior accuracy and reduced computational overhead compared to a classical one, for example YOLOv4 or EfficientDet, which is why it will be able to operate in low-resource agricultural settings in real time.

Conversely, Khatri (2022) provided a Markov decision processes high-level abstraction of quantum networks and suggested a beautiful policy optimization framework to entanglement distribution. This theoretical contribution is important to the development of network-level strategies of decentralized data sensing in remote rural regions.

In addition, Lopetegui et al. (2022) presented a convenient Fisher information-based witness protocol to detect non-Gaussian quantum steering, which would be a practically applicable tool to verify the entanglement even in noisy conditions without complete state tomography. This supplements the cryogenic infrastructural issues that Park et al. (2022) discussed, which was the use of pulse-tube-based dilution refrigerators to maintain superconducting qubits. Cumulatively, the studies indicate that quantum systems can be resilient and field-deployable, on condition that the proper verification and cooling facilities exist.

8.2.4 Frugal quantum technologies

The necessity to bring QT out of the exclusive labs has led to the emergence of FQT, which are systems that balance quantum depth with economic and infrastructural limitations.

As an example, Yon et al. (2025) designed an independent CNN-based charge autotuning algorithm to spin qubit arrays. With model uncertainty estimation and closed-loop optimization, the system achieves 95% success in identifying charge regimes and this showed that automated calibration, enhanced by ML, can significantly reduce overhead costs in the operation.

To this, de Souza et al., (2024) listed the applications of quantum computing in agriculture in the categories of bioinformatics, climate modeling, and smart farming. Authors claimed that quantum systems, because of their high energy efficiency, can be more precise than traditional supercomputers in precision agriculture applications, which has strategic implications in developing economies.

Collectively, these works help to reveal an important development: QT ceased to be hypothetical instruments of elite computation, but a multiscalar innovation paradigm, that is, one that requires both profound theoretical knowledge and context-specific deployment strategies and integration into socio-technical imaginaries. To policy and innovation researchers, working with QT does not only imply monitoring technological diffusion but designing socio-political structures in which quantum systems will be integrated.

8.3 Infrastructure, policy, and capability gaps

With QT moving out of the conceptual frontier into practical applications, the need to achieve equitable preparedness among the different countries has become more urgent. Nevertheless, there are still glaring infrastructural access, institutional support, and strategic capabilities in the Global South.

8.3.1 Current challenges in the Global South

QT, especially quantum computing and quantum communication, require strong digital infrastructure: high-bandwidth connectivity, access to quantum simulators, and cloud-based quantum processing units. However, in most of the Global South, these basics are either immature or nonexistent. The conceptualization of emerging information and communication technologies infrastructure developed by Sarangi et al. (2023) is that the latter is a multidimensional system, the discontinuities in technologies, infrastructural fragmentation, and inadequate regulatory scaffolding intersect to hinder innovation. This is also complicated by asymmetric bandwidth penetration, sporadic computational access, and inadequate hardware integration, a fact that was confirmed by Hussain et al. (2024) who reported the geopolitical fault lines that define digital sovereignty and access to satellite and fiberoptic networks.

Computationally, Sivan and Priya (2025) pointed out that, even in sectors where quantum amplitude estimation (QAE) algorithms are starting to take off, like risk management in IT, a stable and high-fidelity computational ecosystem is necessary to support successful deployment, which is based on secure cloud access, strong encryption standards, and expert digital operators. As demonstrated in their study, the two aspects of infrastructure and predictive accuracy are symbiotically connected: the absence of the computational resources directly undermines the predictive value of the quantum tools.

Digital skills/literacy turn out to be equally constraining. In their cross-national policy analysis, Shim and Suh (2025a, 2025b) unveil that, although countries such as South Korea have implemented a tiered approach to investment, namely, prioritizing quantum education and infrastructure, the emerging economies tend to have no coordinated plans regarding quantum literacy. Seskir et al. (2024) also stated that quantum education is also affected by epistemic gatekeeping, as curricula are too abstract, physics-biased, or culturally incompatible with most non-Western students. This has led to the talent pipeline not just being a thin one but also discontinuous with minimal defined routes between the basic STEM education and the advanced quantum skills. Table 8.3 provides a comparative snapshot of infrastructure, skills, and institutional readiness across selected LMICs, highlighting both the diversity of contexts and the shared gaps that shape quantum deployment pathways.

The literature in its totality highlights a vital observation, namely, that infrastructure and human capital are only part-technical issues but systemic gaps in capabilities determined historically by asymmetries, in industrial and policy settings, and by market disincentives. To deal with them, it is not enough to make gradual improvements but to introduce radical changes that redefine access, ownership, and involvement in quantum ecosystems.

8.3.2 Strategic pathways to quantum readiness

Preparation to quantum transformation is rapidly becoming less a matter of yes or no, and more a matter of degree based on the capacity of that organization to absorb, internalize, and utilize emerging knowledge, absorptive capacity. Using the original construct by Cohen & Levinthal (1990) and Sivan and Priya (2025) go about to show how the current levels of data capabilities, security provisions, and algorithmic literacy in firms determine the extent to which it can adopt QAE in IT-related sectors. Authors discovered that technological availability was intermediated by the absorptive capacity interventional relationship with the application-level innovation, which is also true in the prospect of the quantum-secure digital signatures (Mohanty et al., 2025).

Table 8.3 Infrastructure comparison across low-income countries.

Country	Bandwidth and connectivity	Computational access	Quantum infrastructure	Digital skills and talent pipeline	Policy and institutional support
India	Expanding urban bandwidth; rural gaps persist (Sivan & Priya, 2025).	Moderate access via academic-industry collaboration and cloud simulators (Mohanty et al., 2025).	Early adoption in quantum-enhanced risk modeling and cryptographic research (Mohanty et al., 2025; Sivan & Priya, 2025).	Strong in engineering/STEM; emerging quantum talent nodes (Mohanty et al., 2025).	Institutional support is fragmented; CSR-driven collaboration emerging (Mohanty et al., 2025).
South Africa	High-bandwidth capacity demonstrated through satellite communication (Li et al., 2025).	Portable quantum ground stations tested; operational experimentation ongoing (Li et al., 2025).	Enabled satellite-based QKD with China, crossing 12,900 km (Li et al., 2025).	Skills development emerging via joint R&D; not yet formalized (Li et al., 2025).	Strategic international partnership (esp. China); limited local anchoring (Li et al., 2025).
China	Extensive terrestrial and satellite quantum networks; strong fiber backbone (Li et al., 2025; Hussain et al., 2024).	Among the world's most advanced quantum computing infrastructure (Minbaleev et al., 2024).	Leading quantum satellite constellation; operational QKD (Li et al., 2025).	Government-led talent initiatives and tech diplomacy drive capacity (Minbaleev et al., 2024).	Centralized quantum mission agency; synchronized national strategy (Hussain et al., 2024; Minbaleev et al., 2024).

(continued on next page)

Table 8.3 Infrastructure comparison across low-income countries—cont'd

Country	Bandwidth and connectivity	Computational access	Quantum infrastructure	Digital skills and talent pipeline	Policy and institutional support
South Korea	Nationwide high-speed connectivity; strong digital infrastructure (Shim & Suh, 2025a).	Institutional access to national supercomputing facilities (Shim & Suh, 2025a).	Quantum science designated a national strategic tech; targeted funding (Shim & Suh, 2025a).	Policy-directed investment in quantum education and services (Shim & Suh, 2025a).	Coherent roadmap integrating quantum education, infrastructure, and governance (Shim & Suh, 2025a).
United States	Global leader in infrastructure; some rural limitations (Shim & Suh, 2025b).	Industrial and academic HPC access widespread (Baseri et al., 2025).	Active in QKD, post-quantum crypto, and hybrid networks (Baseri et al., 2025).	Deep STEM base; tech firms driving workforce development (Shim & Suh, 2025b).	NSCI emphasizes cybersecurity and competitive edge; federal investment high (Shim & Suh, 2025; Baseri et al., 2025).
Egypt	Urban digital infrastructure progressing; rural disparity noted (Elmansy, 2024).	Moderate computing access in regulated sectors like Islamic finance (Elmansy, 2024).	Conceptual use of quantum for financial compliance and modeling (Elmansy, 2024).	Financial tech skills improving; quantum skills not yet institutionalized (Elmansy, 2024).	Legal frameworks supportive; technical implementation still lacking (Elmansy, 2024).

Institutional coherence is also crucial besides being absorbed by the absorptive capacity. However, according to Baseri et al. (2025) and Minbaleev et al. (2024), the problem is that most governments tend to do it in terms of fragmentation, or the implementation of siloed policy mechanisms regarding quantum readiness. Science ministries, defence ministries, and commerce ministries can work independently and result in similar work or conflict of regulations. Successful national efforts, exemplified by China's coordinated investments in quantum infrastructure, cryptography, and space-based quantum networks, demonstrate the value of centralized yet adaptive innovation governance.

Nonetheless, the research gap is still significant: the combination of institutional gaps and social absorptive capacity. Although the literature is involved in the firm- or state-level readiness, little attention is paid to the way civil society, local universities, and subnational governments can cocreate quantum capacity through grassroots innovation, or via a decentralized sandbox or open-source platform in the Global South. This omission risks reproducing a technocratic model of readiness that excludes alternative epistemologies and indigenous knowledge systems.

8.3.3 Policy roadmap: Toward inclusive quantum futures

A new body of literature is now emerging that is interested in policy mechanisms, which shift away from techno-solutionism toward inclusive quantum ecosystems. There are three new models that should be looked at more carefully: quantum agri-tech sandboxes, CSR-based infrastructure projects, and readiness benchmarking tools.

To start with, Quantum Agri-Tech Sandboxes, a notion that is still very conceptual in the literature, are conceived regulated experiment areas where quantum algorithms will be used to maximize agricultural output, soil forecasting, and climate reaction. Based on the similarities of De Bartolomeo et al. (2025) in optimizing railways and Liu et al. (2025) in managing microgrids, such sandboxes may use quadratic unconstrained binary optimization (QUBO) and quantum approximate optimization algorithm (QAOA) models to find solutions to resource-constrained, multi-variable agricultural decision-making. Notably, these sandboxes should be localized and include local farmers in them and should be controlled by hybrid public-private frameworks to prevent the pitfall of techno-elitism.

Second, the concept of using corporate social responsibility (CSR) as a way of incentivizing the diffusion of quantum infrastructure is picking up. According to Elmansy (2024) and Baseri et al. (2025), large technological companies have started investing in quantum encryption, diagnostics, and financial modeling. Nevertheless, these investments are still dominated in the Global North. A policy instrument that incentivizes global tech firms to open-sourcing quantum APIs, invest in infrastructure in the under-served regions, and setting up talent fellowships in low-income areas might redefine CSR as a redistributive innovation driver. This involves innovation in regulation, including tax credits on quantum-for-development investment or multilateral CSR agreements either within the World Trade Organization or within the UN.

Last, but not least, a quantum infrastructure readiness index (QIRI) is urgently needed, a benchmarking tool that does not only evaluate technological maturity, but systemic preparedness. Although there is no formal QIRI at present, Pittaluga et al. (2025) and Vermeer et al. (2024) provided the initial ones: the density of quantum devices, access to quantum education, efficiency of policy coordination, and governance ethics. The trick is that epistemic diversity and non-Western representation are to be guaranteed in index design, and the one-size-fits-all approach to readiness should be avoided because it marginalizes local innovations and nonlinear adoption patterns.

The reviewed literature confirms that the quantum transformation in the world is not only a technological competition, but a distributed governance problem. The road to preparedness is not through universal standards and fixed road maps, but through plural, adaptive and equity-oriented approaches. This requires the move beyond the quantum supremacy to quantum solidarity, in which infrastructural, institutional, and policy innovations are cocreated across boundaries.

To this end, the future research should be able to cross disciplinary boundaries, incorporate marginalized geographies, and bring inclusion as a first-order design principle to quantum futures, not as a post hoc solution. The decade ahead provides a slim, yet critical opportunity not only to construct QT, but the justice ecosystems through which they will become relevant to who, where, and how. These considerations are synthesized in Fig. 8.1, which outlines a policy roadmap for quantum agriculture, connecting technological readiness with equity, inclusivity, and systemic cocreation across geographies.

8.4 Ethical and socioeconomic implications

As with any disruptive technology, quantum computing has a duality: a potential to change the world in ways we can hardly imagine and a potential to cause ethical, economic, and governance problems that are equally difficult to imagine. With the pace of its incorporation into other areas such as agriculture, health, education, and security, a multiperspective evaluation is very necessary.

8.4.1 Automation and labor displacement

Quantum-accelerated systems will allow automation, especially in supply chains, optimization tasks, and robotics, which is likely to increase employment vulnerability, especially in agri-fresh and informal sectors. Monlezun et al. (2025) emphasized the fact that, although digital and QT will enhance the scalability of health care and the efficiency of decision-making in LMIC, they will worsen the socioeconomic asymmetries unless accompanied by corresponding investments in the reskilling of the workforce and the construction of the inclusive infrastructure. The same applies to the agri-tech where AI and quantum systems used to predict crops, optimize supply chains, or kill pests might substitute the jobs that are done by informal labor (Balu et al., 2025).

Bongurala et al. (2025) highlighted this contradiction in medical oncology: although AI and quantum computing can lead to a higher level of personalization and diagnostics than ever before, it also displaces lower-skilled healthcare workers in the field of diagnostics, logistics,

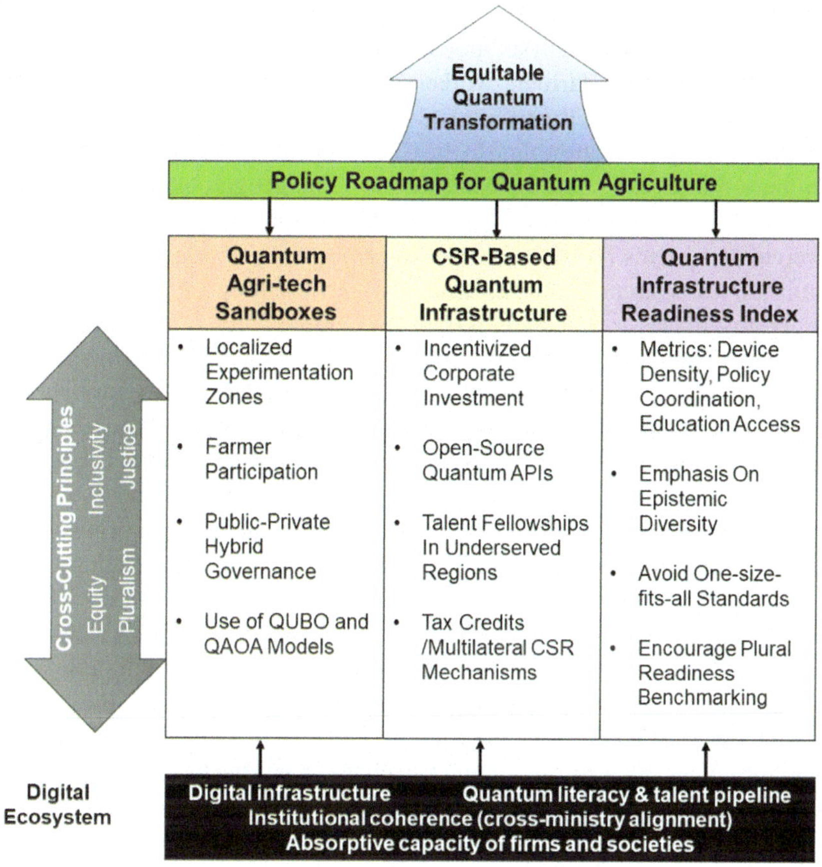

FIGURE 8.1 Policy roadmap for quantum agriculture.

and data collection. Similar trends are expected in agri-fresh industries, which tend to employ informal labor and where predictive analytics, drone-based interventions, and climate modeling algorithms will eliminate labor-intensive work but provide few new jobs in their place.

This follows a larger historical trend of technological innovation: the increase in efficiency exceeding the fair transition of the workforce. However, unlike in the case of previous technological transformations, quantum computing accelerates the pace and intricacy of the change and necessitates that displacement risks should be preemptively managed in the design and implementation plans.

8.4.2 Ethical concerns in quantum-enabled agri-tech

The ethical issues of quantum agri-tech are related to data justice, asymmetries of power, and the lack of transparency of quantum-enhanced decision systems. Umbrello et al. (2024) used the value-sensitive design (VSD) framework to draw the constellation of stakeholders who are impacted by QT. Based on their analysis, they see that the actors in the upstream (developers and institutions) and downstream (farmers, workers, and citizens) are not both equally empowered in the design and governance of the technologies offered.

This imbalance is especially troubling in the Global South, where the tools of decision-making based on quantum analytics could be researched in one geography (e.g., multinational businesses in the Global North), but used to operate in another in the absence of ethical supervision or data sovereignty systems (Zornetta, 2024). The complexity of quantum systems adds to the issues of consent, transparency, and explainability because they are not amenable to the logic that is interpretable in classical AI systems (Qin et al., 2025).

The taxonomy of machine consciousness proposed by Qin et al. (2025) takes the argument further, stating that systems that are close to autonomous decision-making, even in specific fields of application, such as agriculture, should be ethically tested not only in terms of performance but also in terms of proxies of agency, accountability, and self-awareness. In agricultural fields, this could involve algorithms taking real-time irrigation decisions or suggesting pesticide use based on quantum-enhanced weather models. The moral consequences of such choices particularly when they are erroneous stretch to the environmental consequences and the livelihoods of the farmers.

Carbajal-Pina and Acur (2025) claimed that these risks are not addressed sufficiently because there are no proactive, principle-based RI frameworks. They promote the integration of RI in agri-tech design, as a stakeholder cocreator, foresight analysis, and ethical stress-testing, to make sure that technologies are based on local situations and values. Table 8.4 provides an ethical risk matrix for quantum agri-tech, summarizing key challenges, affected stakeholders, root causes, and actionable mitigation strategies to support ethical deployment.

8.4.3 Responsible governance framework

A well-built governance architecture is required to traverse the socio-technical complexity of quantum integration. The concept of human-in-the-loop decision systems, which is already prevalent in the sphere of AI, will have to be applied to the quantum environment to ensure that complete automation does not render human supervision secondary (Joshi, 2025). In such areas as agriculture where local knowledge and tacit expertise are still useful, quantum models will need to be used in hybrid systems that augment, rather than substitute, human judgment.

Holter Ten et al. (2023) stated that it is necessary to enshrine the principles of RI in quantum research consortia. Based on empirical research in the UK quantum network, authors recorded how scientists themselves frequently have neither the means nor the motivation to assess the social consequences of their own inventions. Authors said that to implement RI, structural changes are necessary: special ethics boards, interdisciplinary research, and long-term funding of ethics research, not just compliance checklists.

Table 8.4 Ethical risk matrix for quantum agri-tech.

Ethical risk	Description	Stakeholders affected	Source/cause	Potential mitigation strategies
Labor displacement	Replacement of informal or low-skill workers in agri-fresh supply chains due to quantum-enabled automation and predictive analytics.	Informal workers, local communities, smallholder farmers.	Automation bias, productivity-focused deployment (Balu et al., 2025; Monlezun et al., 2025).	Workforce reskilling programs; inclusive technology design; labor impact assessments.
Algorithmic opacity	Lack of explainability in quantum decision-making models, limiting user understanding and trust in agri-tech interventions.	Farmers, agricultural cooperatives, extension officers.	Complexity of quantum models, lack of XAI tools (Carbajal-Pina & Acur, 2025; Qin et al., 2025).	Explainable AI (XAI) layers; participatory model design; user training for interface transparency.
Data sovereignty and justice	External ownership or control of local farm-level data by multinational firms using quantum infrastructure.	Smallholder farmers, national governments.	Cloud-based quantum systems, foreign IP ownership (Umbrello et al., 2024; Zornetta, 2024).	Local data governance laws; farmer cooperatives for data control; open-source agri-tech frameworks.
Power asymmetry in innovation	Uneven global access to quantum agri-tech systems, entrenching North–South technological divides.	LMIC governments, NGOs, rural populations.	Concentration of quantum R&D in the Global North (Monlezun et al., 2025; Roberson et al., 2023).	South–South collaboration; open innovation platforms; capacity-building funding via CSR or international development programs.

(*continued on next page*)

Table 8.4 Ethical risk matrix for quantum agri-tech—cont'd

Ethical risk	Description	Stakeholders affected	Source/cause	Potential mitigation strategies
Lack of accountability mechanisms	Absence of clear liability structures for decisions made by quantum-enhanced agri systems, particularly in yield prediction or crop failure scenarios.	Farmers, insurers, policymakers.	Gaps in regulation; black-box models (Joshi, 2025; Holter Ten et al., 2023).	Establish Quantum Ethics Boards; formalize human-in-the-loop governance; enforce impact traceability protocols.
Digital literacy and exclusion	Inaccessibility of quantum agri-tech interfaces to users with low digital literacy, increasing dependency on external consultants or vendors.	Smallholder farmers, women farmers, marginalized rural groups.	Complex UIs; lack of localization; poor infrastructure (George & Wooden, 2023; Nita et al., 2023).	Community-based training; localized language interfaces; embedding extension officers in digital rollout programs.
Sustainability trade-offs	Environmental or energy costs of quantum infrastructure not considered in agri-tech deployment, undermining sustainability goals.	Governments, civil society, future generations.	High energy demand of quantum systems (Singh et al., 2024; Veit & Thatcher, 2023).	Green quantum infrastructure mandates; life cycle assessment for agri-tech; alignment with UN SDG targets (especially SDG 12 and 13).

Roberson et al. (2023) made the point more complex by questioning the hype helix that exists around quantum. Although enthusiasm attracts funding and attention, it also poses a danger of shutting down ethical consideration and making promises that exceed what can be delivered. This forms a circular reasoning whereby ethical discrepancy is justified as a sacrifice to innovation and puts off the institutionalization of safe-guards.

The geopolitical and ethical urgency of quantum was determined by Zornetta (2024) who was writing in an encryption policy perspective. The absence of international regulations and common ethical principles would leave the vast advantages of quantum capabilities to be enjoyed by powerful countries to an unfair degree, leaving the rest of the world in the position of the least-trusted-country where distrust, as opposed to cooperation, becomes the norm. It may take the form of data-sharing regimes that are too restrictive, biased models of algorithms, or predatory trade practices based on quantum-accelerated predictive tools in the agricultural setting.

Last, a normative anchoring is that of alignment with the UN sustainable development goals (SDGs). Nita et al. (2023) put forward the concept of quantum literacy as a prior condition to equal quantum access, especially SDG 4 (Quality Education). On the same note, Monlezun et al. (2025) outlined responsible quantum deployment in global health using the principles of SDG, namely equity, accessibility, and sustainability.

Ethical and socioeconomic consequences of quantum computing, especially in the agri-tech and related fields, require a multidimensional response. Unless there are systems to counteract the displacement of labor, guarantee data justice, and incorporate responsible governance, the advantages of quantum can do more harm than good in terms of global inequalities. Such undertakings in the future should not then only give attention to technical efficacy but also to ethical foresight, participatory design, and policy coherence, so that quantum innovations are made to lead to the sustainable and inclusive futures. Fig. 8.2 illustrates a responsible quantum agri-tech governance framework that integrates ethical foresight, inclusive design, and policy alignment to address the sociotechnical risks identified across the quantum innovation lifecycle.

8.5 Innovation ecosystems and startup landscapes

8.5.1 Emerging quantum agri-tech players

The intersection of QT and agriculture is gradually turning into a promising environment to innovate, especially in economies with limited resources. Although the direct quantum application startups in the agricultural sector are still few, the overall innovation ecosystem is developing in related fields, which include nanotechnology, AI, and biotechnology, all of which overlap with agri-tech (El-Ebiary et al., 2024; Jiang & Chen, 2021). This hidden cross-disciplinary basis is setting the stage of future quantum agri-tech entrepreneurship boom.

As an example, the application of carbon quantum dots (CQD) in agriculture, as described by Tavan et al. (2025), showed how nanomaterial innovation may increase the crop yield and stress tolerance. Even though these efforts are not necessarily indicative of the classic start-up, they set the stage to spin-offs and applied ventures at the nexus of materials science and agri-innovation.

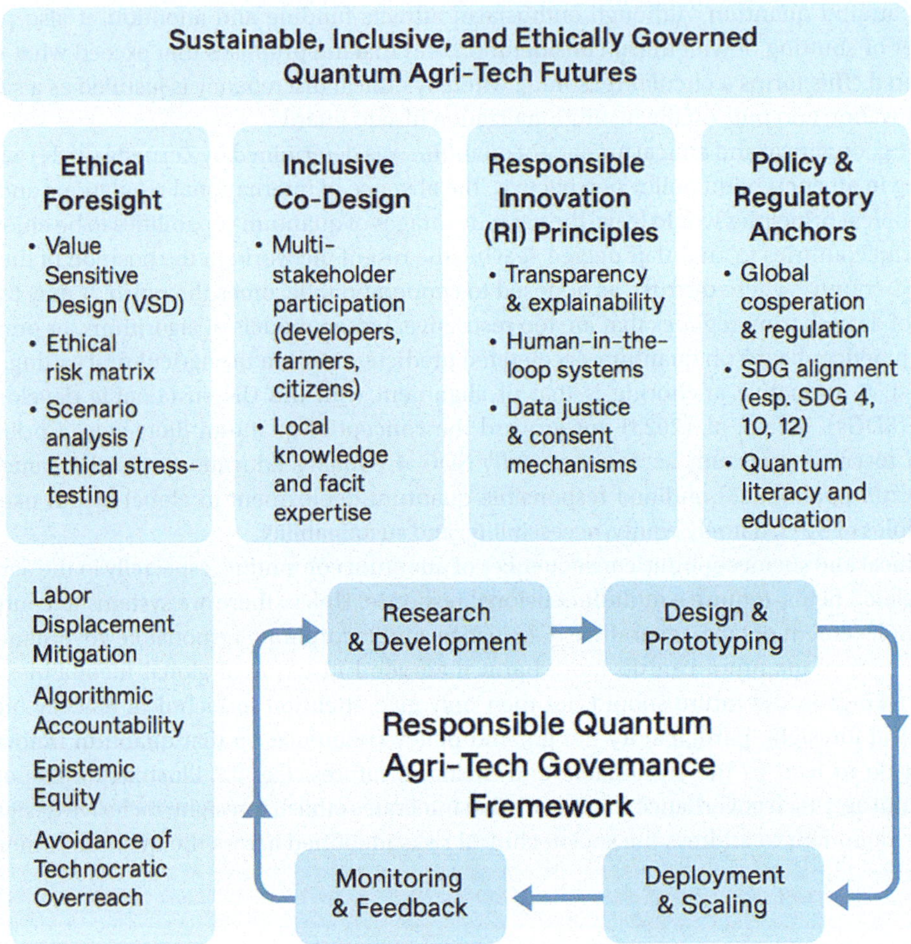

FIGURE 8.2 Responsible quantum agri-tech governance framework.

Moreover, the investigation of quantum ML (QML) in resource economics (Vaz, 2024) offers vital infrastructure to spatial optimization problems that are the core of precision farming, and this is an exciting venture into entrepreneurial experimentation.

Furthermore, the current case studies of biopharmaceutical industries indicate that such latecomer countries as South Korea have managed to take quantum leaps into latecomer industries by means of concentrated business model innovations and orchestration of ecosystems supported by the state (Jeong et al., 2023). The examples are informative to quantum agri-tech ecosystems in the Global South where entrepreneurship can grow rapidly due to a successful coordination of both the public and the private sector.

8.5.2 Enabling frugal quantum ecosystems

A more recent phenomenon, though, is the increased importance of frugal innovation, which is a characteristic of entrepreneurial endeavors in developing environments, as QT gradually finds its way into areas that have long been underserved by high-tech ecosystems. Kadir et al. (2024) noted that grassroots innovation models in related agricultural sectors had already been proven by rural micro-entrepreneurial networks, including Self Help Groups (SHG), made possible by microfinance. These financial inclusion processes are important facilitators of bottom-of-the-pyramid absorptive capacity.

Such austerity-based outlook is consistent with the findings of Adigun et al. (2024) that libraries and other educational establishments in the Fifth Industrial Revolution need to reposition themselves to promote community connection and digital inclusion. With QT entering the mainstream discourse, such institutions can become micro-innovation centers, providing the local stakeholders with quantum literacy and exploration capabilities.

The case of distributed teacher education in modern physics and quantum mechanics (Michelini et al., 2013) in Italy shows how personalized e-learning environments can make the access to advanced technological fields democratic. The blueprint of such frugal, inclusive innovation models, which can be replicated across geographies, particularly in low-infrastructure areas, is provided by this decentralized capacity-building model, which relies on uPortal and collaborative experimentation.

The concept of the grassroots quantum ecosystems also has a basis in Nthaga et al. (2025), who presented a quantum pavement condition Index through drone-GIS integration to assess the state of the road. This research highlights how low-cost, high-impact quantum-inspired tools can be deployed to monitor the public infrastructure, once again, indicating a fertile ground of startups that fit the agri-infrastructure efficiency model. Table 8.5 synthesizes open innovation initiatives focused on bottom-of-the-pyramid contexts, highlighting their alignment with quantum and digital technologies as enablers of inclusive development and grassroots entrepreneurship.

8.5.3 Ecosystem design proposal

In order to establish an inclusive and viable quantum agri-tech innovation ecosystem, it is important to adopt systemic orchestratable stakeholder approach. Based on the concept of multilevel sociotechnical innovation ecosystems developed by Jiang and Chen (2021), the section will suggest a design that will have three nested layers:

- **Core Innovation Actors:** The technological core is composed of startups, research universities, and quantum hardware suppliers. They need to be facilitated by the open innovation platforms and IP-sharing systems to collaborate in a symbiotic manner (Svirina et al., 2016).
- **Supportive Intermediaries:** Connective tissue is in the form of accelerators, NGOs, and inclusive financial institutions. Systemic approaches, such as those provided by participatory

Table 8.5 Open innovation for the bottom billion.

Initiative/Model	Region / context	Innovation focus	Actors involved	Quantum/Tech linkage	Reference
Self-Help Group -Driven Agri Entrepreneurship	North-East India (Tripura)	Microcredit-based grassroots agri-enterprise development	SHGs, Rural Banks (e.g., Tripura Grameen Bank), farmers	Potential to integrate precision agri-tech at micro scale	Kadir et al. (2024)
Teacher-Centric Quantum Literacy Platforms	Italy (National)	Open-access e-learning for physics teachers on quantum content	Universities, teacher networks	Quantum learning modules via uPortal platform	Michelini et al. (2013)
Metaverse-Enabled Livestock Monitoring	Turkey	Frugal livestock wearables selected through participatory fuzzy logic in virtual environments	Farmers, designers, MCDM experts	Neuro-quantum fuzzy MCDM in Metaverse	Ecer et al., (2024)
Systemic Communication Interventions	Jordan (Humanitarian Clinics)	Mental health and well-being through quantum complexity-informed frameworks	NGOs, clinics, refugees, researchers	Quantum systems lens for resilient human-tech interfaces	Parrish-Sprowl et al. (2020)
Open Innovation in Social Enterprises	Emerging Markets (Global South)	Collaborative grassroots innovation models in emerging information economies	Social enterprises, NGOs, government	Open vs. closed innovation pathways with tech enablement	Svirina et al. (2016)
IR4.0 Technologies for Equity	Global South/Libraries	Equitable digital access, education, and inclusion via AI, IoT, and quantum convergence	Libraries, educational institutions, policy networks	Quantum computing framed within 5IR inclusion models	Adigun et al. (2024)
Inclusive Fintech Ecosystems	Indonesia/India/Global	Bottom-up financial innovation and tech integration in grassroots platforms (e.g., GOJEK)	MSMEs, startups, tech incubators, governments	Quantum and AI-driven fintech innovation pathways	Priya et al. (2019)

health innovation ecosystem in Jordan (Parrish-Sprowl et al., 2020), can create transformative results despite the limited resources even with the use of community-based platforms.

- **Governance and Policy Nodes:** Based on the South Korean biopharma example (Jeong et al., 2023), this level includes ministries of science and technology, national innovation councils, and regional planning authorities. This is to allow regulatory sandboxing and policy co-design to allow experimentation and control risk.

Such multi-actor setups should be placed in a RI framework where there is ethical governance and inclusive access as pointed out by Vaz (2024) and Mosteanu and Faccia (2021). Furthermore, in-built capacity-building activities similar to the IDIFO Master model (Michelini et al., 2013) would generate a long-term knowledge creation and sharing system on the ground.

In conclusion, it is possible to say that quantum agri-tech is still a young and developing special field, but the combined power of frugal innovations, friendly intermediaries, and active governance may create the foundations of an inclusive and robust startup ecosystem. Table 8.6 outlines a selection of emerging startups and applied initiatives that demonstrate the diversity of quantum agri-tech use-cases, spanning material science, spatial analytics, and quantum-informed livestock management. Such results do not only indicate the possibility of transformational applications in the Global South but also demand the concerted policy and investment activities, which can connect local entrepreneurship with the global quantum innovation wave. Fig. 8.3 presents an ecosystem stakeholder map that visualizes the layered structure of innovation actors, intermediaries, and governance institutions necessary to support a resilient quantum agri-tech startup environment.

8.6 Challenges, research agenda, and future pathways

Quantum agriculture and the ecosystems that make it possible are coming to the fore at the nexus of computational innovation, sustainability science, and global development. Nevertheless, the road to inclusive quantum change is multidimensional and disjointed.

8.6.1 The threefold barrier: Infrastructure, institutions, and imagination

The single most immediate limitation in LMIC is technological deficits, in particular bandwidth, cloud access, and hardware provisioning of quantum simulation and sensing (Hussain et al., 2024; Sarangi et al., 2023). Such gaps make most of the quantum innovation inaccessible outside of urban academic-industry clusters (Sivan & Priya, 2025).

Institutionally, governance systems are usually divided between ministries and agencies, and not as coherent as the national roadmap of South Korea (Shim & Suh, 2025a, 2025b). The lack of cross-sectoral forums hinders matching the agricultural demands and quantum R&D investments.

Quantum literacy is epistemically very exclusionary. Curriculum is abstract and physics-intensive (Seskir et al., 2024) and the majority of farmers, local entrepreneurs, and rural institutions have no exposure to even low-level concepts, which threatens to form a new quantum

Table 8.6 Emerging startups and use-cases in quantum agri-tech.

Startup/Initiative	Region	Use-case/Application	Technology focus	Stage
Carbon Quantum Dots – Agri Applications	Global/Academic-Industry	Plant growth enhancement, stress resistance, pollutant detection (Tavan et al., 2025)	Quantum Nanomaterials	Early Applied Research
Quantum Pavement Condition Index	Kenya	Road condition and agri-logistics infrastructure optimization (Nthaga et al., 2025)	Quantum-inspired GIS	Pilot Implementation
Communication for Whole Health	Jordan	Systemic mental health platform for agricultural/frontline workers (Parrish-Sprowl et al., 2020)	Quantum-Informed Complexity Science	Deployed Intervention
Quantum Spatial ML for Urban Agriculture	Europe	Urban agri-resource mapping and regional planning (Vaz, 2024)	Quantum Machine Learning	Proof-of-Concept
Wearable Smart Collars for Livestock	Turkey	Livestock monitoring with AI and quantum fuzzy logic for stress/disease detection (Ecer et al., 2024)	Quantum Spherical Fuzzy Decision Systems	Prototype and Field Testing

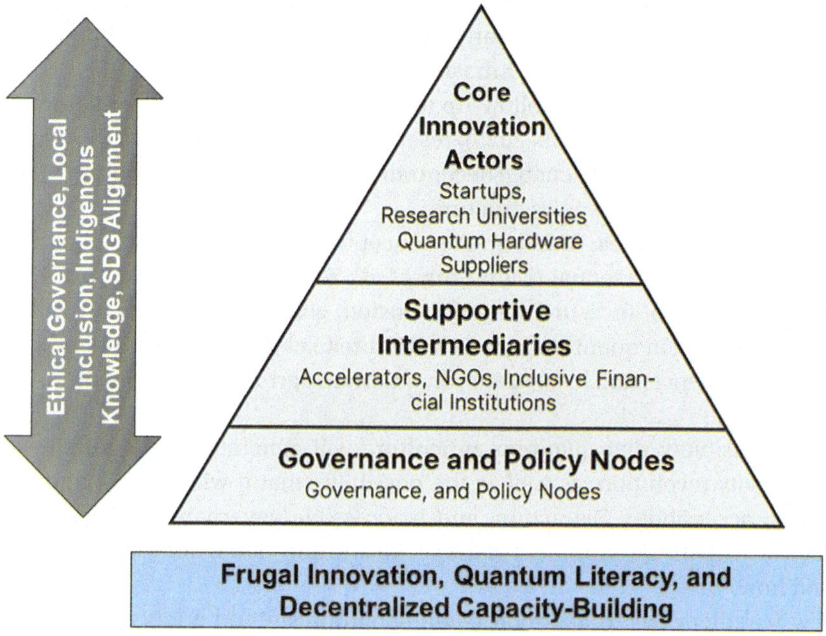

FIGURE 8.3 Ecosystem stakeholder map.

Table 8.7 Threefold barrier to inclusive quantum transformation.

Barrier type	Description	Illustrative challenges	Key references
Technological	Infrastructure deficits in bandwidth, cloud access, and quantum hardware	Limited access in rural LMIC areas	(Hussain et al., 2024; Sarangi et al., 2023)
Institutional	Fragmented governance frameworks and weak inter-ministerial coordination	Disconnect between agri-needs and R&D	(Shim & Suh 2025a, 2025b; Sivan & Priya, 2025)
Epistemic/ Imaginative	Lack of quantum literacy, exclusionary curricula, absence of public discourse on quantum futures	Technological alienation in rural societies	(Nita et al. 2023; Seskir et al., 2024)

divide as well as the digital divide. Table 8.7 summarizes the threefold barrier to inclusive quantum transformation (technological, institutional, and epistemic) highlighting the specific constraints and systemic challenges faced by LMICs in quantum agri-tech adoption.

8.6.2 A frugal and just quantum research agenda

Instead of pursuing quantum supremacy, the future will be quantum sufficiency, or tools that are context-sensitive and that have equitable access:

- **Frugal-Quantum Hybrids:** Low-resource settings require further design and field-testing of hybrid quantum-classical tools. Research should be able to come up with modular tools that can operate in the current agri-IoT infrastructures.
- **Localized Policy Models:** As a follow-up to the idea of the so-called quantum agri-tech sandboxes (De Bartolomeo et al., 2025), researchers are encouraged to develop prototype models of governance, which enable decentralized innovation and experimentation without the need to build a full-scale infrastructure.
- **Ethics and Impact Studies:** Although the conceptual argumentation of labor displacement or data sovereignty is essential (Monlezun et al., 2025; Umbrello et al., 2024), more longitudinal aspects of measurement of inclusion, going beyond presenting theorized barriers to inclusion in quantum agri-tech (Umbrello et al., 2024), are urgently needed that track equity, resilience, and local agency in quantum agri-tech deployments.

There is a possibility that quantum agriculture will generate a new climate-smart, data-driven productivity revolution, as well as the possibility that it will contribute the greatest inequity through accessibility, extraction, and inaccessible governance (Roberson et al., 2023; Zornetta, 2024). With the development of the sector, the question is not whether quantum works, but who and how.

This new paradigm has to abandon techno-solutionism and adopt quantum solidarity, a pledge to participatory, ethical, and place-based innovation. The level of success will not be gauged on the quantum advantage but how it has been merged with resilience, accessibility, and justice.

8.6.3 Research gaps and future research questions

Although the quantum innovation is picking up momentum, there are still gaps in theory, empirical practice, methods, and translation, especially when looking at the application of quantum in agriculture and sustainability in LMICs.

Most frameworks are constructed at the theoretical level around high-resource, hardware-intensive quantum systems, with little concern to the so-called frugal quantum paradigm that would balance resource constraints with technological adequacy. This vacuity indicates that there are no models that theorize quantum sufficiency, where the ability of the contextual appropriateness, and modular quantum justice, which values equitable access and adaptability over predominance.

The narratives of governance are usually centralized and technocratic and leave out other epistemologies or governance systems based on indigenous logics, local autonomy, or participatory pluralism (Holter Ten et al., 2023; Umbrello et al., 2024). The majority of policy models reflect top-down models of the Global North, and they do not provide much advice to LMICs with different institutional realities or decentralized innovation requirements (Seskir et al., 2024).

Empirically, the literature is biased and unbalanced with high-income environments, and little elaborated case studies of quantum integration in LMIC agriculture. Among the most promising frameworks, such as frugal quantum sensing or blockchain-based IoT systems

(Sezer et al., 2023), the application in the real world of rural or underserved communities is still occasional, and hardly ever longitudinally assessed. In addition, very little research presents quantifiable measures of inclusion, equity, or distributive justice in quantum innovation outcomes (Monlezun et al., 2025; Zornetta, 2024).

Methodologically, such research methods as the co-design based on simulation (e.g., entanglement games, federated QML) are conceptually abundant, yet operationally disjunctive. The tools have not been used consistently in participatory governance or testing policy situations. At the same time, although much has been said about RI, in practice, the integration of the RI principles into startup ecosystems, ethics boards, or national innovation councils is still rather superficial (Carbajal-Pina & Acur, 2025; Holter Ten et al., 2023).

Finally, the route between quantum laboratory advances and community tools is not well known in translational research. The localization of QT, be it in the language, pedagogy, interface design, or co-ownership models, has not been systematically studied, especially with regard to digitally marginalized, non-English-speaking groups (George & Wooden, 2023; Nita et al., 2023). That presents the danger of a techno-colonial path, in which even well-meaning-sounding quantum instruments create real-world exclusions on a systemic level.

In order to fill these gaps and influence inclusion into meaningful scholarship, the following research questions are suggested:

A. **Theoretical Gaps**
   a. What are the best theoretical frameworks that can describe the socio-technical dynamics of frugal quantum ecosystem in agriculture?
   b. How would quantum governance regimes support epistemic diversity and locally based value systems?
B. **Empirical Gaps**
   a. What are the quantitative and qualitative measures that can help to measure the distributive effect of quantum agri-tech in LMICs?
   b. Which are the success factors and failure points of the real-world pilots of quantum-enabled agriculture in the Global South?
C. **Methodological Gaps**
   a. What would it take to make the design of agri-policies more universal through simulation-based co-design (e.g., quantum game theory, digital twins)?
   b. How might RI frameworks be applied into the quantum startup ecosystems, especially to frugal innovation settings?
D. **Translational Gaps**
   a. Which are the most efficient localizing quantum innovation in agri-tech mechanisms: language, pedagogy, interface, or co-ownership?

By addressing these questions directly, research in the future will be able to go beyond the technical curiosity to influence a context sensitive and just quantum agri-tech agenda. Closing these divisions will be critical to make quantum innovation a positive rather than a negative force in terms of equity, sovereignty, and resilience in food systems and rural livelihoods. The second stage of investigation should make participatory ethics, social-political consciousness, and interdisciplinary innovation its central pillars and change quantum into an architecture of solidarity, rather than a field of exclusion.

8.7 Conclusion and Future Scope

This chapter has already formed an equity-based approach to the role of quantum and quantum-inspired technologies in low- and middle-income country agricultural systems. It asserts that the agri-tech quantum power may be revolutionary in frugality, situational versatility, and inclusive system design in lieu of accounts about quantum supremacy and obtrusive experimentation. The chapter presents quantum agriculture as a socio-technical transition, as opposed to a computational advance, by incorporating the wisdom of QT, frugal innovation, RI, and the socio-technical systems theory.

The discussion shows that inefficient resource allocation, climate vulnerability, fragile supply chains, and asymmetries in data governance are some of the current issues that can be resolved using modular and affordable quantum and quantum-inspired tools. Meanwhile, it also highlights that quantum agri-tech cannot be considered a value-neutral technology, without proper governance, institutional capacity, and ethical foresight, these technologies can actually strengthen the current inequalities. The suggested frugal quantum agri-tech structure and policy alternatives consequently reposition quantum innovation along the distributive justice, resilience, and sustainability policy.

In the future, some research and policy priorities become apparent. First, empirical and longitudinal trials should be done to measure actual implementations of frugal quantum agri-tech in LMIC settings, both in terms of technical performance and social performance. Second, equity, inclusion, and data sovereignty should be operationalized in future work to determine distributive effects. Third, to achieve responsible translation of instantiations of localized governance, such as quantum agri-tech sandboxes and community-owned data platforms, experimentation with those can help.

Finally, computational advantage is not the only factor that should be used to measure the success of quantum agri-tech, but rather its role in promoting equitable, resilient, and inclusive agricultural changes. This chapter also puts quantum innovation in a perspective of enabling architecture of sustainable food systems and rural livelihoods instead of the source of new technological exclusion by putting the discussion on quantum supremacy into quantum sufficiency and quantum solidarity.

References

Adigun, G. O., Ajani, Y. A., & Enakrire, R. T. (2024). The intelligent libraries: Innovation for a sustainable knowledge system in the fifth (5th) industrial revolution. *Libri, 74*(3), 211–223. De Gruyter Saur https://doi.org/10.1515/libri-2023-0111 .

Amato, L., Grimm, M., & Müller, M. (2025). Enhanced quantum sensitivity and coherence of symmetric magnetic clusters. *Physical Review B, 111*(13). American Physical Society https://doi.org/10.1103/PhysRevB.111.134406 .

Attri, I., Awasthi, L. K., & Sharma, T. P. (2025). EQID: Entangled quantum image descriptor an approach for early plant disease detection. *Crop Protection, 188*. Elsevier Ltd https://doi.org/10.1016/j.cropro.2024.107005 .

Balu, S., Raghabu, S. M. A., Mahalingam, P., Krishnasamy, J., Kadambarajan, J. P., & Pattar, D. S. (2025). Advanced optimized quantum learning for robust water quality assessment with an IoT sensor data. *Ingenierie des Systemes d'Information, 30*(2), 359–366. International Information and Engineering Technology Association https://doi.org/10.18280/isi.300207 .

Baseri, Y., Chouhan, V., Ghorbani, A., & Chow, A. (2025). Evaluation framework for quantum security risk assessment: A comprehensive strategy for quantum-safe transition. *Computers and Security, 150*. Elsevier Ltd https://doi.org/10.1016/j.cose.2024.104272 .

Bongurala, A. R., Save, D., & Virmani, A. (2025). Progressive role of artificial intelligence in treatment decision-making in the field of medical oncology. *Frontiers in Medicine, 12*. Frontiers Media SA https://doi.org/10.3389/fmed.2025.1533910 .

Carbajal-Pina, C., & Acur, N. (2025). From principles to practice: Responsible implementation of emerging technologies through innovation dilemmas. *IEEE Transactions on Engineering Management, 72*, 18–28. Institute of Electrical and Electronics Engineers Inc https://doi.org/10.1109/TEM.2024.3479413 .

Chan, M. L., Bell, T. J., Pettersson, L. A., Chen, S. X., Yard, P., Sørensen, A. S., & Paesani, S. (2025). Tailoring fusion-based photonic quantum computing schemes to quantum emitters. *PRX Quantum, 6*(2). American Physical Society https://doi.org/10.1103/PRXQuantum.6.020304 .

Cohen, W. M., & Levinthal, D. A. (1990). Absorptive capacity: A new perspective on learning and innovation. *Administrative Science Quarterly, 35*(1), 128–152.

De Bartolomeo, M., Cavalleri, G., & De Nicola, A. (2025). Quantum computing in the railway sector; [Quantum computing nel settore ferroviario]. *Ingegneria Ferroviaria, 80*(2), 111–148. CIFI Collegio Ingegneri Ferroviari Italiani https://doi.org/10.57597/IF.02.2025.ART.2 .

de Souza, K. X. S., Bolfe, É. L., de Andrade Leite, M. A., Bambini, M. D., Visoli, M. C., Júnior, A. L., da Silva, F. R., Esquerdo, J. C. D. M., & de Carvalho Teixeira Yassitepe, J. E. (2024). Quantum computing: Current and potential applications in digital agriculture; [Computação quântica: Aplicações atuais e potenciais na agricultura digital]. *Pesquisa Agropecuaria Brasileira, 59*. Embrapa https://doi.org/10.1590/S1678-3921.PAB2024.V59.03753 .

Ecer, F., Ögel, İ. Y., Dinçer, H., & Yüksel, S. (2024). Assessment of Metaverse wearable technologies for smart livestock farming through a neuro quantum spherical fuzzy decision-making model. In *Expert Systems with Applications: 255* (p. 124722). Elsevier. https://doi.org/10.1016/j.eswa.2024.124722.

El-Ebiary, Y. A. B., Mjlae, S. A., Ahmad, H., Rababa'h, S. Y., Rababah, M. A., & Arabiat, O. G. (2024). IR4.0 and internet of things: Future directions towards enhanced connectivity, automation, and sustainable innovation. *Telkomnika (Telecommunication Computing Electronics and Control), 22*(6), 1469–1477. Universitas Ahmad Dahlan https://doi.org/10.12928/TELKOMNIKA.v22i6.25487 .

Elmansy, A. M. (2024). Quantum computing and islamic banking: Legal framework for transforming Egypt's financial sector. *Manchester Journal of Transnational Islamic Law and Practice, 20*(3), 369–374 Electronicpublications.org Ltd.

George, B., & Wooden, O. (2023). Managing the strategic transformation of higher education through artificial intelligence. *Administrative Sciences, 13*(9). Multidisciplinary Digital Publishing Institute (MDPI) https://doi.org/10.3390/admsci13090196 .

Golter, D. A., Clark, G., El Dandachi, T., Krastanov, S., Leenheer, A. J., Wan, N. H., Raniwala, H., Zimmermann, M., Dong, M., Chen, K. C., Li, L., Eichenfield, M., Gilbert, G., & Englund, D. (2023). Selective and scalable control of spin quantum memories in a photonic circuit. *Nano Letters, 23*(17), 7852–7858. American Chemical Society https://doi.org/10.1021/acs.nanolett.3c01511 .

Holter Ten, C., Inglesant, P., & Jirotka, M. (2023). Reading the road: Challenges and opportunities on the path to responsible innovation in quantum computing. *Technology Analysis and Strategic Management, 35*(7), 844–856. Routledge https://doi.org/10.1080/09537325.2021.1988070 .

Hussain, F., Hussain, Z., Khan, M. I., & Imran, A. (2024). The digital rise and its economic implications for China through the digital silk road under the belt and road initiative. *Asian Journal of Comparative Politics, 9*(2), 238–253. SAGE Publications Ltd https://doi.org/10.1177/20578911231174731 .

Jeong, H., Song, C., Kim, S., & Shin, K. (2023). Quantum jump in biopharmaceutical industry: A case of Korea's catching up with Europe and US. *European Planning Studies, 31*(5), 885–904. Routledge https://doi.org/10.1080/09654313.2022.2132781 .

Jiang, S.-Y., & Chen, S.-L. (2021). Exploring landscapes of quantum technology with Patent Network Analysis. *Technology Analysis and Strategic Management, 33*(11), 1317–1331. Routledge https://doi.org/10.1080/09537325.2021.1928056 .

Joshi, H. (2025). Emerging technologies driving zero trust maturity across industries. *IEEE Open Journal of the Computer Society, 6*, 25–36. Institute of Electrical and Electronics Engineers Inc https://doi.org/10.1109/OJCS.2024.3505056 .

Kadir, A., Majumder, S. H., & Mondal, B. (2024). Analyzing self-help group-bank linkages through volume of savings and credit: Insights from the north eastern states of India. *Agricultural Research Journal, 61*(1), 75–85. Punjab Agricultural University https://doi.org/10.5958/2395-146X.2024.00011.3 .

Khatri, S. (2022). On the design and analysis of near-term quantum network protocols using Markov decision processes. *AVS Quantum Science, 4*(3). American Institute of Physics Inc https://doi.org/10.1116/5.0084653 .

Kumar, A., & Agrawal, S. (2023). Challenges and opportunities for agri-fresh food supply chain management in India. *Computers and Electronics in Agriculture, 212*, 108161.

Kumar, A., & Agrawal, S. (2024). A quality-based sustainable supply chain architecture for perishable products using image processing in the era of industry 4.0. *Journal of Cleaner Production, 450*. Elsevier Ltd https://doi.org/10.1016/j.jclepro.2024.141910 .

Kumar, A., & Agrawal, S. (2025). Enhancing quality-based classification of perishable products: A convolutional neural network approach with statistical hyperparameter optimization. *Multimedia Tools and Applications, 84*(15), 15429–15452. Springer. https://doi.org/10.1007/s11042-024-19700-z.

Kurmapu, M. K., Tiunova, V. V., Tiunov, E. S., Ringbauer, M., Maier, C., Blatt, R., Monz, T., Fedorov, A. K., & Lvovsky, A. I. (2023). Reconstructing complex states of a 20 -qubit quantum simulator. *PRX Quantum, 4*(4). American Physical Society https://doi.org/10.1103/PRXQuantum.4.040345 .

Li, Y., Cai, W.-Q., Ren, J.-G., Wang, C.-Z., Yang, M., Zhang, L., … Pan, J.-W. (2025). Microsatellite-based real-time quantum key distribution. *Nature, 640*(8057), 47–54. https://doi.org/10.1038/s41586-025-08739-z.

Liu, M., Liao, M., Zhang, R., Yuan, X., Zhu, Z., & Wu, Z. (2025). Quantum computing as a catalyst for microgrid management: Enhancing decentralized energy systems through innovative computational techniques. *Sustainability, 17*(8). Multidisciplinary Digital Publishing Institute (MDPI) https://doi.org/10.3390/su17083662 .

Lock, E. H., Lee, J., Choi, D. S., Bedford, R. G., Karna, S. P., & Roy, A. K. (2023). Materials innovations for quantum technology acceleration: A perspective. *Advanced Materials, 35*(27). John Wiley and Sons Inc https://doi.org/10.1002/adma.202201064 .

Lombe, P., Carvalho, E., & Rosa-Santos, P. (2024). Drought dynamics in sub-Saharan Africa: Impacts and adaptation strategies. *Sustainability, 16*(22), 9902. https://doi.org/10.3390/su16229902.

Lopetegui, C. E., Gessner, M., Fadel, M., Treps, N., & Walschaers, M. (2022). Homodyne detection of non-gaussian quantum steering. *PRX Quantum, 3*(3). American Physical Society https://doi.org/10.1103/PRXQuantum.3.030347 .

Michelini, M., Santi, L., & Stefanel, A. (2013). E-learning in teacher professional development in innovation and formative guidance on modern physics: The case of IDIFO master's programs. *Journal of E-Learning and Knowledge Society, 9*(2), 47–79. https://doi.org/10.20368/1971-8829/833.

Michigan State University. (2025). Ethiopia: economy. https://globaledge.msu.edu/countries/ethiopia/economy.

Minbaleev, A., Zenin, S., & Evsikov, K. (2024). Prospects for legal regulation of quantum communication. *BRICS Law Journal, 11*(2), 11–54. University of Tyumen https://doi.org/10.21684/2412-2343-2024-11-2-11-54 .

Mohanty, T., Srivastava, V., Debnath, S. K., Roy, D., Sakurai, K., & Mukhopadhyay, S. (2025). An experimentally validated feasible quantum protocol for identity-based signature. *Sadhana—Academy Proceedings in Engineering Sciences, 50*(1). Springer https://doi.org/10.1007/s12046-025-02676-3 .

Monlezun, D. J., Omutoko, L., Oduor, P., Kokonya, D., Rayel, J., Sotomayor, C., Girault, M. I., De los Ríos Uriarte, M. E., Sinyavskiy, O., Aksamit, T., Dugani, S. B., Garcia, A., & Gallagher, C. (2025). Digitalization of health care in low- and middle-income countries; [Numérisation des soins de santé dans les pays à revenu faible et intermédiaire]; [Digitalización de la atención sanitaria en países de ingresos bajos y medios]. *Bulletin of the World Health Organization, 103*(2), 148–154. World Health Organization https://doi.org/10.2471/BLT.24.291643 .

Mosteanu, N. R., & Faccia, A. (2021). Fintech frontiers in quantum computing, fractals, and blockchain distributed ledger: Paradigm shifts and open innovation. *Journal of Open Innovation: Technology, Market, and Complexity, 7*(1), 1–19. Multidisciplinary Digital Publishing Institute (MDPI) https://doi.org/10.3390/joitmc7010019 .

Muzammul, M., Assam, M., & Qahmash, A. (2025). Quantum-inspired multi-scale object detection in UAV imagery: Advancing ultra-small object accuracy and efficiency for real-time applications. *IEEE Access, 13*, 2173–2186. Institute of Electrical and Electronics Engineers Inc https://doi.org/10.1109/ACCESS.2024.3523405 .

Nikmehr, N., Zhang, P., & Bragin, M. A. (2022). Quantum distributed unit commitment: An application in microgrids. *IEEE Transactions on Power Systems, 37*(5), 3592–3603. Institute of Electrical and Electronics Engineers Inc https://doi.org/10.1109/TPWRS.2022.3141794 .

Nita, L., Mazzoli Smith, L., Chancellor, N., & Cramman, H. (2023). The challenge and opportunities of quantum literacy for future education and transdisciplinary problem-solving. *Research in Science and Technological Education, 41*(2), 564–580. Routledge https://doi.org/10.1080/02635143.2021.1920905 .

Nthaga, T. M. E., Nyomboi, T., & Mwaniki, M. (2025). An integrated remote sensing and GIS road condition assessment framework applying geospatial techniques to improve pavement condition analysis. *Engineering, Technology and Applied Science Research, 15*(2), 21021–21028. Dr D. Pylarinos https://doi.org/10.48084/etasr.9785 .

Park, J., Kim, B., & Jeong, S. (2022). A review on a 4 K cryogenic refrigeration system for quantum computing. *Progress in Superconductivity and Cryogenics (PSAC), 24*(2), 1–6. Korea Institute of Applied Superconductivity and Cryogenics https://doi.org/10.9714/psac.2022.24.2.001 .

Parrish-Sprowl, S., Parrish-Sprowl, J., & Alajlouni, S. (2020). Innovations in addressing mental health needs in humanitarian settings: A complexity informed action research case study. *Frontiers in Communication, 5*. Frontiers Media S.A https://doi.org/10.3389/fcomm.2020.601792 .

Parvini, T. S., Römling, A.-L. E., Sharma, S., & Kusminskiy, S. V. (2025). Cavity-enhanced optical manipulation of antiferromagnetic magnon pairs. *Physical Review B, 111*(1). American Physical Society https://doi.org/10.1103/PhysRevB.111.014416 .

Pettit, R. M., Karasahin, A., Von Den Driesch, N., Jansen, M. M., Pawlis, A., & Waks, E. (2022). Correlations between cascaded photons from spatially localized biexcitons in ZnSe. *Nano Letters, 22*(23), 9457–9461. American Chemical Society https://doi.org/10.1021/acs.nanolett.2c03527 .

Pittaluga, M., Lo, Y. S., Brzosko, A., Woodward, R. I., Scalcon, D., Winnel, M. S., Roger, T., Dynes, J. F., Owen, K. A., Juárez, S., Rydlichowski, P., Vicinanza, D., Roberts, G., & Shields, A. J. (2025). Long-distance coherent quantum communications in deployed telecom networks. *Nature, 640*(8060), 911–917. Nature Research https://doi.org/10.1038/s41586-025-08801-w .

Priya, K., Ganesh, N., & Balaraman, P. (2019). Basics of business model, emerging fintech & case insights on Gojek business model. *International Journal of Engineering and Advanced Technology (IJEAT), 8*(5).

Qin, R., Zhou, C., & He, M. (2025). A comprehensive taxonomy of machine consciousness. *Information Fusion, 119*. Elsevier B.V https://doi.org/10.1016/j.inffus.2025.102994 .

Rao, B. D., & Jayaraman, R. (2023). A novel quantum identity authentication protocol without entanglement and preserving pre-shared key information. *Quantum Information Processing, 22*(2). Springer https://doi.org/10.1007/s11128-023-03832-6 .

Roberson, T., Raman, S., Leach, J., & Vilkins, S. (2023). Assessing the journey of technology hype in the field of quantum technology; [Bewertung der Entwicklung des Technologie-Hypes im Bereich der Quantentechnologie]. *Zeitschrift fur Technikfolgenabschatzung in Theorie und Praxis /Journal for Technology Assessment in Theory and Practice, 32*(3), 17–21. Oekom - Gesellschaft fuer Oekologische Kommunikation mbH https://doi.org/10.14512/tatup.32.3.17 .

Rocha, J. C. A., Pires, D. G., Motta, M. R. L., Litchinitser, N. M., Vianna, S. S., Fonseca, E. J. S., & Ferraz, J. (2022). Controlled conversion of transverse symmetries in a four-wave mixing process. *Journal of Optics (United Kingdom), 24*(4). IOP Publishing Ltd https://doi.org/10.1088/2040-8986/ac57b2 .

Sarangi, A. K., Pradhan, R. P., & Misra, M. (2023). ICT infrastructure and innovation: Determinants, potentials, and policy implications. *World Review of Science, Technology and Sustainable Development, 19*(4), 314–348. Inderscience Publishers https://doi.org/10.1504/WRSTSD.2023.133888 .

Seskir, Z. C., Goorney, S. R., & Chiofalo, M. L. (2024). Educating to the "Culture" of quantum technologies: A survey study on concepts for public awareness. *European Journal of STEM Education, 9*(1). Lectito https://doi.org/10.20897/ejsteme/14193 .

Shibayan, D. (2025). Nigeria has a food security problem as water for crops is harder to find. https://apnews.com/article/nigeria-climate-rivers-farming-agriculture-drought-0985473d47de9b752cd4b07f6df99c22

Sezer, B. B., Turkmen, H., & Nuriyev, U. (2023). PPFchain: A novel framework privacy-preserving blockchain-based federated learning method for sensor networks. In *Internet of Things: 22* (p. 100781). Elsevier. https://doi.org/10.1016/j.iot.2023.100781.

Shim, H., & Suh, M. (2025a). Comparative study on national HPC core strategies for AI adaptation-focusing on South Korea and the United States. *Edelweiss Applied Science and Technology, 9*(3), 689–696. Learning Gate https://doi.org/10.55214/25768484.v9i3.5283 .

Shim, H., & Suh, M. (2025b). Enhancing quantum science and technology in South Korea: A strategic investment framework. *International Journal of Innovative Research and Scientific Studies, 8*(2), 2543–2550. Innovative Research Publishing https://doi.org/10.53894/ijirss.v8i2.5733 .

Singh, M. P., Singh, J., Ravi, V., Gupta, P., Alahmadi, T. J., Singh, P., … Verma, M. (2024). Impact and implications of quantum computing on blockchain-based electronic health record systems. *The Open Bioinformatics Journal*: 17. Bentham Science Publishers https://doi.org/10.2174/0118750362316814240820051945.

Sivan, A., & Priya, K. (2025). Quantum computing and risk prediction accuracy: An analysis of IT companies' risk appetite. *International Journal of Business and Systems Research, 19*(2), 111–139. Inderscience Publishers https://doi.org/10.1504/IJBSR.2025.145483 .

Svirina, A., Zabbarova, A., & Oganisjana, K. (2016). Implementing open innovation concept in social business. *Journal of Open Innovation: Technology, Market, and Complexity, 2*(4). Multidisciplinary Digital Publishing Institute (MDPI) https://doi.org/10.1186/s40852-016-0046-8 .

Tatlock, C. W. (2006). Water stress in Sub-Saharan Africa. https://www.cfr.org/backgrounder/water-stress-sub-saharan-africa

Tavan, M., Yousefian, Z., Bakhtiar, Z., Rahmandoust, M., & Mirjalili, M. H. (2025). Carbon quantum dots: Multifunctional fluorescent nanomaterials for sustainable advances in biomedicine and agriculture. *Industrial Crops and Products, 231*. Elsevier B.V https://doi.org/10.1016/j.indcrop.2025.121207 .

Trading Economics. (2025a). Ethiopia—employment in industry (% of Total Employment). https://tradingeconomics.com/ethiopia/employment-in-industry-percent-of-total-employment-wb-data.html

Trading Economics. (2025b). Benin—Agriculture value added (% of GDP). https://tradingeconomics.com/benin/agriculture-value-added-percent-of-gdp-wb-data.html

Ulandssekretariatet. (2025). Labour Market Profile 2025: Benin. https://www.ulandssekretariatet.dk/wp-content/uploads/2025/01/Benin-LMP-2025-final1.pdf

Umbrello, S., Seskir, Z. C., & Vermaas, P. E. (2024). Communities of quantum technologies: Stakeholder identification, legitimation and interaction. *International Journal of Quantum Information, 22*(7). World Scientific https://doi.org/10.1142/S0219749924500126 .

Vaz, E. (2024). Quantum machine learning in spatial analysis: A paradigm shift in resource allocation and environmental modeling. *Letters in Spatial and Resource Sciences, 17*(1). Springer Science and Business Media Deutschland GmbH https://doi.org/10.1007/s12076-024-00374-y .

Veit, D. J., & Thatcher, J. B. (2023). Digitalization as a problem or solution? Charting the path for research on sustainable information systems. In *Journal of Business Economics: 93* (pp. 1231–1253). Springer. https://doi.org/10.1007/s11573-023-01143-x.

Vermeer, M. J. D., Heitzenrater, C., Parker, E., Moon, A., Lumpkin, D., & Awan, J. (2024). Evaluating cryptographic vulnerabilities created by quantum computing in industrial control systems. *Journal of Critical Infrastructure Policy, 5*(2), 88–110. John Wiley and Sons Inc https://doi.org/10.1002/jci3.12024 .

Yon, V., Galaup, B., Rohrbacher, C., Rivard, J., Morel, A., Leclerc, D., Godfrin, C., Li, R., Kubicek, S., Greve, K. D., Dupont Ferrier, E., Beilliard, Y., Melko, R. G., & Drouin, D. (2025). Experimental online quantum dots charge autotuning using neural networks. *Nano Letters, 25*(10), 3717–3725. American Chemical Society https://doi.org/10.1021/acs.nanolett.4c04889 .

Zornetta, A. (2024). Quantum-safe global encryption policy. *International Journal of Law and Information Technology, 32*(1). Oxford University Press https://doi.org/10.1093/ijlit/eaae020 .

9

Quantum agriculture and its ethical futures: A multidisciplinary management inquiry

Harshita Gupta[a], Rajib Roy[b] and Sonali Narbariya[c]

[a]INDEPENDENT RESEARCHER, GURUGRAM, HARYANA, INDIA [b]JAIPURIA INSTITUTE OF MANAGEMENT, NOIDA, UTTAR PRADESH, INDIA [c]T A PAI MANAGEMENT INSTITUTE, MANIPAL ACADEMY OF HIGHER EDUCATION, MANIPAL, KARNATAKA, INDIA

9.1 Introduction

Recent studies have identified that the adoption of frontier technologies in ecological fields of application, particularly in agriculture, should be done conceptually and contextually discerningly (Ahlborg et al., 2019; Anderies, 2014; Birhane, 2021; Redman & Miller, 2015). It is warned by scholars that an unsystematic perspective of human–environment–technology entanglements can result in unintentional ecological or social impacts of well-intended innovation (Mallinger & Baeza-Yates, 2024). In this changing scene, the agricultural sector is on the edge of a transformational revolution never seen before. The adoption of advanced technologies has ceased to be an option, as the world community urges to shift to sustainable innovation and sound food systems. Some of the new frontiers in technology include quantum computing and instrumentation and have the prospects of changing the manner in which we model biological systems, predict and optimize agricultural operations radically. In addition to improving the speed of simulations of crop responses, and creating more effective climate prediction models, quantum technologies are said to offer not just speed, but a new form of complexity handling that has proven challenging to classical systems. This convergence is not occurring alone. With the global agri-value chains developing under the burden of climate change, population pressures, and geopolitical changes, we are seeing the dawn of what some refer to as Agriculture 5.0, which is a paradigm of autonomous decision-making, hyper-connected devices and artificial intelligence (AI)-enhanced farming implements (Bergier et al., 2024; Saiz-Rubio & Rovira-Mas, 2020; Taha et al., 2025). The instrumentation of quantum is directly placed in the context of this development, which could allow a single-ecosystem-wide, fast, secure, and accurate data processing. The particular applications, including pest detection and soil profiling as well as irrigation optimization and market shock prediction, had been described by scholars like

Quantum Computing and Instrumentation for Smart Agriculture. DOI: https://doi.org/10.1016/B978-0-443-44247-6.00008-7

Polymeni et al. (2023). Qayyum et al. (2023) examined how technological inventions can be used to improve sustainable agricultural practices, especially through improved environmental systems modeling. Naveh (2022) also suggested that quantum technology is becoming a strategic asset of companies and countries in general, which indicates its increased economic and geopolitical importance. Scalability is another emerging concern related to this fast growth, as well as the interoperability and the role of public–private partnerships in the driving of responsible innovation. An example of this is the integration of quantum sensors in agriculture supply chains, not only in terms of speed and precision but reevaluating the feedback between soil, crop, climate, and consumer demand (Maraveas et al., 2025). In a world where nations are competing to gain quantum advantage, pilot projects in smart farming have already proven that the hybrid AI-quantum platform can produce insights based on the high-dimensional agricultural data and complex farming conditions (Pasupuleti, 2025). These features have special promise in the case of smallholder farmers and agri-startups, who tend to be at the periphery of the technological revolution. Quantum-driven insights can provide a level playing ground with sufficient support and a reasonable design, presenting customized interventions on-the-fly, whether to predict diseases or to keep track of nutrients.

In addition, one should not dismiss the geopolitical aspect of quantum development. With governments in North America, Europe, and Asia pouring investments into quantum R&D, agriculture stands as one of its areas of experimentation and strategic power (Raj, 2025). The conceptualization of quantum infrastructure as a national resource, such as space or nuclear technologies, is an indication of a tectonic plate change in the conceptualization of agricultural innovation, which is becoming less about incremental development and more about the ability to be sovereign. The redefinition has far-reaching consequences on food security in the world, intellectual property, and the morality of data sovereignty, especially on transnational agri-trade networks. The field of agriculture is more and more considered not only as a victim of the new technology but also a subject of experiments associated with the extreme frontiers of computation and instrumentation. The intensity, magnitude, and push of such endeavors underscore a new epoch of techno-centric optimism, an era that tends to lay emphasis on efficiency, predictability, and control over local demands, stakeholder heterogeneity, and societal viability in the long term. However, in the scramble toward technical superiority, one of the most important aspects which in many cases is not thoroughly researched is ethics-oriented management. This chapter starts by putting the focus on that same dimension, not as an appendix of the technological dialogue, but as its foundation stage. We hypothesize that there is no quantum leap in agri-tech without a corresponding change in organizational thinking, alignment of stakeholders, and value-based decision-making. Similarly to the way in which Feynman (1982) imagined quantum computers to be less scientifically realistic than classical systems, we propose that the ethical and organizational scaffolds surrounding such systems should also be redesigned in order to be more real to the aspects of human ecosystems, which are heterogeneous, uncertain, and intertwined. Thus, the chapter is written in an integrated prism, relying on the fields that are traditionally outside the scope of engineering: Marketing to understand stakeholder perception and bottlenecks of adoption; organizational behavior and

HR to discuss the transition of workforce, its resistance, and reskilling; and strategy to redefine missions and governance systems in the face of quantum transformation. Although quantum computing introduces entanglement and superposition to machines, this chapter is the one where quantum computing is bound to the human and ethical future, not to inquire of what can be done, but of what must be done and how. This way, the chapter will not only add to the literature in digital agriculture or quantum systems, but also to a larger epistemic shift, that of techno-centric solutions to techno-ethical synthesis.

Objectives of the Chapter

The initial goal of the chapter is to analyze the ethical issues that arise due to the introduction of quantum technologies into modern agri-business ecosystems. This chapter also describes how the high rate of technological acceleration can destabilize agency, trust, and inclusion in the agricultural value chains by locating quantum-enabled agriculture in the wider context of socio-technical and institutional systems. The second goal is to redefine ethics in quantum agriculture by using an interdisciplinary managerial view based on the perspectives of marketing, strategy, and organizational behavior. This strategy demonstrates the manner in which ethical considerations are integrated into not only technological design but also in strategic decision-making, stakeholder communication, as well as organizational preparedness. The third goal is to build and define the EPQ-AgriMgmt Framework as a model of integration of ethical preparedness of quantum agriculture. The framework conceptualizes the interdependency of strategic ethics, people ethics, and communication ethics and makes foresight, inclusivity, and trust to be important mechanisms of ethical alignment. This chapter can help to implement good governance and application of quantum technologies in agriculture by translating ethical principles into practical considerations. In this respect, this chapter is structured in such a way that it establishes conceptual background and moral enlightenment of quantum-based agricultural systems. After the introduction, the literature review synthesizes available literature on advanced agricultural technologies, ethical governance, and socio-technical change and finds gaps in the ways ethics is theorized at the current agri-technological environments. This chapter further redefines ethics in quantum agriculture by a variety of management lenses, including marketing ethics, formulating ethical strategies, and developing human agency in organizational behavior to illustrate how ethics tensions occur in the decision-making process, the communication process, and the practice. Through these views, a combined model is presented to show how the strategic, people, and communication ethics are integrated to form ethically prepared quantum agriculture ecosystems and managerial implications to the practitioners and policymakers. The last part of this chapter is the reflection on the main findings and defines the direction of the future research to solve the emerging ethical issues of quantum technologies in agri-business.

Organization of the chapter

The rest of the chapter is organized as: Section 9.2 enlightens literature review. Section discusses reframing ethics in quantum agriculture. Section 9.4 highlights integrated framework and managerial takeaways. And, finally Section 9.5 concludes the chapter with future scope.

9.2 Literature review

Since the time of the earliest gatherers, agriculture has experienced significant transformation. It has served as a driving force for scientific progress, a global economic powerhouse, and a channel for technological innovations, fundamentally altering human society and preventing famines by increasing food production. But goodness has its own evil. At present, agricultural areas used for growing crops and raising livestock cover 50%–60% of the planet's land surface and the numbers are ever increasing. The practices associated with farming, along with land use, contribute approximately 10%–14% to the annual greenhouse gas emissions (Pinto et al., 2010), which is very detrimental for this planet. Other than easily measurable environmental risks, technological progress in agriculture has resulted in the extensive use of harmful substances containing chemicals that interfere with hormones. These substances can hinder human cell development and damage the watersheds within ecosystems.

In the pretext of such arrangements where economic, social, as well as human aspects of life have tainted the legacy of farming and agri-business quite significantly, the arrival of the newest beast in the technological front, i.e., AI can work as boon or bane depending upon the whole sector's adaptability to it. As agribusiness had already shifted from promoting a traditional farming culture, henceforth, now supports advanced scientific and technological innovations such as AI, which farmers need to adopt in order to remain competitive with multinational farms in the international market (Amini, 2008; Penson et al., 2006).

Although the manufacturing and service sectors have occupied the spots for major contributors in the national GDPs across states, still agriculture as a sector holds substantial importance due to its comparatively lower economic productivity henceforth, deeper linkages with the well-being of the lower strata of the society. Considering this kind of elongated and to and fro equations between the society and the agri sector where multifaceted issues such as soil health, water resources, carbon emissions, greenhouse effects, climate change, food security, international trade, human migration, and local social values get twisted deserves greater public reviews when the policymakers began designing AI policies (Hoffman, 2011; Penson et al., 2006; Weart, 2011). Three major constituents in this policy making are public policymakers, various stakeholders of agri-business, and the science and technology communities who are invariably concerned whether the arrival of AI would force changes in their prevailing conducts. Where the public policymakers are more concerned about the greater sociopolitical issues, the agri-business partners are worried about whether the new policies would force them to significantly change their successfully established corporate behavior, and finally, the science and technology communities are apprehensive of whether the broader concern in the society about AI's socio-ethical aspect would impede the free development and adoption of quantum technology in agri-business. This chapter explores how agri-business stakeholders are responding to the integration of quantum technologies and their transformative implications for everyday decision-making. It introduces a triadic model where people ethics and communication ethics support strategic ethics, interconnected by the guiding values of foresight, trust, and inclusivity.

Former researchers used to justify individual ethics, taking into consideration a three-dimensional ethical model in which the overlap of people and ethics forms a complex model that is determined by three major traditions of ethical approaches that affect business decision-making activities (DesJardins, 2011). These are the ethical traditions namely the utilitarian tradition, deontological tradition, and virtue ethics tradition. The tradition of utilitarianism gives vehement emphasis on making business decisions that maximize the general good through the calculation of consequences of decisions to all the stakeholders. On the other hand, the deontological approach concentrates on the actions of morality and emphasizes applications of truth and openness in doing businesses without covering up on what occurred. Virtue ethics, in its turn, stresses the value of moral character and integrity in creating moral corporate cultures.

Contemporary research suggests that ethical issues in agri-business are particularly complex, affecting current market dynamics and future generations. Typically, the literature emphasizes that unethical behavior ultimately undermines company objectives, as evidenced in cases such as Takata Corporation or Cadbury Schweppes' business practices (Cadbury Schweppes Corporate and Social Responsibility Report, 2002). Now, adjoining the AI integration in agri-business, we first need to train different stakeholders how to participate in this gigantic tech-shift. As the stakeholders include hugely diverse partners such as agri-tech firm workers as well as poorly educated, oftentimes rurally based farmers, henceforward, it requires balancing multiple considerations: transparency in environmental sustainability concerns, product information, job protection of the farmers, AI's technology diffusion timeframe, as well as end consumers rights, while acknowledging that ethical behavior often must originate from within organizational culture rather than being externally imposed. The addition of wider ethical considerations into agri-business would lead to the development of more comprehensive approaches to the frameworks for "People Ethics" that extend beyond mere profit considerations to encompass broader societal values, environmental responsibilities, and finally, promoting sustainable AI integration in agri-tech development.

The investigation into communication ethics within the workplace has expanded considerably. Communication ethics, defined as the established moral values and corporate standards that govern communication, play a crucial role in influencing decision-making. The literature indicates that communication ethics, whether documented or not, are crucial to the effective operation and motivational dynamics of any work environment through specific cultural perspectives, operational structures, and strategic initiatives; henceforth, the whole ecosystem can reap negatively or positively (Kolk, 2016). Among many communication ethics frameworks often surveyed by erstwhile researchers, the Seven Cs Ethical Model of Communication represents a comprehensive framework for ethical communication practices, particularly in agricultural development contexts (McCann, 2012). Differing significantly from the traditional Seven Cs of effective communication (clarity, conciseness, concreteness, correctness, coherence, complete, and courteous), this model could be best suited for AI integration in agri-tech development, which encompasses seven key components: Collaboration, culture, community, conservation, capacity, care, and a seventh C component (varies by source). Considering the inherent uncertainties of a disruptive technology such as AI, this ethical model of communication emphasizes sustainable development and stakeholder engagement, with a structured approach to ethical decision-making in diverse communication scenarios, particularly in addressing AI concerns in agri-business.

In the realm of agri-business, strategic approaches extend beyond the scientific domain to include communication efforts, awareness initiatives, and community involvement. This chapter explores the diverse strategies used to boost the adoption of agri-tech, tackling issues like hesitancy toward AI adoption, differences in public awareness, and navigating various sociocultural contexts. Any strategic campaigns revolving around AI adoption in agri-business must tailor their messages through culturally competent, linguistically appropriate content, which includes important stakeholders' inclusivity by participatory decision-making processes. Here, especially the foresight of the leadership plays a significant role in determining the speed of AI adoption. Because agriculture has always been a localized community activity, henceforth, without the involvement of trusted opinion leaders who can easily align with the sociocultural fabric of the particular community while addressing issues regarding socialcultural-economic barriers, the full potential of agri-business's AI acceptance may fail to leverage global partners well-being (Vveinhardt et al., 2019).

9.3 Reframing ethics in quantum agriculture: Perspectives across management disciplines

9.3.1 Marketing ethics in the age of quantum agriculture ecosystems

It is now widely recognized in academic literature that the ethical implications of emerging technologies are inseparable from how they are perceived, communicated, and integrated into existing sociotechnical systems (Jakku et al., 2019; Liao & Sundar, 2022). In agricultural domains specifically, the success of smart and quantum innovations often hinges on how well they align with stakeholder dynamics and trust relationships (Bronson & Knezevic, 2016; Sonka, 2016; Wolfert et al., 2017). As quantum technologies permeate the agricultural domain, marketing practitioners are confronted with a complex ethical terrain. The very features that make quantum systems powerful, namely, granular personalization, predictive modeling, and high-dimensional data synthesis, also raise profound questions about transparency, inclusion, and consumer agency. Fig. 9.1 illustrates the constellation of stakeholders in the agricultural data ecosystem. From a marketing perspective, ethical responsibility does not end with the consumer, it must account for upstream stakeholders such as farmers and ATPs, whose data are often leveraged without adequate compensation or transparency (Kaur et al., 2022).

The main issue of marketing ethics in this new area is the management of agricultural data as a marketable item. Platforms powered by quantum computing can monitor the health of the soil, the behavior of crops, the weather, and even consumer preferences in a way never seen before. Although this makes supply–demand matching, it is likely to turn farmers and consumers into passive data points in a high-frequency feedback loop. Who controls this data? Who is the beneficiary of its insights? These are no longer the marginal questions, but they have to be in the center of marketing discourse (Kaur et al., 2022).

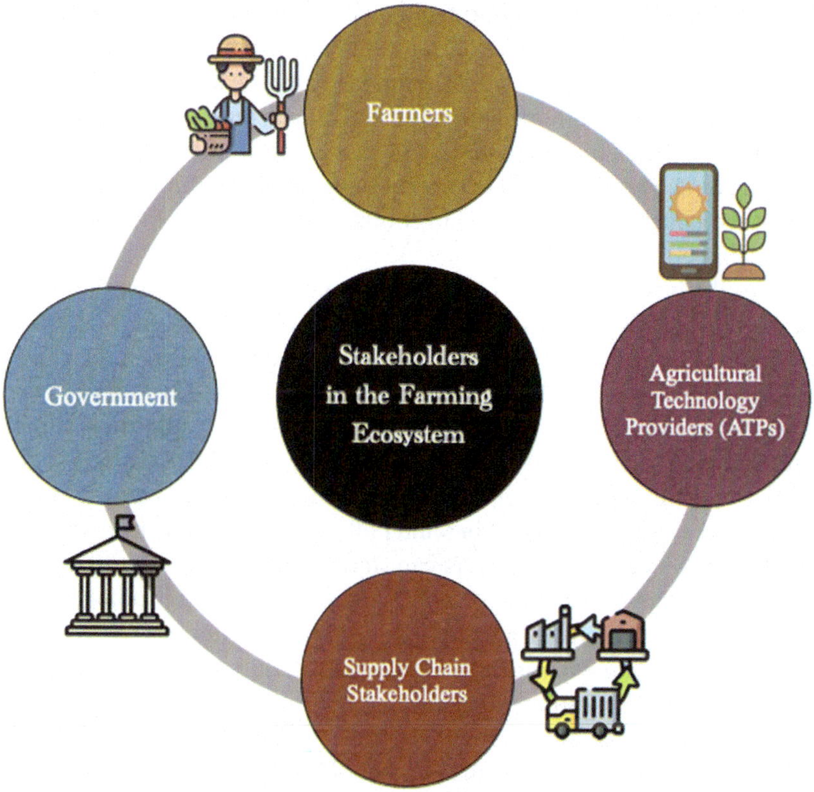

FIGURE 9.1 Stakeholders in the farming ecosystem[1]. *Source: Modified from Kaur et al. (2022).*

For example, smallholder farmers, who are often the intended beneficiaries of precision agriculture, may find themselves disadvantaged if quantum-powered agri-input recommendations are filtered through proprietary algorithms owned by large agritech firms. The ethical tension emerges not only from data asymmetry but also from how control over digital infrastructure shapes whose interests are prioritized in agri-tech ecosystems (Carbonell, 2016).

Also, the use of quantum-enhanced insights in agriculture introduces ethical risks tied to algorithmic bias, information asymmetry, and opacity. While tailored recommendations based on soil-genomic combinations can boost precision, they also risk deepening inequities if certain farmer groups are underrepresented in training data or lack the resources to interpret or challenge the outputs. Without adequate explainability and equitable access, such targeting can shift from personalization to manipulative influence, especially when decision-making power is concentrated among a few dominant agritech players (Sustainability Directory, 2025). These ethical concerns, particularly those related to data opacity, information asymmetry, and centralised control, also carry direct implications for marketing practices in agriculture. When quantum-powered insights are used to target farmers with personalized inputs or services, the lack of

[1] While originally presented in the context of data privacy and governance, these groups also represent traceable nodes in quantum agriculture systems, raising considerations of ethical marketing and data responsibility

transparency, and the imbalance in information can be leveraged to steer choices, potentially compromising farmer autonomy and reinforcing dependency on dominant agritech platforms.

The marketing ethics should evolve to take into account the intergenerational and environmental impact. Will quantum-influenced wisdom encourage long-term sustainability decisions or encourage profit-maximizing behavior in the short term in the name of accuracy? Ethical challenge is not about being able to sell better, but rather about communicating in a responsible manner, so that the value propositions do not conflict with the good of the ecosystem as well as community (Pasupuleti, 2025).

In addition to the immediate issues of data governance and consumer agency, a new ethical perspective emerges that is particularly important, which is posthuman ethics. Quantum agriculture dictates that posthuman ethics compels marketers to rethink the conventional anthropocentric models. Posthumanism proposes a new approach to farmers and consumers, as opposed to seeing them as a series of data points or market niches; rather, they should be seen as part of the messy interactions between humans, technology, and the environment. This school of thought focuses on the interrelationship of all stakeholders of the agricultural system including nonhuman organisms such as soil microbes and pollinators. Upon relying on posthuman ethics as developed by De Freitas, 2017, marketers can start to re-conceptualize strategy as an ethical practice that acknowledges the relational entanglement of people, technologies, and ecological factors and creates more sustainable and equitable agricultural futures.

Considering these aspects, we suggest that marketing ethics in quantum agriculture should be based on a set of operational principles going beyond traditional promotional reasoning. *Data stewardship* is the accountable management of high-dimensional farming and behavioral information, such as the consent procedure, information minimization measures, clarifiable analytics, and protection against secondary or extractive uses of data in farmer-focused applications. *Transparent value communication* involves communicating, in nonopaque forms that are context dependent and predictive of uncertainty and algorithm assumptions of quantum-enabled systems, instead of converting them into opaque performance claims. *Inclusive messaging* entails the precise planning of the communication strategies that will acknowledge heterogeneity among the farming communities in terms of language, literacy, scale of operation, and technological preparedness, so that quantum propositions will not favor digital elite stakeholders. Lastly, *co-created insights* emerge through participatory marketing and feedback architectures where farmers are not just sources of data or objects of persuasion, but epistemic participants that define value discourse, adoption processes, and system improvements. Such technically based practices are the only ones that can bring quantum marketing beyond precision and toward provision, i.e., the provision of trust, equity, and long-term resilience within the food ecosystem.

9.3.2 Ethical strategy formulation for quantum-enhanced agri-futures

The ethical strategy formulation for quantum technology can be approached through the intersecting lenses of technology perspective, human perspective, and managerial perspective.

From a technological standpoint, before this current generation (The Sustainable and Smart Agriculture: A. 05) of agricultural evolution, the civilization has already experienced another four stages of technological progression. They can be summarized as (1) The Pre-Industrial Era (A. 01); (2) The Industrial Era (A. 02); (3) The Green Revolution (A. 03); and (4) The Digital Revolution (A. 04). In each of these evolutionary juggernauts, all stakeholders directly partnering agri-business's technological breakthroughs have not only taken care of personal and community well-being but have brought human civilization ahead. The technology partners encompassing the sustainable and smart agriculture can be plentiful in numbers, and each of them will play a magnificent role in the ongoing quantum technology adoption. The most significant technology strategy requires implementation of cloud computing, blockchain technology, big data analytics, IoT, smart sensors, agricultural robotics and smart drones, geo-tagging and satellite imaging, agri-biotech, and AI based smart-agri apps (Ali et al., 2024). Now the application strategy for these agri-techs to develop a wholesome quantum technology framework will require an impressive modus operandi at the diverse collaborators end. Further, to make this framework functional, the mammoth infrastructure development requires not only the closely linked partners to invest generously but the government at large to participate altruistically. These multi-partner agri-tech strategy frameworks, if accomplished successfully, the outcomes at the direct participants' end will resolve decades long environmental, biological, and economic productivity related unpredictability. The shared framework for consequence and management will help the parties with instant information, directions and continuous guiding on climate conditions, field conditions, crop health, precision agriculture, livestock and poultry monitoring, greenhouse automation, smart irrigation, pest and disease control, smart fertilizer infusion, intelligent agricultural machine, smart harvesting, soil management, crop yield, weed management, water management, agri-products' quality, and safety traceability (Arya et al., 2023).

Viewed from a human perspective, irrespective of the types of agri-tech partners, eventually the strategic implications of quantum technology adoption will impact human life at large. Now the broader social good will reflect through the various parties' genuineness toward ethically sound strategic intent design. There could be multitude of concerning topics surrounding the ethical issues but as outlined and visualized by (Jobin et al., 2019; Ryan & Stahl, 2020), respectively, they can be themed as transparency, justice and fairness, nonmaleficence, responsibility, privacy, beneficence, freedom and autonomy, trust, sustainability and dignity, and solidarity. Although many of these ethical concerns will impact agri-business partners via ecological apprehension, the direct impact of some of them such as privacy, freedom and autonomy, and transparency will be obvious through social and personal implications. Major ramifications of nonethical strategy formulations surrounding agri-techs, while adopting quantum technologies can be listed as data privacy, breaches and leakages, malware injection attack (ransomware and botnet), social engineering (phishing attack), denial-of-service (DoS), and man-in-the-middle (MitM) attacks, replay, eavesdropping, and insider attacks, supply chain and side-channel attack, radio-frequency jamming, rogue device deployment, node tampering, spoofing, agroterrorism, false data injection, signature wrapping, reconnaissance attacks, SQL injection, sybil and black-hole attack, evasion attack, AI data poisoning, tracing and protocol attack, infusion forgery, buffer overflow, and RFID-based attacks. Henceforward, the onus of developing ethically sound strategic pathways for quantum technology infrastructure spread, where a socially responsible humanoid robot works freely or alongside human-mates without any obstacles (informational and physical) (Stock & Gardezi, 2021) is the minimum viable skeleton (MVS) for sustainable quantum technology fusion in agri-business management.

Looking at the agri-business realm through the classical strategic management lens encourages the various stakeholders to quantify and compare the real economic outcomes of participating or nonparticipating in this tectonic quantum-tech led disruption. Presently, there exists a lot of skepticism around the mission, vision, goals, and objective settings, in the pursuit of developing an economically and environmentally sustainable basis for each of their long-term foundations. Major issues in the failure of developing managerially sound quantum technology strategies could be the inability to define the right target markets, desirable competitive positions, actionable growth strategies, and deriving suitable marketing mix. Henceforth, the outcomes could be combined miscalculation in both the perceived as well as real cost of quantum technology adoption in agri-business. The new-age tech-focused agri-products' market placements may face varied hindrances linked to market offerings, communication channels, supply chain modelling etc. Moreover, due to the sustained arm-twisting between the investors of quantum technology in agri-business and the agri-entrepreneurs may impede further fundraising (colossally sized investment numbers required) due to the initially poor return on invested capital (ROIC). Finally, enticing the customers of quantum-tech linked agri-products with a compelling value proposition, for example, quality and safety traceability attributes may face reasonable barriers in the competitive markets if the "The Battle Between Good and Better" doesn't have a clear winner.

9.3.3 Human agency and organizational behavior in quantum agriculture

As quantum technologies enter the agricultural sector, conversations have largely focused on infrastructure and technical capacity. Yet, a deeper shift is underway and one that reconfigures human agency, work identities, and behavioral norms. Quantum agriculture does not merely enhance tools; it transforms how decisions are made, how labor is valued, and how individuals relate to both land and data. This section explores how quantum systems challenge embodied expertise, reshape psychological contracts, and recast roles across all organizational levels. It considers the evolving role of HRIS in managing skills, well-being, and equity, and asks: What does agency mean when machine-led precision guides practice, and when human judgement is filtered through quantum logic?

Fig. 9.2 summarizes how the introduction of quantum-AI systems reshapes decision-making and knowledge authority, with implications for human agency and epistemic balance, mediated by organizational enablers and reflected in outcomes such as trust, adaptation, and ethical practice.

Framed through *the lens of empowerment, authority, and epistemic balance,* agency as a belief in the possibility of action lies at the basis of the adoption of agricultural innovations, especially with the support of self-efficacy and peer influence (Chandra et al., 2022). The concept

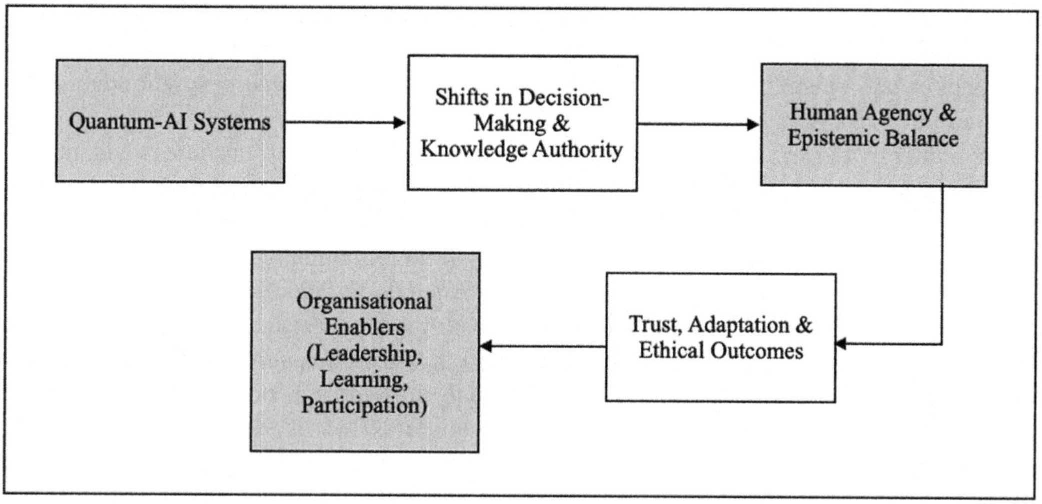

FIGURE 9.2 Key organizational and behavioral dimensions shaping human agency in quantum agriculture.
Source: Developed by the authors.

of agency in quantum agriculture is that it happens at an individual level but also at the level of collectives or intermediaries that enable access to technologies (Patel & Mehta, 2023). The implementation of quantum-AI decision systems can however move the power of cognisance away from traditional knowledge custodians to information-based managers, which can marginalize them epistemically. This happens when local, experiential knowledge is put to the back burner in defavor of algorithmic reasoning (Birhane, 2021). To guarantee epistemic balance (Jain and Banerjee, 2024), participatory forms of decision-making must be put in place so that field-level wisdom is funded legitimately. Organizations have an obligation to inculcate the systems of knowledge pluralism so that quantum advancements supplement human action and not replace them.

Quantum Leadership is a new leadership model that is best suited to the complexity, ambiguity, and fast changes of organizations that dominate the modern-day business environment. Quantum leadership is based on one or more of the principles of quantum physics that include uncertainty, entanglement, and nonlinearity (Porter-O'Grady & Malloch, 2015). It also does not support linear, command-and-control models but rather creates an atmosphere where the leadership can come out at any organizational network node. Quantum leadership is, in essence, about distributed decision-making, facilitative communication, and constant continuous learning, and it is an environment that encourages and promotes innovation and organizational resilience. These leaders do not perform as central forces, but rather as a sensemaker and co-producers of meaning and adaptive power across teams (Uhl-Bien et al., 2007). In those areas where there are volatility and interdependence like the agri-food business- quantum leadership has been empirically found to be pivotal in achieving organizational ambidexterity—the ability to exploit existing resources in a way that is efficient and explore and harvest the potential novel opportunities in a way that is innovative (Rosing et al., 2011). To be more accurate, research discovered that the combination of the quantum leadership and the culture of knowledge sharing helps the organization better negotiate between the short-term organization needs or the demands of daily operation (exploitation) and the long-term strategic changes or the adaptability (exploration) (Turner et al., 2013).

In addressing *trust, transparency, and human–AI co-design,* it becomes evident that quantum-AI systems cannot be designed, deployed, and then engineered to deserve trust; it must be designed from beginning to end. Wang et al. (2025) recommended a triadic model of trust-ability, benevolence, and integrity adopted within reinforcement learning structures to serve agricultural decision systems. This according to their findings indicates that farmers have a stronger tendency to implement AI systems that not only work but also hold the priorities of the local values and agronomic principles. This backs the shift to participatory human-AI co-design, in which the farmers and field experts become not only the end-users but also the co-developer of data inputs, model explainability, and situational standards (Li et al., 2024). These contributors form the epistemic frame of the AI system, grounding it on the relationships of trust instead of the technical abstraction. Evidence indicates that transparent and challenge-able decision criteria on algorithmic systems increase the legitimacy of such systems (Saatçı et al., 2024). It is not necessary to skip human agency, and instead, successful AI systems are used by employing algorithm-in-the-loop approaches, where users tune their dependence on contextual and confidence signals (Lee & Park, 2024). Such systems neither prescribe action nor offer or remove judgment, actively leaving the task of exerting judgment to people but presenting uncertainty, collaborative judgment making, and supporting a distributed paradigm of intelligence.

When it comes to *learning cultures, resistance, and sense-making dynamics,* the success or failure of quantum change depends on the learning elasticity of the host culture; preparation is anchored on self-efficacy, value fit, and leadership support (Armenakis et al., 1993). It is important to note that the resistance that is usually witnessed during pilot stages is usually an indication of the rational defence of tacit craft knowledge as opposed to the obstinacy (Nonaka & Krogh, 2023). Organizations have to organize adaptive learning frameworks so that this tension can be directed into the productive channel of energy. The sense-making model proposed by Weick avails iterative action of enact–select–retain cycles; expression of this model can be viewed in terms of peer-moderated after action huddles, scenario planning sprints to unveil buried assumptions and replacement of liaison roles, rotated between agronomists and data scientists (Weick, 1995). Psychological safety studies indicate that trust in inability to question algorithms can be removed (where coders can ask inquiries without the fear of reprisals), which increases reporting of errors and creates a more reliable system (Edmondson, 1999). Such leaders do not perform as core managers but as creators of sense and co producers of meaning and adaptive power within teams (Uhl Bien et al., 2007). The quantum ready firms solve this by focusing the core performance currency on knowledge exchange as raw prediction accuracy, rather than the other way around; they structure knowledge exchange so that human insight iterates and improves machine intelligence and vice versa and convert resistance to continuous learning.

In the realm of *psychological adaptation and quantum anxiety,* even to the hardened technologist, the probabilistic, counter-intuitive quantum-AI logics are discomforting; to frontline agronomists, the sentiment usually takes the form of what I call quantum anxiety, a mix of technophobia and identity crisis. Indeed, according to change management research, failure to express such anxiety results in it crystallizing as passive resistance or learned helplessness. The theory of psychological safety thus recommends the climates to which doubt, naive inquiries, and minor errors are not punished, rather they serve as the learning catalyst (Edmondson, 1999). That climate in quantum agriculture should be built based on dual-track capability building. First, there is the classical upskilling, sensor dashboards, RL-based recommendations, simple quantum concepts, and it offers mental scaffolding to the employees. Second, the development of emotional-intelligence (EI) prepares them to learn cues of anxiety, empathize with their peers, and self-regulate. The sense-makers are the leaders who normalize uncertainty, i.e., everyone is a learner here and explain failures through data but not by failure (Uhl-Bien et al., 2007). Quantum anxiety is also reduced by narrative practices. The storytelling of the knowledge development that involves veteran farmers sharing their evidence-based knowledge and data scientists explaining how an algorithm works builds a so-called shared mental model and respects both epistemes. Sessions of value clarification, where work groups explain why quantum technology is beneficial to the community, i.e., saving soil, fair market access, tie the technology to purpose, not to hype. These kinds of interventions can assist the stakeholders in redefining professional identity: once single guardians of land they are now liaisons between the cyber and physical world to combine both intuition and machine intelligence. Lastly, frequent sense-making checks, short and scaffolded reviews of what took people unawares or puzzled them, does not allow background worries to solidify into distrust. When these reviews input design adjustments, employees feel that their concerns have been converted to concrete improvement of systems, thereby completing the loop of anxiety. In this sense, psychological adaptation is not an HR enhancement, but a strategic need that nurtures engagement and ethical agency through the growth of quantum agriculture.

Overall, the ethical progression of quantum agriculture requires that organizational decisions respect the same values as we require of the land: equity, mutuality, and stewardship. Human-centered design should then be fused with agrarian ethics, where local voices are empowered, data sovereignty achieved, and benefits spread along gender, class, and farm yard boundaries. Contradictorily, ethics is the practice of leadership that brings the diverse stakeholders around the table as opposed to telling them what to do; the algorithms are inculcated with trust that comes with transparency and responsibility. Preservation of psychological well-being is an ethical responsibility, and high velocity of change should not corrupt dignity or identity. The aim of flexible and feedback-rich systems that are desired to balance yield optimization and sustainability on the basis of soil health, biodiversity, and intergenerational equity should be guided by ethics of stewardship. The presence of such pillars of ethics working in tandem can make quantum innovation stop being merely a productivity lever, but a pathway to democratic inclusion, social energy, and successive regeneration of agriculture in the society.

9.4 Integrated framework and managerial takeaways

Quantum agriculture cannot be easily split into its ethical environment. It requires a subtle, mutually supportive system that will combine strategic foresight, community-based preparedness, and open communication into a single ethical system. By being complex and predictive,

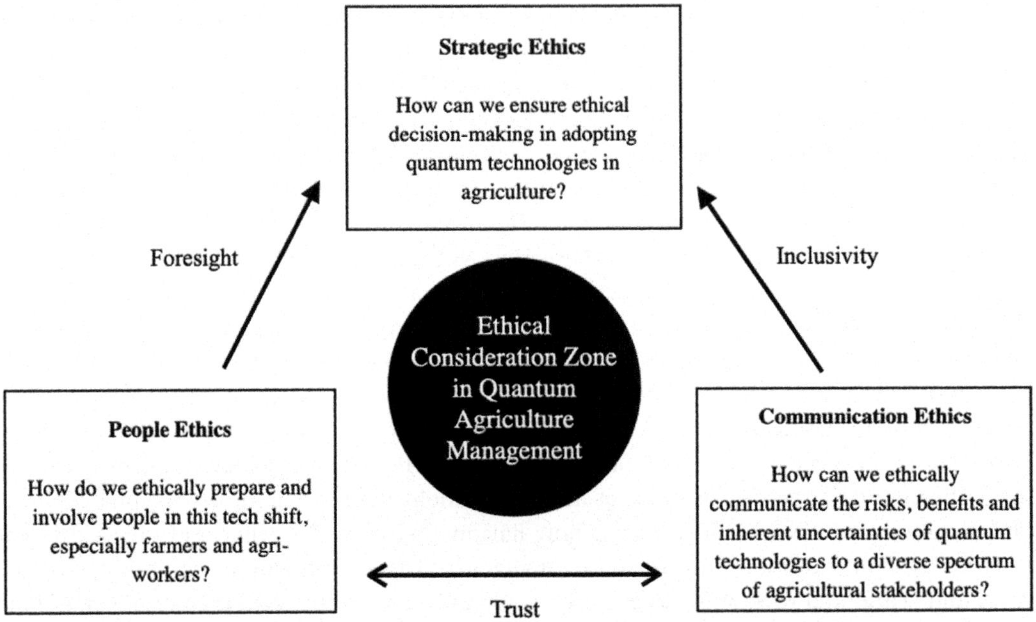

FIGURE 9.3 EPQ-agri-mgmt. framework. (Ethical Preparedness for Quantum Agriculture Management). *Source: Developed by the authors.*

quantum technologies will have a revolutionary impact on the agri-sector, but its social and ethical implications are yet to be charted (Ahlborg et al., 2019; Anderies, 2014; Redman & Miller, 2015; Wolfert et al., 2014, 2017). Unless a human–environment–technology approach is taken in the conceptualization of smart farming solutions, Mallinger and Baeza-Yates (2024) assert, unintended systemic disruptions are liable to occur. This will require a restructuring of ethics as fixed codes to dynamic processes of alignment.

To cope with such interruptions, the proposed Ethical Preparedness for Quantum Agricultural management (EPQ-AgriMgmt) framework (Fig. 9.3) is a combination of experience in various areas of management in a flowing, interactive ethical model to provide the fluidity of quantum agricultural systems. Instead of taking Strategic, People and Communication Ethics as different silos, the framework insists on their interconnection, where one area reinforces, informs and in some cases, limits the other. This interdependence resembles the relationality of quantum theory and the agricultural practice where the actions resonate across the scales, communities and ecosystems. The main element of this model is the idea of ethical alignment, framed not as a one-time compliance benchmark but as a recursive, adaptive process shaped by changing contexts and competing demands. The triangulation between foresight, trust and inclusivity mobilizes ethical alignment and they all serve as stabilizing vectors to ethical reasoning in an otherwise uncertain and constantly changing field. Strategic ethics, which are supported by accelerating agri-technological innovation, require actors to use anticipatory governance, mapping not just immediate efficiencies but long-term social, environmental and cultural impacts

(Eastwood et al., 2019). But vision of air without foundation on earth may fail. The translation of anticipation into action has to be ethical by using inclusive and context-sensitive design in accordance with the systems, values, and constraints of those on the ground (Stitzlein et al., 2020). The technology is also likely to be perceived as a foreign force instead of a locally based intervention when the concept of strategic foresight is no longer tied to the reality of farmers (Garnett & Godfray, 2012; Higgins et al., 2017; Jones-Garcia & Krishna, 2021; Mahon et al., 2017). This association leads us to the field of people ethics where the human implications of a technological transformation are most apparent and immediate. Vulnerability is a structural aspect of the agricultural experience, particularly among the smallholders and marginalized farmers who regularly operate within the limits of unpredictable weather patterns, unforeseeable market dynamics and unpredictable access to resources (Cabas et al., 2010; Felkner et al., 2009; Patil & Veettil, 2024). In such communities, the implementation of new farming techniques including those stimulated by quantum technologies is hardly a linear and rational choice. Rather, it is frequently a balancing act which is precarious, mediated by uncertainties as well as being circumscribed by material conditions. The existence of such coping mechanisms as diversification of income, selling assets, migration or dependence on kinship groups is an adaptive reaction but also an ethical indicator that the infrastructural and institutional framework surrounding these mechanisms is still weak (Aryal et al., 2020; Zimmerman & Carter, 2003). Top-down dissemination would not be sufficient to bring about ethical alignment. It demands the co-invention of participatory infrastructures, systems, which involve communities not just as receivers of innovation, but as co-creators of its conditions and pace. It means developing not only awareness but also agency making sure that individuals are able to influence, oppose or even refuse technological interventions. Ethical preparation in this sense is not only educational outreach but also a structural necessity, which calls us to redesign the models of distribution, reinforce the systems of extension, and redesign the institutional logics that inform the making of agricultural decisions.

The third axis on the model is communication that runs across strategy and participation that serves as the glue holding and tearing ethical coherence. Communication in this case mediates relationships, expectations, and trust in quantum agriculture rather than passing information. The epistemic asymmetries are great because of the representational gap between developers of quantum systems and the users of quantum systems, which are in many cases rural communities with their own cultural, linguistic, and experiential knowledge. Such asymmetries are not technical alone but rather ethical, since they decide who has the privilege of knowing, doubting as well as shaping the course of technological implementation. Trust in AI-driven systems, according to Liao and Sundar (2022), is a factor of affordances of transparency and user interaction and semantic clarity, which, as the researchers claim, must be consciously established in the communicative architecture initially. In farming, where the new tools tend to touch on old embedded practices, and local epistemologies, there is necessarily more than clarity or simplification in ethical communication, namely the development of mutual intelligibility and respect. The acceptance and deployment of quantum and smart farming instruments, in its turn, are contingent upon the understanding, adoption, and adaptation of the systems to the reality by farmers and other decision-makers and thus is an inseparably coupled process associated with current trust relations and familiarity with the institutions (Bronson & Knezevic, 2016; Sonka, 2016; Wolfert et al., 2017). As Jakku et al. (2019) pointed out, trust-based advisory networks are essential in the rural setting. Such networks are interpretive bridges, sieving, transposing, and harmonizing quantum-era propositions to the beats of everyday farming life. Not only is communication ethics, then, about messaging, but also about co-narration, i.e., building up of plural, credible, and context-sensitive understandings of what quantum agriculture is, and who it is aimed at. The directional flow of the EPQ-AgriMgmt Framework is conscious and indicative of the manner in which ethical considerations trickle down the domains of managerial functions. Whereas trust incorporates the sense of a two-way relationship between People Ethics and Communication Ethics, the recognition of the co-constitutive aspect of stakeholder engagement, foresight and inclusiveness are made bidirectional vectors toward strategic ethics. This organization acknowledges that enlightened strategic choices on ethical grounds should be informed by ground-level readiness and involving communication despite the lack of inevitability of the reflexively manifested influence. Accordingly, the model favors bottom-up ethical congruence, according to which strategy is developed out of, as opposed to down out of, the lived conditions of agrarian societies. People ethics leads into strategic ethics via foresight to indicate that no one-looked-to-the-future planning is ethically true unless it is based on the vulnerabilities, adaptation strategies and lived lives of the farming communities. In the same spirit, the paradigm of strategic ethics transpires communication ethics to underline the impossibility of an ethically coherent strategy without initially incorporating varied narratives, values, and interpretive frames. These unilateral directions are not designed to imply inflexibility, but only to emphasize the ethical sequence of managerial action, strategy does not get tough when it dictates; it gets tough when it listens. Combined, these vectors serve to remind us that the ethical future of quantum agriculture is going to have to be told in a ground-up manner. Conversely, two way flow of trust between People Ethics and Communication Ethics highlights its relational aspect in that trust is established through clear and culturally resonant communications and is maintained through the response and reciprocating nature of communities. This reciprocally confirms that an ethical engagement is not a message that is delivered in a single occasion but a continuous dialogue of presence, listening, and a back and forth feedback. The EPQ-AgriMgmt Framework provides a unified ethical zone with interlocked domains as opposed to isolated checkpoints. Ethical alignment is defined as a shared practice: strategic responsibility is enhanced by foresight, inclusion by participation, and legitimacy by communication which is co-constructed. These three spheres are interdependently dynamic and can be used to respond to the ethical turbulence of quantum changes in the agricultural industry. As an agricultural manager, policy architect, and technology designer, the EPQ-AgriMgmt Framework can be used in a diagnostic and design capacity to clarify what they should be aware of and how they have to formulate an intervention capable of being ethically based. Scenario planning that is ethical, strategically, can allow long-term ecological and socio-political changes, which will require managers to think outside the cost-benefit matrices and inside the impact-forecasting that incorporates the voices of the marginalized. It necessitates stakeholder mapping at an early stage that puts into focus, stakeholders that tend to be left out in the quantum innovation processes so that their positionalities can affect the vision and implementation. When dealing with vulnerable groups whose livelihoods are tied to the land, human-based design becomes something that is not negotiable since it is based on participatory research. This involves tailoring the technologies to local demands and creating adaptive infrastructures, such as extension networks and training hubs, that change in response to the local demands with time. Communication can no longer be considered a peripheral part of innovation cycles but rather be viewed as a stable development process. This will mean harmonizing the technological accounts with the local expertise, the development of deliberative spaces and epistemic justice between specialists and professionals. The managers should actively create and sustain knowledge-sharing ecosystems in which trust and a conversation are operational requirements instead of the occasional results. Institutions have to develop feedback loops that are more listening than informative, and make learning processes that make community-based understanding into iterative governance practices. Lastly, this combined framework forces decision-makers to go beyond dichotomous views of efficiency vs. equity. It advocates a framework where quantum agricultural technologies not only get evaluated on their technical output but also on their ethical resonance, that is how well they celebrate the complexity of the communities, ecosystems and stories into which they enter. It is not merely, however, the managerial task, to lead change, but to serve alignment, an alignment in which progress is marked by mutual trust, rooted foresight and significant inclusion.

9.4.1 Illustrative case—Ethics-aligned quantum marketing in an agri-forecasting ecosystem

Consider an example of a quantum-powered agrarian forecasting project implemented by a farmer group in a climate-sensitive agrarian area. The system integrates quantum optimization models with real-time satellite data and agronomic data on the ground to offer crop-specific advisories on sowing, irrigation and risk mitigation. Although the potential of such systems to be effective in terms of predictive accuracy is the technology, the combination of strategic, communicative, and human considerations is what will determine their success ethically and as a manager. At the *people ethics* level, the experience of living with the volatile climatic conditions, exclusion of farmers throughout the history of high-technology decision systems, and the digital literacy of different degrees, influence the perception of the system and its adoption. Notably, the *foresight to strategize does not arise only from calculative ability* only but in the awareness of such human conditions as moral feeding. The leadership of the cooperative considers such insights in its strategic planning by giving primary attention to adaptive timelines, low-risk trial stages, and shared accountability structures, where the long-term strategic decisions are informed by the agency of farmers and the experience. On the *communication ethics* level, the project focuses on open value communication which is based on nontechnical explanations, interfaces which are in a vernacular and ongoing communication and not the dissemination of information in one direction. Such communication practices bring out asymmetries in the access, trust and understanding between groups of farmers. *Ethical communication results directly influence the strategic decisions*, which subsequently result in inclusive design decisions including the use of modular systems, collective dashboards, and opt-out features. In such a manner, inclusivity is a communication ethic that transfers to strategic ethics, as opposed to retrofitting, once implemented. *Trust is a two-way ethical ground* between individuals and communication ethics. Open communication helps farmers to be willing to provide sensitive information and interact with probabilistic recommendations, and the continued feedback of farmers helps to make information framing and presentation a reconstruction of the same. This is to the effect that trust becomes an outcome of marketing, not in isolation as such but is a relational ethical process. Overall, this case demonstrates that ethics-aligned quantum marketing in the agricultural sector will demand transitioning beyond precision-based optimization to the domain of *provision,* which is the provision of trust, inclusiveness, long-term sustainability and foresight in food ecosystems. Strategic ethics in this regard is not a managerial imposition that is developed top-down but rather a meeting point that is formed through ethical communication and human agency.

9.5 Conclusion and Future Scope

From quantum possibility to ethical practice, the opening key question in this chapter has been: how can the transformative potential of quantum technologies in agriculture be tapped in an ethically and inclusively oriented and not an efficient only way? It is stated that ethics cannot be an afterthought or a burden on innovation, instead it serves as the scaffolding keeping the social, technological, and ecological aspects of quantum transitions. The fundamental objective was to remodel the conventional interpretation of ethics as conformity to co-constructive, under which trust, vision, and inclusiveness define the premises of quantum agricultural frameworks. A cross-disciplinary investigation into marketing, organizational behavior, human resource systems, and strategic ethics, this chapter attempted to address the question that has been posed above: how can we make certain that quantum farming does not overshadow the human agency, the environmental responsibility, or the democratic process? This chapter has also achieved its purpose by analyzing quantum agriculture under the three very important lenses of management, which include marketing, strategic thinking, and organizational behavior. Marketing-wise, it proved that there is a necessity to prioritize marketing ethical dealings, which implies making data transparent, being intergenerationally responsible and practicing posthuman ethics, so as to consider farmers and consumers as the co-value-creators of algorithmic systems rather than data points per se. Strategically, it promoted an alternative to its technocratic efficiency, and that is value-based foresight, which centers on communication, inclusion, and collective governance as the aspects of effective ethical decision-making. This led to a deeper perspective through the organization behavior lens that viewed the psychological and behavioral aspects of change. It demonstrated the methods through which the culture of adaptive learning, emotional intelligence, and the frameworks of human and AI co-design may help reduce the number of quantum anxiety and facilitate inclusive, future-oriented working conditions. In combination, these views assisted in the operationalization of the EPQ-AgriMgmt Framework, a triadic model including strategic ethics, people ethics, and communication ethics. Instead of existing separately, these dimensions can serve as the reinforcing lever of one another, which allows agricultural innovation to happen with not only technological advanced properties but with a social and ethical maturity as well. This chapter has a number of conceptual as well as practical contributions. It provides, in the first place, a transdisciplinary synthesis to span fields that are typically drawn into silos-quantum engineering, HR systems, marketing ethics, and rural change management in order to create a shared vocabulary of quantum agriculture. Second, it posits EPQ-AgriMgmt Framework as an effective instrument of an ethical preparedness evaluation in terms of strategy, workforce development, and stakeholder communication. Third, it brings to the fore the notion of a specific pandemic: quantum anxiety, a unique psychological phenomenon in technologically mediated transformations to farming, offering such interventions as storytelling, EI-building, sensemaking ceremonies. The conclusion of the chapter reframes leadership and organizational design as a morally imperative act - in which decentralized decision-making, trust building and epistemic pluralism become a core means of dealing with uncertainty and complexity. The weakness of this chapter lies mainly in the fact that it makes up an idea. Most of the insights provided are not necessarily a result of the case-studies themselves (in behavioral and organizational contexts), but on the existing literature and knowledge on AI and digital transformation in general. EPQ framework, though logically developed, has failed to be tested in the field or using large-scale empirical research. Some other factors that may affect the cross-contextual application of these principles are regional and cultural differences in using agri-tech. Quantum technologies offer new possibilities to model biological systems with high accuracy never seen before. In the process, however, they challenge us to think about the systems of meaning, value, and belonging, which support agricultural life. Getting from possibility to practice, we ask not just what can we do with quantum agriculture, but what is it anyway that quantum agriculture should do and to whom? Ethical, inclusive, and trust-driven innovation makes quantum agriculture not only the instrument of optimization but regenerative justice, the instrument of democratic self-governance, and the means to care about the planet at the scale of generations.

Future Scope

This work has some significant prospects to be followed in further research and practical experimentation. The first suggestion is systematic empirical research into the effects of trust-integrated quantum decision support tools on the adoption behavior of farmers, autonomy and well-being in the long run productivity of farmers in various agrarian settings. These investigations may adopt mixed-method methodologies, that is, field experiment and longitudinal ethnographic investigation, to be able to observe both measurable changes and lived experiences. Second, policy sandboxes are an encouraging means of experimenting with data ethics, benefit-sharing terms, and algorithmic accountability in the conditions of large-scale farming, where regulators and practitioners can experiment with unintended outcomes over time before rolling out large-scale applications. Third, quantum-based agri-technologies should be implemented with systematic education and reskilling programs that should not only be linguistically but also culturally localized with familiar metaphors and practice-based learning to reduce digital stratification and marginalization. Fourth, co-creative positions can be used in the design, testing, and fine-tuning of context-sensitive quantum applications based on local knowledge regimes through interdisciplinary living labs that include farmers, ethicists, engineers, agribusiness managers, and rural youth. Lastly, future studies will need to create additional assessment metrics that transcend yield and efficiency metrics in order to include, through this means, indicators of ecological regeneration, labor dignity, epistemic inclusion, and knowledge equity, and, by so doing, open up a wider spectrum of evaluation of value creation in quantum-enabled agricultural ecosystems.

References

Ahlborg, H., Ruiz-Mercado, I., Molander, S., & Masera, O. (2019). Bringing technology into social-ecological systems research—Motivations for a socio-technical-ecological systems approach. *Sustainability*, (7), 11. https://doi.org/10.3390/su11072009.

Ali, G., Mijwil, M. M., Buruga, B. A., Abotaleb, M., & Adamopoulos, I. (2024). A survey on artificial intelligence in cybersecurity for smart agriculture: State-of-the-art, cyber threats, artificial intelligence applications, and ethical concerns. *Mesopotamian Journal of Computer Science*, 53–103. https://doi.org/10.58496/MJCSC/2024/007.

Amini, F. (2008). Eco views: An interview with Dr. Vandana Shiva. *International NGO Journal, 3*(4), 77–83.

Anderies, J. M. (2014). Embedding built environments in social–ecological systems: Resilience-based design principles. *Building Research & Information, 42*(2), 130–142. https://doi.org/10.1080/09613218.2013.857455.

Armenakis, A. A., Harris, S. G., & Mossholder, K. W. (1993). Creating readiness for organizational change. *Human Relations, 46*(6), 681–703. https://doi.org/10.1177/001872679304600601.

Arya, S., Tripathi, S., Srivastava, A., Aggarwal, S., Soni, N., & Ansar, S. A. (2023). Double-edged agriculture 4.0: Hodiernal expedient technologies and cyber-security challenges. In *2023 6th International Conference on Contemporary Computing and Informatics (IC3I): 6* (pp. 313–320).

Aryal, J. P., Sapkota, T. B., Rahut, D. B., Krupnik, T. J., Shahrin, S., Jat, M. L., & Stirling, C. M. (2020). Major climate risks and adaptation strategies of smallholder farmers in coastal Bangladesh. *Environmental Management, 66*(1), 105–120.

Bergier, I., Barbedo, J. G. A., Bolfe, É. L., Romani, L. A. S., Inamasu, R. Y., & Massruhá, S. M. F. S. (2024). Framing concepts of agriculture 5.0 via bipartite analysis. *Sustainability, 16*(24), 10851. https://doi.org/10.3390/su162410851.

Birhane, A. (2021). Algorithmic injustice: A relational ethics approach. *Patterns, 2*(2).

Bronson, K., & Knezevic, I. (2016). Big data in food and agriculture. *Big Data & Society, 3*(1), 2053951716648174. https://doi.org/10.1177/2053951716648174.

Cabas, J., Weersink, A., & Olale, E. (2010). Crop yield response to economic, site and climatic variables. *Climatic Change, 101*(3), 599–616. https://doi.org/10.1007/s10584-009-9754-4.

Cadbury Schweppes. (2002). *The importance of ethics in business.*

Carbonell, I. M. (2016). The ethics of big data in big agriculture. *Internet Policy Review, 5*(1), 1–13. https://doi.org/10.14763/2016.1.405.

Chandra, R., Singh, V., & Joshi, M. (2022). Agency and adoption: Understanding farmer readiness in digital agriculture. *Regional Environmental Change, 22*(1), 45–59.

De Freitas, E. (2017). Karen Barad's quantum ontology and posthuman ethics: Rethinking the concept of relationality. *Qualitative Inquiry, 23*(9), 741–748. https://doi.org/10.1177/1077800417725359.

DesJardins, J. (2011). *An introduction to business ethics* (4th ed.). McGrew-Hill International.

Directory, S. (2025). *What are the ethical considerations of using quantum computing in agriculture? Sustainability Directory* https://sustainability-directory.com/question/what-are-the-ethical-considerations-of-using-quantum-computing-in-agriculture/.

Eastwood, C., Klerkx, L., Ayre, M., & Dela Rue, B. (2019). Managing socio-ethical challenges in the development of smart farming: From a fragmented to a comprehensive approach for responsible research and innovation. *Journal of Agricultural and Environmental Ethics, 32*(5), 741–768. https://doi.org/10.1007/s10806-017-9704-5.

Edmondson, A. C. (1999). Psychological safety and learning behavior in work teams. *Administrative Science Quarterly, 44*(2), 350–383. https://doi.org/10.2307/2666999.

Felkner, J., Tazhibayeva, K., & Townsend, R. (2009). Impact of climate change on rice production in Thailand. *American Economic Review, 99*(2), 205–210. https://doi.org/10.1257/aer.99.2.205.

Feynman, R. P. (1982). Simulating physics with computers. *International Journal of Theoretical Physics, 21*, 467–488. https://doi.org/10.1007/BF02650179.

Garnett, T., & Godfray, C. (2012). Sustainable intensification in agriculture. Navigating a course through competing food system priorities. (p. 51). University of Oxford, UK: Food climate research network and the Oxford Martin Programme on future of food.

Higgins, V., Bryant, M., Howell, A., & Battersby, J. (2017). Ordering adoption: Materiality, knowledge and farmer engagement with precision agriculture technologies. *Journal of Rural Studies, 55*, 193–202. https://doi.org/10.1016/j.jrurstud.2017.08.011.

Hoffman, A. J. (2011). Talking past each other? Cultural framing of skeptical and convinced logics in the climate change debate. *Organization & Environment, 24*(1), 3–33. https://doi.org/10.1177/1086026611404336.

Jain, M., & Banerjee, P. (2024). Repositioning knowledge hierarchies in smart agriculture. *Journal of Rural Studies, 104*, 80–91. https://doi.org/10.1016/j.jrurstud.2023.11.008.

Jakku, E., Taylor, B., Fleming, A., Mason, C., Fielke, S., Sounness, C., & Thorburn, P. (2019). If they don't tell us what they do with it, why would we trust them?" Trust, transparency and benefit-sharing in Smart Farming. *NJAS-Wageningen Journal of Life Sciences, 90*, 100285. https://doi.org/10.1016/j.njas.2018.11.002.

Jobin, A., Ienca, M., & Vayena, E. (2019). The global landscape of AI ethics guidelines. *Nature Machine Intelligence, 1*, 389–399. https://doi.org/10.1038/s42256-019-0088-2.

Jones-Garcia, E., & Krishna, V. V. (2021). Farmer adoption of sustainable intensification technologies in the maize systems of the Global South. A review. *Agronomy for Sustainable Development, 41*(1), 8. https://doi.org/10.1007/s13593-020-00658-9.

Kaur, J., Hazrati Fard, S. M., Amiri-Zarandi, M., & Dara, R. (2022). Protecting farmers' data privacy and confidentiality: Recommendations and considerations. *Frontiers in Sustainable Food Systems, 6*, 903230. https://doi.org/10.3389/fsufs.2022.9032.

Kolk, A. (2016). The social responsibility of international business: From ethics and the environment to CSR and sustainable development. *Journal of World Business, 51*(1), 23–34. https://doi.org/10.1016/j.jwb.2015.08.010.

Lee, S., & Park, J. (2024). Balancing AI transparency: Trust, certainty, and adoption. *Journal of Information Science*. https://www.researchgate.net/publication/392230169_Balancing_AI_transparency_Trust_Certainty_and_Adoption.

Li, J., Yang, Y., Zhang, R., & Lee, Y. C. (2024). *Overconfident and unconfident AI hinder human-AI collaboration* https://arxiv.org/abs/2402.07632.

Liao, Q. V., & Sundar, S. S. (2022). Designing for responsible trust in AI systems: A communication perspective. In *Proceedings of the 2022 ACM conference on fairness, accountability, and transparency* (pp. 1257–1268). https://doi.org/10.48550/arXiv.2204.13828.

Mahon, N., Crute, I., Simmons, E., & Islam, M. M. (2017). Sustainable intensification: "Oxymoron" or "third-way"? A systematic review. *Ecological Indicators, 74*, 73–97. https://doi.org/10.1016/j.ecolind.2016.11.001.

Mallinger, K., & Baeza-Yates, R. (2024). Responsible AI in farming: A multi-criteria framework for sustainable technology design. *Applied Sciences, 14*(1), 437. https://doi.org/10.3390/app14010437.

Maraveas, C., Arvanitis, K. G., Bartzanas, T., & Loukatos, D. (2025). Potential applications of quantum sensors in agriculture: A review. *Computers and Electronics in Agriculture, 235*, 110420. https://doi.org/10.1016/j.compag.2025.110420.

McCann, E. L. (2012). *The seven CS ethical model of communication: Environmental communication and indigenous knowledge management strategies in international agricultural development.* Texas A&M University https://hdl.handle.net/1969.1/ETD-TAMU-2012-08-11702.

Naveh, Y. (2022). Quantum is strategic to companies and countries, upside and downside alike. *Forbes.* https://www.forbes.com/councils/forbestechcouncil/2022/06/29/quantum-is-strategic-to-companies-and-countries-upside-and-downside-alike/.

Nonaka, I., & Krogh, G. (2023). Tacit knowledge and knowledge conversion: Controversy and advancement in organizational knowledge creation theory. *Organization Science, 34*(2), 261–280. https://doi.org/10.1287/orsc.2023.1676.

Pasupuleti, M. K. (2025). Quantum AI for food security and sustainable agriculture. In *Food security and sustainable agriculture with quantum AI* (pp. 101–110). National Education Services. https://doi.org/10.62311/nesx/97953.

Patel, R., & Mehta, S. (2023). Collective capacity and technology mediation in Indian agriculture. *Agricultural Systems, 208*, 103623. https://doi.org/10.1016/j.agsy.2023.103623.

Patil, V., & Veettil, P. C. (2024). Farmers' risk attitude, agricultural technology adoption and impacts in Eastern India. *Agriculture & Food Security, 13*(1), 50. https://doi.org/10.1186/s40066-024-00497-x.

Penson, J., Capps, O., Rosson, C. P., & Woodward, R. (2006). *Introduction to agricultural economics* (4th ed.). Pearson.

Pinto, A., Maghalaes, M., & Ringler, C. (2010). Potential of carbon markets for small farmers. International Food Policy Research Institute *(IFPRI).*

Polymeni, S., Plastras, S., Skoutas, D. N., Kormentzas, G., & Skianis, C. (2023). The impact of 6G-IoT technologies on the development of agriculture 5.0: A review. *Electronics, 12*(12), 2651. https://doi.org/10.3390/electronics12122651.

Porter-O'Grady, T., & Malloch, K (2015). *Quantum leadership: Building better partnerships for sustainable health.* Jones & Bartlett Learning.

Qayyum, M., Zhang, Y., Wang, M., Yu, Y., Li, S., Ahmad, W., & Gan, J. (2023). Advancements in technology and innovation for sustainable agriculture: Understanding and mitigating greenhouse gas emissions from agricultural soils. *Journal of Environmental Management, 347*, 119147. https://doi.org/10.1016/j.jenvman.2023.119147.

Raj, M. (2025). *Quantum* crops & sovereign coins*: The* new agri-monetary order https://doi.org/10.2139/ssrn.5262776.

Redman, C. L., & Miller, T. R. (2015). The technosphere and earth stewardship. In *Earth stewardship: Linking ecology and ethics in theory and practice* (pp. 269–279). Springer International Publishing.

Rosing, K., Frese, M., & Bausch, A. (2011). Explaining the heterogeneity of the leadership–innovation relationship: Ambidextrous leadership. *The Leadership Quarterly, 22*(5), 956–974. https://doi.org/10.1016/j.leaqua.2011.07.014.

Ryan, M., & Stahl, B. C. (2020). Artificial intelligence ethics guidelines for developers and users: Clarifying their content and normative implications. *Journal of Information, Communication and Ethics in Society, 19*(1), 61–86. https://doi.org/10.1108/JICES-12-2019-0138.

Saatçı, M., Kaya, R., & Ünlü, R. (2024). Resume screening with natural language processing (NLP). *Alphanumeric, 12*(2). 121–140 https://doi.org/10.17093/alphanumeric.1536577 .

Saiz-Rubio, V., & Rovira-Más, F. (2020). From smart farming towards agriculture 5.0. *A Review on Crop Data Management. Agronomy, 10*(2), 207. https://doi.org/10.3390/agronomy10020207.

Sonka, S. (2016). Big data: Fueling the next evolution of agricultural innovation. *Journal of Innovation Management, 4*(1), 114–136. https://doi.org/10.24840/2183-0606_004.001_0008.

Stitzlein, C., Fielke, S., Fleming, A., Jakku, E., & Mooij, M. (2020). Participatory design of digital agriculture technologies: Bridging gaps between science and practice. *Rural Extension and Innovation Systems Journal, 16*(1), 14–23.

Stock, R., & Gardezi, M. (2021). Make bloom and let wither: Biopolitics of precision agriculture at the dawn of surveillance capitalism. *Geoforum, 122*, 193–203. https://doi.org/10.1016/j.geoforum.2021.04.014.

Taha, M. F., Mao, H., Zhang, Z., Elmasry, G., Awad, M. A., Abdalla, A., Mousa, S., Elwakeel, A. E., & Elsherbiny, O. (2025). Emerging technologies for precision crop management towards agriculture 5.0: A comprehensive overview. *Agriculture, 15*(6), 582. https://doi.org/10.3390/agriculture15060582.

Turner, N., Swart, J., & Maylor, H. (2013). Mechanisms for managing ambidexterity: A review and research agenda. *International Journal of Management Reviews, 15*(3), 317–332. https://doi.org/10.1111/j.1468-2370.2012.00343.x.

Uhl Bien, M., Marion, R., & McKelvey, B. (2007). Complexity leadership theory: Shifting leadership from the industrial age to the knowledge era. *The Leadership Quarterly, 18*(4), 298–318. https://doi.org/10.1016/j.leaqua.2007.04.002.

Vveinhardt, J., Stonkutė, E., & Sroka, W. (2019). Discourse on corporate social responsibility in the external communication of agricultural enterprises. *European J. of International Management, 13*(6), 864. https://doi.org/10.1504/ejim.2019.102816.

Wang, Z., Jang, W., Ruan, B., Wang, J., & Xiao, S. (2025). *Developing and integrating trust modeling into multi objective reinforcement learning for intelligent agricultural management* https://arxiv.org/abs/2505.10803.

Weart, S. (2011). Global warming: How skepticism became denial. *Bulletin of the Atomic Scientists, 67*(1), 41–50. https://doi.org/10.1177/0096340210392966.

Weick, K. E. (1995). *Sensemaking in organizations* https://us.sagepub.com/en-us/nam/sensemaking-in-organizations/book5256.

Wolfert, S., Ge, L., Verdouw, C., & Bogaardt, M.-J. (2017). Big data in smart farming: A review. *Agricultural, Systems, 153*, 69–80. https://doi.org/10.1016/j.agsy.2017.01.023.

Wolfert, S., Goense, D., & Sørensen, C. A. G. (2014). A future internet collaboration platform for safe and healthy food from farm to fork. In *2014 Annual SRII Global Conference* (pp. 266–273). https://doi.org/10.1109/SRII.2014.47.

Zimmerman, F. J., & Carter, M. R. (2003). Asset smoothing, consumption smoothing and the reproduction of inequality under risk and subsistence constraints. *Journal of Development Economics, 71*(2), 233–260. https://doi.org/10.1016/S0304-3878(03)00028-2.

10

Ethical considerations and social impact in quantum agriculture

Sowmiya Velmurugan[a], Ramya Karuppasamy[a], Kumaresan Mayakannan[b] and Arunkumar Munimathan[b]

[a]DEPARTMENT OF AGRICULTURAL ENGINEERING, HINDUSTHAN COLLEGE OF ENGINEERING AND TECHNOLOGY, TAMIL NADU, INDIA [b]DEPARTMENT OF MECHATRONICS ENGINEERING, HINDUSTHAN COLLEGE OF ENGINEERING AND TECHNOLOGY, TAMIL NADU, INDIA

10.1 Introduction

The advent of disruptive technology such as artificial intelligence, quantum computing, and improved instrumentation systems is driving a paradigm-shifting picture within the global agriculture industry. These things are acclaimed as the new big thing of making the environment more sustainable, resilient, and productive of agricultural products. Demands to adopt more viable, dynamic, and data-informed agricultural operations have been increasing since population growth, climate change, and resource shortage aggravate the situation. Considering this fact, smart agriculture evolved to be an end-to-end solution that transforms the traditional mode of agriculture into highly streamlined systems based on automations, real-time data, and predictive analytics models. Along with the transformative potential, however, this also comes with an ethical burden. Besides the difficulties in operations and technology, the fact that the use of digital and quantum technologies in agriculture is being rapidly introduced brings up some relevant moral and social concerns that should be carefully observed. The problems of data ownership, algorithmic bias, the availability of resources, and cultural displacement should be solved as early as possible and carefully so that inequalities are not reproduced, and harm is not done inadvertently. The marginalized farming populations (especially those already being disadvantaged by geography or economic status) may become systematically excluded through ethical blind spots (Keswani et al., 2019). The chapter surrounds with two functions. To begin with, it aims at providing a systematic and thorough overview of the way advanced instrumentation and quantum computing are transforming the agricultural sector. It explores the principles of technology, its newer applications, and the positive effects it may hold in optimizing food systems. Second, and more importantly, the chapter is to act as a wake-up call to agricultural

Quantum Computing and Instrumentation for Smart Agriculture. DOI: https://doi.org/10.1016/B978-0-443-44247-6.00005-1

innovation with an ethical foresight. Specifically, their rights and dignity should be upheld by inclusive frameworks such as smallholding farmers and indigenous communities, and those with reduced access to digital resources. This chapter will serve as a basis for how a more sustainable and inclusive model of agricultural transformation can be developed by integrating technical analysis and ethical reflection. It seeks to ensure that no society is left behind in the digital agricultural revolution and benefits of innovation are fairly spread. Governments, technologists, researchers, and policymakers have to play a role in developing governance mechanisms that uphold the values of justice, transparency, and fairness. Finally, this chapter is a manifestation of a future where a powerful sense of ethical accountability informs the growth of technological advancements in agriculture to ensure the enhancement in agriculture is not only measured on the scale of yield and or efficiency but also integrated in the value of long-term implications in environmental health, social equity, and cultural perpetuation.

10.1.1 Overview of smart agriculture and emerging technologies

Precision agriculture, or smart agriculture, is an interactive collation of the latest digital technologies and traditional farming practices aimed at boosting sustainability, efficiency, and the ability to produce (Saini et al., 2020). At each step of the farming process, including soil preparation, planting seeds, harvest, and market supply chain, the application of this approach in real-time data collection and analysis is used. Smart agriculture is basically converting a farm that relies on experience-based decisions to one that is data-driven and can respond to the changes in markets and in the environment more rapidly. Three major emerging technologies lie at the core of this change. The Internet of Things (IoT) is a very important part of smart farming since it enables the implementation of networked sensors and equipment in agricultural environments. These sensors collect large amounts of information on important parameters such as crop condition, temperature, moisture and humidity, pH reading, soil moisture, and nutrient levels. Artificial intelligence (AI) and machine learning (ML) algorithms can be used to analyze this data and offer real-time actionable data to farmers on cloud-based platforms (Rueda-Delgado et al., 2025). The other all-transforming mechanism is big data analytics, which unites datasets produced by multiple sources, like satellite images or weather predictions and past farm data, in order to discover patterns and make predictions (Madhumathi et al., 2021). Performing tasks in a very precise and efficient way, robotics, automation, i.e., autonomous tractors, drones, and robotic harvesters, is increasingly being used to reduce the cost of manual labor and maximize the use of inputs. These innovations save time, money, and reduce the harmful impact that the utilization of excessive water, pesticides, and fertilizer has on our environment since they minimize their overuse. An agricultural aspect of a very profound task is the application of quantum computing. Even though quantum computing is in its early stages, it has the potential of changing the way modeling and data processing are radically carried out. Quantum computers run on quantum bits, or qubits, and actually can analyze large amounts of information simultaneously, unlike classical computers, which read information sequentially. It also allows modeling complex biological processes, climate, and genetic scenarios in a rapid manner. An example of a quantum algorithm would be to help determine the most suitable crop types under specific environmental conditions or reduce global food chains in situations when demand is fluctuating and the weather is changing. The existence of such technologies can transform the way farming is conducted; however, through integration, important issues concerning both economic security and accessibility, as well as ethical use, are brought forth. Poor digital infrastructure, limited access to power or the internet, and low technical literacy are some of the barriers that affect the majority of smallholder and marginalized farmers in developing countries. Moreover, monopoly or dependencies enabled by proprietary technology platforms are feared to further damage resource-poor farmers. Therefore, although smart agriculture can be a very promising solution, it is necessary that these possibilities are implemented in an inclusive and context-sensitive manner (Mukhopadhyay & Mason, 2013). The assumption of the potential of developing agricultural technologies implies recognizing the digital divide and ensuring equal access to tools and training and promoting the model of participatory innovation. This chapter goes deep into these dimensions, examining the functioning of smart technologies and social structures that affect their success and impact.

10.1.2 Ethical and social dimensions: Why they matter

The combination of the best modern solutions, such as AI, quantum computing, and smart sensors, could result in a significant rise in productivity, efficiency, and sustainability in agriculture. With these changes comes also a complex system of moral and societal concerns that should be addressed in a speedy and transparent manner. There are no adequate ethical safeguards; technological innovation can unknowingly cause even further inequity, human rights violations, and destroy cultural traditions that have been serving communities for many generations. Some of the most pressing ethical concerns are data ownership and privacy. A smart agriculture system is made up of drones, satellites, and field sensors to collect granular soil quality, crop health, water utilization, and even farmer activity data. The issue of who exactly is the owner of this data is brought up by its storage and processing by the private companies or governmental platforms in most cases. In the absence of clear regulations, farmers, especially those who own small holdings, can lose control over their data unwillingly and risk being used as a source of profit, or could miss the benefits of its analysis. Autonomy is as important as informed consent. The purposes of farming can change so much that farmers would be reduced to mere users of technologies due to the increased usage of algorithms in determining how to plant their crops, when to irrigate them, and how to kill the scourges. This dissociates the farmer from their traditional knowledge pool and withdraws their agency. There is a trend that farmers adopt such technologies without having a clear understanding of the technologies, the risks, and the party that gets in control of the outcomes. This lack of informed participation raises major ethical issues as far as accountability, transparency, and trust are concerned. The digital divide is an important social challenge. A lot of smallholder farmers, in particular those located in developing countries, lack access to such basic things like digital literacy, mobile connection or electricity (Senoo et al., 2023). The modernization of commercially large-scale operations and technologically efficient farms threatens to have an unnecessary edge in the implementation of high-tech tools, which will again displace the disadvantaged groups. It may result in more land consolidation, the formation of an agricultural class of technological elites, and an increase in the division between the socioeconomic dissimilarity in rural communities. On top of that, it has severe cultural implications. Traditional farming is not only a farming practice but is also related to ways of life, fondness of lands, and regional effacements. Automated procedures and ML models could lead to the decay of cultures since ancient customs can be replaced with them. As an example, over-reliance on external tools of decision-making can underestimate the value of approaches that may be closely associated with the local climates and ecosystems, e.g., organic composting, saving, seeds, and seasonality planning. The ethical and social considerations of smart agriculture must be accepted because there is no alternative. Ethical issues have to be put into the development and use of agricultural technologies by researchers, developers, and policymakers. This involves developing participative, explainable systems, which should be appropriate to the needs of different communities of farmers (Gondchawar & Kawitkar, 2016). All the stages of technological progress should be oriented on the main principles of transparency, accountability, and inclusivity. By ensuring that these matters are resolved before they become a serious problem, we will be able to ensure that agriculture will be enjoying a smart, just, equitable, and sustainable future. The innovation is not hindered by ethical and social aspects, which are important in legitimizing innovation and making it more sustainable.

Objectives of the chapter

The objectives of the chapter are:

1. To critically analyze the place of highly developed instrumentation and quantum computing in smart agriculture;
2. To examine the ethical issues related to the data-driven and quantum-enabled agricultural technologies;
3. To assess the social consequences of smart agriculture on farmers and the rural population;
4. To investigate governance, policy, and regulatory frameworks of responsible agricultural innovation;
5. And, to recommend avenues of inclusive, ethical, and sustainable agricultural innovation.

Organization of the chapter

The rest of the chapter is organized as follows: Section 10.2 covers the technological change in agriculture, which is advanced instrumentation and the new role of quantum computing. Section 10.3 discusses some of the major ethical implications, such as the governance of data, freedom of the farmer, openness, and the removal of old processes. Some of the social impacts that are analyzed in Section 10.4 include the digital divide, marginalization of smallholder farmers, and technological elitism. Section 10.5 deals with governance, policy, and regulatory frameworks needed in the implementation of ethics. Section 10.6 stresses inclusive and sustainable innovation with the help of digital literacy, participatory design, and ethical standards. And, finally Section 10.7 concludes the chapter with future scope.

10.2 Technological transformations in agriculture

10.2.1 Role of advanced instrumentation

Advanced instrumentation is a core component of the process of changing the traditional, experience-based methods of farming into data-driven and precision-based systems that are indispensable to the concept of smart agriculture. It is the inclusion of modern sensors along with automated observation equipment and analytics that provides real-time data on all aspects of farming processes, which is causing this shift. The technologies have transformed the way in which farmers care for their fields, respond to any change in the environment, and make informed decisions that influence sustainability, profitability, and production. One of the critical components of advanced instrumentation involves the deployment of field-portable sensors to monitor soil contents on a continuous basis, such as salinity, pH, nutrient levels, and moisture content. Through transmitting the details to central systems, these sensors allow reliable fertilizing and irrigation schedules that are devised in accordance with the specific needs of different zones of the field. Besides boosting production and body strength, such a technique is effective at decreasing waste of resources, which leads to stewardship of the latter and minimal pressure on farmers in terms of financial affairs. Weather monitoring is another important application of highly advanced instrumentation. Weather parameters measured at strategically located automated weather stations at key farms include temperature, humidity, speed, and precipitation of atmospheric conditions. The data is used in planning how to control pests, predicting when diseases might occur, and determining the most appropriate times to plant and harvest. Living in the era of increased climate fluctuation, these predictive powers are essential since they enable farmers to adjust their working processes and avoid potential losses. Aerial instrumentation introduces another rank of accuracy, particularly those drones with thermal and multispectral imaging cameras. These drones capture high-resolution photographs of the fields, hence allowing early disease, stress, and crop nutrient deficiency warning signs to be observed. Unlike reconnaissance of the field with an assistant, drones allow quick and accurate scanning of a large space and provide farmers with an aerial view of activities, and help identify issues early in their rise. Those, in addition, complicated advanced tools, such as geographic information systems (GIS) and remote sensing satellites, simplify macro-level farm analysis. The technologies provide spatial information that can be utilized in the mapping of resources, zoning of crops, and land-use planning. The global positioning systems (GPS) coupled with variable-rate technology (VRT) made possible by the use of these instruments make possible site-specific use of inputs such as seeds, water, and agrochemicals. VRT enhances the productivity of the farmland without wasting farm inputs since the square meter of farmland is being managed accordingly. Employment of the advanced instrumentation is not free of challenges; however, despite the benefits. Significant acquisition costs, maintenance requirements, and technical skills necessary to operate and comprehend such systems can be one of the main barriers, especially to small holders and resource-limited farmers. Question marks about affordability, independence, and sustainability in the long-term are posed by the fact that the new cutting-edge technology might lead to the development of dependence on proprietary platforms or third-party service providers. Consequently, the introduction of comprehensive technology models, financial aid systems, and capacity-building measures requires implementation together with an increased implementation of such tools. The government-subsidized procurement projects, schemes of collective access such as community-based equipment hubs, and digital literacy training to enhance the number of digitally literate persons can help the program become adopted among the underserved farming population.

10.2.2 Introduction to quantum computing in agriculture

As a new technological breakthrough, quantum computing can fully change data analysis, problem-solving, and decision-making in many fields, including agriculture (Pandey et al., 2021). Quantum computers utilize qubits, which can be in one state and another state at the same time, because of quantum mechanics' superposition and entanglement effects, instead of using binary bits (0 and 1) as in a type of classical computer. Due to this fact, quantum systems are able to perform complex computations exponentially faster than classical systems. Such capabilities have the capability of transforming the way modelling, analysis, and problem solving are done in agriculture in a drastic manner. The total quantity and complexity of data produced in the agricultural systems have drastically grown with the arrival of smart farming equipment such as sensors, satellites, and IoT devices. This information can cover the growth of crops and the nature of the soil, in addition to market dynamics and climatic factors. This data processing might be extremely demanding, and the insights that will allow us to apply them might not be produced in time, in conditions of an uncertain environment, when fast decisions should be made. Quantum computing makes it possible, and the processing and optimization of data in parallel can be conducted in a quantity successfully unattainable even by the most powerful classic supercomputers. One of quantum computing in agriculture promises to be the most exciting projects: crop genetics and breeding. Traditional programs of crop improvement include cross-breeding and field trials that require much time and material. Quantum algorithms, on the other hand, can precisely forecast by performing simulations of thousands of combinations of genes. It is likely to accelerate the manufacture of vigorous and high-yielding forms of crops by accelerating the discovery of favorable qualities such as drought resistance, pest resistance, or high nutritional value. Another important use is climate-smart agriculture. Quantum computing can raise the resolution and predictive capacity of climate models by joining massive data sets of soil databases, weather forecasts, and satellite observations together. These models enable farmers to make informed choices regarding future crop choices based on various climatic conditions, the best irrigation methods, and the decision concerning plant timing and locations. When aberrations in climate and catastrophically bad weather have become more frequent and threaten the international food supply, precision is essential. Quantum-enhanced optimization can also be applicable in supply chain management and in logistics. Presence of algorithms, such as the ability to recognize the presence of bottlenecks within agricultural distribution systems, real-time matching of available supply and fluctuating demand, and optimization of routes followed when transporting perishable items. These improvements facilitate sustainable food systems, more profits, and fewer wastes. Quantum computing in the field of agriculture is just beginning,

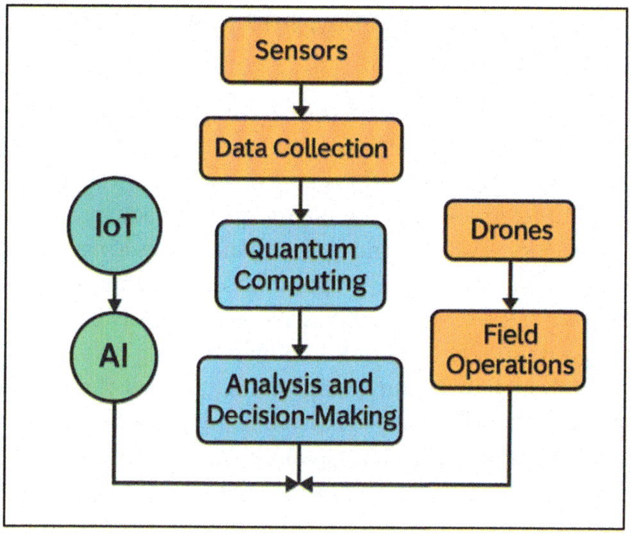

FIGURE 10.1 Technological ecosystem in smart agriculture.

Table 10.1 Comparison of classical vs. quantum computing in agriculture.

Feature	Classical computing	Quantum computing
Data handling	Sequential processing	Parallel processing using qubits
Use case	Basic farm analytics	Genetic modeling, optimization
Speed and efficiency	Limited with large data	Extremely high for complex problems
Current status	Widely adopted	Emerging and experimental

even though it is revolutionary in its nature. The presence of such problems as the expensive nature of quantum hardware, the lack of skilled employees, and the need to implement special infrastructure means that current applications remain mostly limited to research environments and pilot projects. Moreover, quantum computing is only available to large technology firms, well-endowed startups, or esteemed research institutions. To ensure that quantum computing can make a meaningful contribution to agriculture, governments, research institutions, and even the stakeholders in the private sector have to collaborate to democratize access. The open-source platforms, cloud quantum simulators, and collaborative research ventures are just some of the ways to close the gap between theory and practice. Fig. 10.1 depicts that smart agriculture is not an isolated tool driven by a technology ecosystem that acts as a team to achieve the best outcomes.

This can be well depicted in Table 10.1 that compares classical and quantum computing, wherein the quantum model performs better in parallel interaction of data, hence can be used in complex tasks such as genetic simulations, climate modeling, and so on.

10.2.3 Predictive capacity, efficiency, and precision: Opportunities ahead

AI and quantum computing, as well as the latest equipment, are making possible hitherto unimaginable precision, operational efficiency, and predictive capacity in agriculture. Besides modifying how agricultural system operations are carried out, these abilities are also enabling the existence of food production models that are resilient, proactive, and sustainable in nature. This shift lies in the possibility to predict and optimize outcomes at each stage of the agricultural value chain, including a choice of crop, preparation of the ground, delivery to the market, and adapting to the climate. Predictive analytics allows seeing insights into several agricultural variables in real time when integrated with ML models and big data. As an example, monitors in the fields collect data about temperature, crop health, the level of nutrients, and soil moisture repetitively. This data could be effectively predicted using predictive algorithms to forecast the risk of disease, pest epidemics, and yield anticipations. These projections help farmers to reduce losses and maximally produce by putting a minimal amount of treatments, scheduling harvest properly, and intervening before things get out of hand. Since quantum computing can provide the possibility to analyze huge complex data very fast, it increases the threshold of predictive ability. Quantum systems are particularly effective in solving multifactor problems, which exhibit excessive levels of off variation and interdependence, since they perform operations in parallel, as they do not perform operations sequentially as classical systems do. This would involve modeling interactions on soil carbon, simulating and predicting impacts of distinct climate scenarios on crop production, and optimal genetic selection methods toward climate-resilient crops. These insights are available to formulate both short-term decisions made within an on-farm environment as well as long-term planning and policy-making. The other notable advantage of these predictive technologies is precision agriculture. Precision methods such as site-specific fertilization, variable-rate irrigation, and pesticide application are impossible without forecasting models and specific data that tie to a very definite location. As an example, farmers will have the opportunity to apply inputs on the farms when they are needed by integrating satellite imagery with AI and quantum-driven simulation. This will reduce environmental degradation and overexploitation, and reduce expenses. This will ensure every intervention is resource-effective, ecologically friendly, and science-based. Livestock industry is also improving predictive efficiency. Wearable sensors on animals can be used to track physiological sensors such as body temperature, locomotion, and eating behaviors. Predictive models analyze this data in order to identify early signs of disease, stress, or poor nutrition. Prevention measures will ensure that all livestock quality is enhanced with limited expenditure on veterinary services by addressing the issue when it is still at an early stage. Moreover, the agricultural economy is being transformed with predictive tools by enhancing market intelligence. By considering historic pricing, weather conditions, and interruptions within the supply chain, predictive models can be used to forecast what the market will want, along with when it should plant, harvest, and sell, among other applications. This helps to achieve maximum profits and minimal wastage because farmers can then adjust their production cycles to the market conditions. That aspect of trying to come up with these technologies is not easy, but also ensuring that one is able to use as well as enjoy the benefits of the same. To have farmers, particularly in resource-constrained environments, trust and then make good use of the predictive tools, they need to be transparent, allow flexibility so they adapt to local conditions, and be backed by capacity-building efforts. Fair access and ethical usage would be the only way to actualize the promise of precision, efficiency, and predictive capacity and turn farming into an inclusive, resilient, and sustainable framework.

10.3 Ethical considerations

10.3.1 Data ownership and privacy in agricultural systems

Digital transformation in agriculture has resulted in the creation and harvesting of tremendous data using different sources; the examples include soil sensors, drones, satellites, weather stations, and smart machinery. This information is very sensitive yet valuable data that should be linked to the land conditions, crop performance, estimate of crop yields, inputs consumed, patterns of pests, as well as even financial parameters that relate to the operation choices of a farmer. Questions of ownership of such data, control of data, and use of the data when it comes to conducting agricultural activities have become dominant ethical issues as agricultural activities increasingly depend on the ability to manage such data through decision-making. Data ownership is a legal, ethical, and moral right that a person or organization should have access to, manage, control, and profit from the data they produce or add to the environment. The case study of the farmer in agriculture usually produces most of the data, but they may not own the legal properties or control them absolutely. Most contemporary agri-tech devices gather information through the cloud-based networks and software systems that are privately owned. By signing a list of terms and conditions that can generally give the technology provider the right to make wide access, farmers leave themselves at risk of being without control over information about what they are operating. Such a loss of control may result in a number of dangers. To begin with, there is a possibility of commercialization of data without the consent or knowledge of the farmer; in this way, the companies earn profits, but the farmer gets no profit or reward. Second, it exposes the possibility of competitive disadvantage since big companies will combine their data on various farms and possibly devise better analytics or market plans, a situation that a smallholder may struggle to solve. Third, sensitive information can be abused to extrapolate values of the land, determine yield results, or measure financial risk, which obviously can result in exploitation, land acquisition pressures, or credit system exclusion. Privacy is an aspect of privacy that is directly connected, which concerns the circumstances under which data is emulated, stored, or shared. Farmers, in most of the cases, are not sufficiently educated about who can have access to their data, the period that will expire, and whether they can give the information to third parties like the insurers, providers of agri-inputs, or the government. Such a lack of transparency and absence of informed consent is a contravention of the basic ethical principles of autonomy and trust. Violation of privacy might even result in disclosure of personal data, such as geolocation, financial records, and land tenancy information, which can be used to spy or intimidate an individual. In order to arrest such situations, there is a need to have well-defined and implementable data governance models that respect the rights of farmers. These

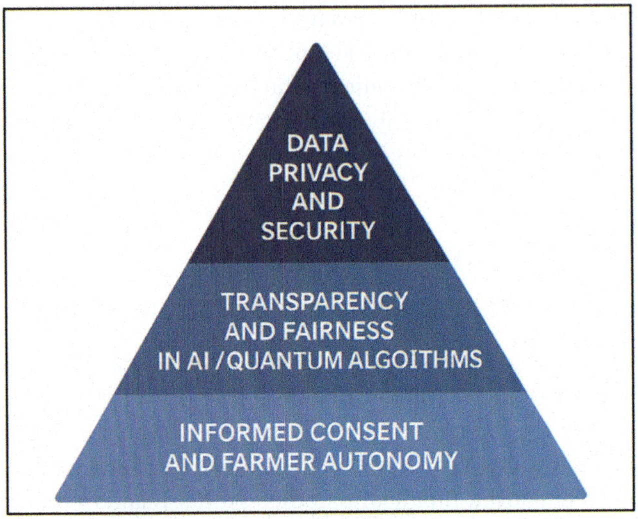

FIGURE10.2 Ethical decision-making framework in smart agriculture.

Table 10.2 Ethical challenges and mitigation strategies.

Ethical concern	Description	Suggested mitigation
Data ownership	Lack of farmer control over data	Clear policies, consent mechanisms
Algorithm transparency	Opaque AI decisions	Explainable AI, audits
Cultural displacement	Loss of traditional practices	Integrate local knowledge
Consent and privacy	Farmers unaware of data use	Local training, opt-in protocols

frameworks must outline ownership mechanisms, put data usage limits, and guarantee transparent consent procedures. Among the strategies, it is possible to distinguish the development of farmer-observed or society-managed data cooperatives, which may provide producers with the collective bargaining power and control over their data monetization or sharing. Blockchain solutions and open-source platforms can make agile industries transparent, trackable, and share data in an equity way, which government agencies and institutions should strive to establish. Additionally, training courses are necessary to focus the attention of farmers on digital rights and privacy protection and ethical uses of data. Fig. 10.2 can assist in understanding the structure that should be built to make ethical decisions in smart farming, starting with informed consent and ending with benefit-sharing.

Table 10.2 summarizes some important ethical issues (e.g., ownership of data and transparent algorithms) and mechanisms facilitating their mitigation, including local consent protocols and algorithm audits (Morchid et al., 2024). To make these complicated matters easy, the ethical challenges associated with data governance are outlined in the table given below, with proposed mitigating measures giving priority to the empowerment of farmers and transparency in such systems.

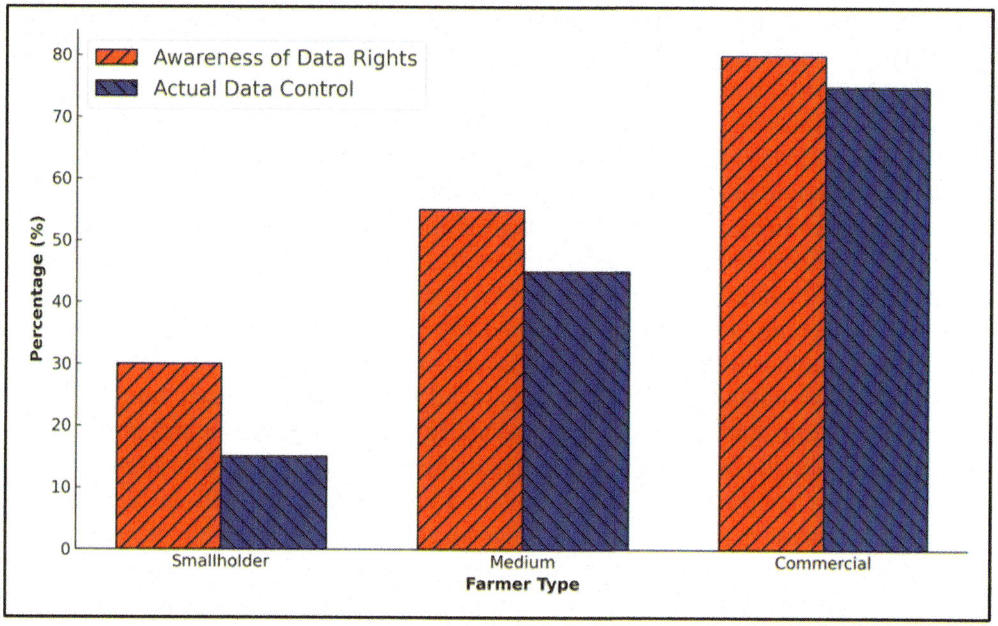

FIGURE 10.3 Awareness of data rights vs. actual data control.

The present capability of the farmers to control their data statistically is vastly disparate from their knowledge of the rights of data-access, as it is represented in Fig. 10.3. An infographic that gives a visual representation of the divide existing between what farmers know can happen with their data and what it actually does.

10.3.2 Informed consent and farmer autonomy

Informed consent and autonomy are major ethical concepts preserving the rights of an individual in opposition to the power of technologies and institutions. Such principles have particular applicability to the setting of smart agriculture since a growing share of farming choices is being made on the back of digital platforms, data-driven tools, and AI-based recommendations (Lin et al., 2019). Farmers, particularly those belonging to rural and marginal communities, must retain the right of choice and be able to decode the consequences of adopting new technologies because it will help them to ensure that innovation becomes their tool rather than demeans them or benefits others. A user agreement does not consist of merely a checkbox in informed consent in its most ethical form. That is, the farmers are well aware of what information is being gathered, how the information will be utilized, who will access the information, and the benefits or risks involved in its utilization. Unluckily, this rate of comprehension is not attained in most instances. The agricultural software or sensor-based system's terms of service are commonly expressed in legal ways, which even technologically savvy users cannot interpret, not to speak of farmers, who not only may lack literacy but also may be without access to a local-linguistic version of the document. Further, most farmers are exposed to smart technology either by government schemes or vendor demos or institutional pilots, where the focus lies more on adoption than knowledge. Consequently, farmers can use tools or share their data that are not aware of the degree of surveillance, commercialization of data, or bias in their choices through algorithms. This contravenes the concept of autonomy by making farmers engage in digital ecosystems even without their knowledge and consent into all the implications. Autonomy in agriculture would also mean the freedom of the farmers to make up their own minds based on their knowledge, context, and preferences. Although the AI algorithms and predictive modelling can provide helpful recommendations, they need to be backed up by the farmer, who should interpret, question, and reject them. The danger is that excessive dependence on digital solutions may lead farmers to become users of preprogrammed and unchangeable advice, something that will render them unfamiliar with their land, experience, and traditional activity. This dependence does not just make individual agency less powerful but threatens to have long-term effects of the erosion of indigenous agricultural wisdom. In efforts to maintain and remain ethical, technology providers/developers and policymakers should move to enhance educated choices and freedom. This involves making the terms and conditions simplified, including those in visual or spoken plane expression in local lingo, and also incorporating the decision to opt-in at each phase of the user honchie spaces. The use of digital tools needs to be accompanied by farmer training programs at the center of its deployment, where users can learn the usage of the technology, but also how to question and comprehend data and its outputs. Ownership and trust are also increased by participatory design processes, making actors codesigners of the technology they use. Also, institutional protection measures like the independent review boards, grievance redress systems, and ethical oversight bodies should also be implemented to supervise the process of consent taking and honor. These systems will make sure that they are active, and are accountable, and provide a remedy in case of breach of ethics. It is an honor to help farmers in protecting their rights, but making sure that the consent and autonomy of farmers is not a niche, but a business made possible by trust, empowerment, and cocreation of a culture of digitization in agriculture. Technology can be as good as it can only be when the farmers are partners in innovation.

10.3.3 Transparency and accountability in AI/quantum algorithms

The motivation behind transparency and accountability in AI and quantum computing is even more acute in agriculture, as these technologies become more involved than ever as part of the fundamental decision-making in the agriculture sector. Such technologies affect the fundamentals of farming, including its choice of crops, planting periods, irrigation systems, and supply chain facilities. They provide remarkable computing capabilities and the capacity to maximize performance, but due to basic complexity, their inner operations may remain mysterious, particularly to final consumers like farmers, agricultural employees, and even to policymakers. Their black box character can be viewed as an incredibly serious ethical issue that should be resolved to promote responsible innovations. Transparency in algorithmic systems is the extent to which the operations, decision reasoning, and data inputs of an algorithm are couched, explainable, and challengeable. Sadly, most of the available AI and quantum technologies already used in agriculture are not open source but are proprietary and built without a close participation of the users. This renders the manner in which recommendations are formulated, the assumptions upon which the models are made as well as the biases in the output incomprehensible to the stakeholders, particularly farmers and even local institutions. To use an example, an AI model can propose a particular pesticide routine or propose crop rotation plans on the basis of correlations of the statistics, yet without indicating why these proposals should be made. In case the farmer loses crop after carrying out the recommendation applied, it might not be easy to understand where the problem might be occurring, i.e., the algorithm or input data or other factors outside. When there is no transparency, then not only does the trust break, but the farmer does not have the option to accept accountability or compensation. Another dimension is brought in by quantum algorithms. It becomes even more difficult to understand their operations by an ordinary user, as their work depends on superposition, entanglement, and probabilistic reasoning. Although these technologies have the potential to make sense of huge amounts of data and provide highly streamlined solutions, the abstract mathematical background may render it almost impossible for individuals to trace how a particular outcome was derived, even amongst the developers themselves. This sends waves of crucial issues regarding explainability, reproducibility, as well as fairness in decision-making. Accountability refers to the statement that developers of technologies, vendors of technologies, institutions, and governments should be answerable for the consequences and effects of the technologies they develop and put in use. Lacking explicit accountability measures, the mistakes or biases in the AI/quantum algorithms will translate into physical harm in the real world, e.g., false deployment of fertilizers, price discrimination in market algorithms, or lack of access to financial systems by particular farming communities. There are a number of steps to be taken in order to promote transparency and accountability. First, explainable AI (XAI) is a system aimed at explaining the working principles of an algorithmic system. This implies producing models that can explain their arguments in a comprehensible fashion, or at least a reasoned route to their findings. Reliability and objectivity would also be achieved through independent audits and reviews of peers of the algorithmic systems. Second, technology output should be better matched with realities on the ground, through participatory algorithm development: farmers, agronomists, and local stakeholders should take part in the development and testing of models. Governments also put in place regulation mechanisms that would place an obligation on ethical review of algorithmic systems, disclosure of assumptions made in various models, and liability in the event of harm (Kaewmard & Saiyod, 2014). Using the measure of increased transparency, as shown in Fig. 10.4, increased trust and adoption coincide with increased confidence by farmers. Black box systems actually raise suspicion, whereas one that can be explained, i.e., the method of receiving decisions, is apparent gains more acceptance and reliability in the long run.

10.3.4 Ethical implications of replacing traditional practices

Modern technologies, including the application of AI, IoT devices, drones, and quantum computing in agriculture, have implemented effective changes facilitating an increase in productivity, a reduction of resource waste, and the predictability of choices. Nevertheless, such a digital

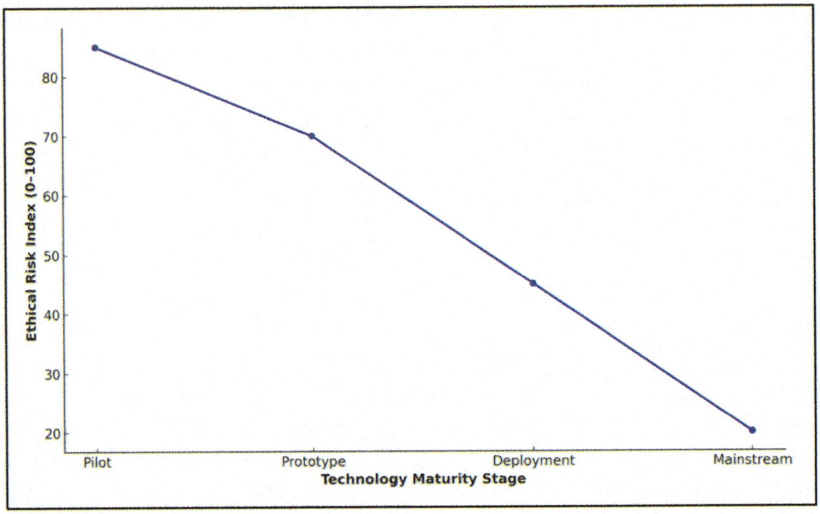

FIGURE 10.4 Ethical risk index vs. technology maturity.

revolution comes along with the gradual substitution of the traditional agricultural practices that are often strongly rooted in local cultures, ecosystems, and knowledge systems in a community. Due to this trend, multiple ethical issues are vital, and they are more than technological efficiency, as issues and questions include identity, fairness, sustainability, and future resiliency of agricultural communities (Rayhana et al., 2021). Traditional agriculture is not simply an old-fashioned way of farming; it is rather a living tradition of indigenous knowledge that has developed through the ages. Among such methods, some of the most common are intercropping, saving seeds, natural pest management, organic composting, rainwater collection, and livestock-crop combinations. They are generally connected to the surroundings and are guided by intensive knowledge of soil, weather, and biodiversity patterns in a specific area. The culture of agriculture is also incorporated with rituals and beliefs, including spirituality and communal practices, in most rural and indigenous societies. In such a manner, the replacement of these practices with modern technologies under certain circumstances may trigger the loss of not only the agronomic diversity but even of the cultural heritage and social cohesion. One of the ethical aspects of this process is the chance of ecological nudity. Conventional agricultural practice usually focuses on the biodiversity, strength, and less disturbance of the environment. The modern mechanized and data-driven systems, on the contrary, though efficient, may occasionally grow monocultures, chemical dependence, and unrestrained use of natural resources. Take, for example, accurate pesticide broadcasting using drones, where workers are also replaced, but destroying insects and pollinators could be interrelated with the alienation of the ecological peculiarities. The agency of farmers is yet another issue. Even in cases when there is no alternative to knowledge accumulated by experience, the latter is subject to underestimation or avoidance as decision algorithms proliferate. Machines that tell farmers what, how, and when to farm may sideline farmers who have perfected the crafts of cultivation over many decades due to their

Table 10.3 Ethical risks vs. technologies.

Technology	Key ethical concern	Risk level (Low/Medium/High)
IoT sensors	Data privacy, surveillance	Medium
Drones	Consent, land rights	Medium
AI algorithms	Bias, lack of transparency	High
Quantum computing	Algorithm opacity, accessibility	High

interaction and understanding of nature. In the long term, this might result in the reliance on outside technologies and market-driven inputs that undermine community self-reliance and resilience to market or climate-related shocks. A homogenization of culture is also a possibility. Corporations or other research institutions that are based either in an urban or globalized environment often design technological solutions to the problem of agriculture. It is possible that such technologies are not congruent with local farming customs or social norms or ecological factors. Consequently, farmers are at risk of abandoning their traditional ways of life in order to accept standardized and commercially approved practices, which limits the number of cultures available in rural societies and harms the richness and diversity of rural cultures. To alleviate these fears, then ethical innovation in the agricultural sector should also focus on including rather than supplanting traditional practices. This entails the codesign process with farmers where new tools are fixed to support radically as opposed to displace native techniques. Community archives, academic collaborations and national levels policies must help in documentation, protection, and revitalization of the traditional knowledge. Also, both conventional agricultural practices and scientific methods must be taught through training and outreach programs, and a dual effort toward the exchange of ideas, being as inclusive and noncontradictory as possible, must be instituted by the agricultural community. Technology is able and needs to develop in the same communion with cultural values and ecological wisdom, not at the expense of the latter. As the category of technologies listed in the Table 10.3 on ethical risks shows, the level of attention to opaque AI models and proprietary quantum platforms has increased. Mentions IoT, drones, AI, and quantum computing and lists particular moral hazards (e.g., surveillance and bias).

10.3.5 Ethical dilemmas in AI-based optimization

The models of AI applied to agriculture may optimize agricultural outcomes like output or land input. This optimization, however, might be in terms of long-term sustainability or equity. To take an example, an AI model can prescribe overwatering or overuse of fertilizer, other than considering long-run effects of soil erosion or loss of biodiversity. In addition, algorithms can underprice crops grown by smallholders or women farmers if they are out of the commercial market standards. Having mathematically optimal relationships is not always obtained ethically. So, there is an increasing necessity of including a multi-objective optimization framework, which will favor production, as well as ecological stability, fairness, and cultural applicability.

10.4 Social impacts and risks

10.4.1 Digital divide and technological exclusion

Digitization of agriculture holds the promise of greater productivity, more accurate and sustainable agriculture, but also has the potential of further disadvantaging the disenfranchised by an all-too-common term today, the digital divide. This is the term that reflects the radical differences in access to technology, digital infrastructure, and digital literacy among various societies, economies, and geographic populations. The digital divide is a problem in agricultural terms that proves to be one of the most crucial in terms of making sure that people equally enjoy the results of smart farming in any department of the population. In most less developed areas (mostly in rural and remote areas), basic infrastructure is either poor or nonexistent, in terms of electricity and availability of reliable internet connection and cellular network. The access to digital devices such as smartphones, tablets, and computers could be restricted even where the infrastructure exists because it may be expensive or have a socio-cultural issue. These infrastructural and economic disparities deny the smallholder agriculture sector (especially those in marginalized settings) an opportunity to engage and be part of the digital revolution in the agricultural sector. Additionally, rural populations also tend to have low digital literacy since it is not only the understanding of digital tools but also how to use them efficiently. The process of working on smart devices, mobile applications, or data interpretation dashboards may not be known to many farmers. The problem is further determined by language issues, a humble educational background, and the absence of the ability to participate in training. This has a dual effect, first of course on farmers being able to actually access weather forecasts, market prices, or crop recommendations, but also in not allowing farmers to access or add input through digital systems, such that they are also marginalized in innovation ecosystems. The outcome of this technological marginalization is vast. It is also possible that farmers without access to digital instruments cannot use resources efficiently, observe the state of the crop, or take action to develop climate-resistant practices. Consequently, this makes them happen to stay at the same level or even become less productive, whereas even more technologically advanced farms will be on the rise in efficiency and profits. Such a growing disparity is a factor of economic inequality and the probability of large agribusinesses taking over the market, which can afford and use the latest technologies. Along with that, digital exclusion may strengthen other social marginalizations. As an example, less access to mobile phones, training programs, and rights to land is characteristic of women farmers, which puts them at a higher risk of being left behind in the smart agriculture movement. Likewise, the people who engage in traditional farming ways might also be overlooked by the digital platforms, which are inclined toward the standardized and commercialized way of farming. This digital divide must be solved in a multi-pronged way. Governments should invest in rural digital infrastructure, such as low-cost broadband connections, mobile towers, and ways of using renewable energy, such as solar-powered device charging stations. Such efforts can be scaled at a reasonable price with the help of public-private partnerships (PPPs). In conjunction with this, digital literacy initiatives should be constructed in the local languages, culturally meaningful content, and active pedagogy. The focus of these programs must not only be on the farmers but also on the extension workers,

Table 10.4 Technological inclusion vs. exclusion.

Factor	Inclusion outcome	Exclusion outcome
Internet access	Real-time decision-making	No access to platforms/tools
Literacy programs	Confident use of AgriTech	Low adoption, increased dependency
Infrastructure support	Efficient operations	Technology gap widens inequality

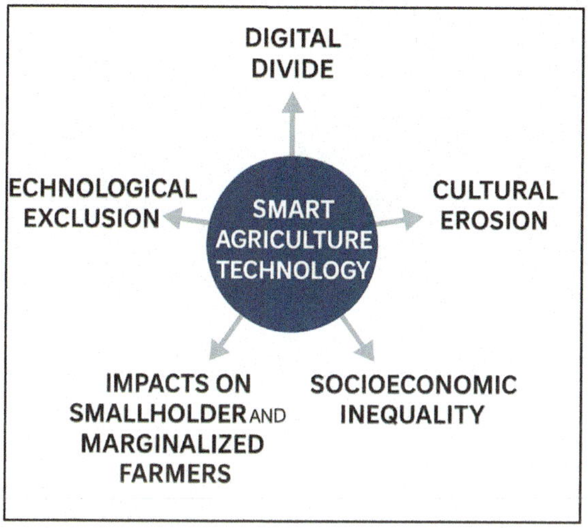

FIGURE 10.5 Social impact mapping of technology effects.

cooperatives, and the younger generation in the far-flung communities, in order to establish a conducive environment for the transfer of digital adoption. Accessibility means that there should be an affordable and accessible technology model for inclusion. Entry barriers can be lowered by means of subsidized devices, shared access centers, and technology-as-a-service (TaaS) platforms. Notably, participatory design should be the essence of innovation, and all the tools should be adapted to the real-life needs of farmers in various settings (Ibba et al., 2021). Contrasts what is gained or what is lost (e.g., real-time decision- making on the one hand, growing inequality on the other) through access conditions to the internet and literacy. Such dichotomy is reflected in Table 10.4, inclusion vs. exclusion, and as Table 10.4 points out, the lack of digital infrastructure can aggravate inequality in agricultural productivity and income.

A table of the social implications of either opposing or giving access to technology. Fig. 10.5 charts the macro effects of technological access on the society correlating the existence of the infrastructure to education performance, gender equality, and food security amongst the larger population.

Fig. 10.6 illustrates clearly that regions with unsupported digital access will always depict a low adoption level of smart genes. Displays the relationship between connectivity and the rate of agri-tech adoption.

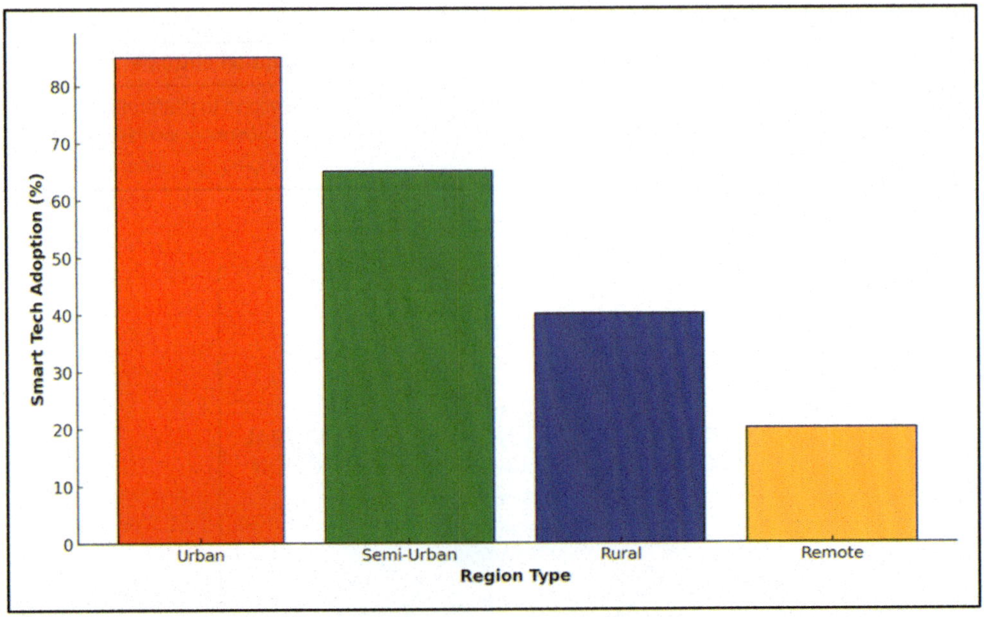

FIGURE 10.6 Digital access vs. adoption of smart agriculture tools.

10.4.2 Impacts on smallholder and marginalized farmers

The farmer groups most exposed to the changes brought about by the fast development of technology are smallholder farmers who form the majority of agricultural producers in many developing nations. The much-hyped smart agriculture technologies, such as AI, IoT devices, drones, and quantum computing, presently being made more integrated into farm systems, are becoming increasingly feared as being especially relegating to smallholder and marginalized farmers, while the larger farmers and agricultural companies are reaping the benefits. These economic and digital marginalizations are of critical ethical and developmental risk and require strategic incision given to the fact that technology needs to be used and that it has to be used equitably. Smart agriculture systems usually demand farms that are exposed to high technological infrastructure, technical assistance, and financial resources. Such systems presuppose the consistency of electricity, connection to the internet, and technical literacy, which are rather common gaps in the context of smallholders. Consequently, most of the tools and services that are being launched, starting with cloud-based data dashboards, continuing onto super-accurate equipment, are unavailable or unworkable to farmers who are working on small parcels that have limited resources at their disposal. Also, the economic model of the majority of agri-tech goods does not correspond to the financial capabilities of smallholders. Most systems need large initial expenses in hardware (sensors, smartphones, or drones), subscription charges to data, and maintenance expenses. Such investments are too costly to the smallholder farmers who run on thin margins and have no access to borrowed money. Unless brought on board into the technology benefits through specific financial means of microcredit, pay-as-you-go, or community-sharing systems, these farmers will not experience the advantages of technology to their production and survival. Further, there is marginalization, which applies to the women farmers, indigenous people, landless workers, and ethnic minorities. There are striking gender gaps in the areas of access to land, financial services, and education, as well as mobile technology. As an illustration, women farmers cannot afford smartphones or digital tool training, even though they do the bulk of the agricultural work, in most regions of the world. Their involvement in decision-making strategies in the area of agriculture, as well as arguments and the adoption of technologies, is restricted by socio-cultural norms and legal rules. Exclusion of this kind poses a threat to normative inequalities in that it can create poverty cycles and entrench vulnerability. It is not only that smallholders or marginalized groups will not get access to new opportunities, but that even their disadvantages will be made worse. Bigger farms are becoming competitive in the marketplace when they install technologies that improve yields and reduce the cost of production. This may result in the suppression of prices, consolidation of land, and the even greater marginalization of the inability to keep up. In the future, the gap between high-tech and modern agriculture and the smallholder ones can turn into an abyss of livelihoods and food security. Inclusive technology design and policy intervention help to curb such risks. Agencies, development agencies, non-governmental organizations (NGOs), and governments should focus on the necessities of the smallholders in the strategies of agricultural innovation. They encompass the participation of farmers in the development (technologies are codeveloped and tested, hence, are relevant and applicable). Extension services ought to be reinforced and redirected toward digital literacy, training, and advisory services to the smallholders and marginalized communities. Moreover, smart farming tools will become more affordable due to specific financial instruments addressed to this purpose, like low-interest loans, farmer cooperatives, and subsidies on technology. Because of this acknowledgment that both smallholder and marginalized farmers represent their own set of significant challenges, the agricultural industry can correct its mistakes and implement solutions that accentuate technological development as a driver of solidarity but not a divide.

10.4.3 Socioeconomic inequality and risk of technological elitism

There is that immense potential that comes with digital transformation in agriculture; however, there is also a great danger of increasing socioeconomic inequality based on a phenomenon called technological elitism. This is because of the incredible benefits of technology, e.g., more yield, access to markets, and optimum utilization of resources to a dominant group of relatively advantaged farmers who are already resource-endowed with financial, educational, and infrastructural capabilities to use and employ the up-and-coming technologies. The consequence is that producers with greater wealth or with corporatized agricultural enterprises can get an inordinate share of the benefits of innovation, with small-scale and resource-constrained producers further disadvantaged. Technological elitism is one of the leading reasons to focus on inequalities, according to access to digital tools and information. However, cutting-edge agricultural technologies, including satellite imaging, autonomous tractors, AI-powered analytics, and quantum computing systems, are made with the idea of having massive commercial-scale farms. Such operations are usually endowed with the capital, manpower, and system to invest and operate on sophisticated technologies. In comparison, smaller farmers grow on small budgets and may not be able to cover the expense of precision farming systems and software licenses, or text messages and surveys. Further, the advantages gained by the digital platforms and the market-based technologies tend to be enjoyed by people who are already networked on formal supply chains, credit institutions, and digital literacy networks. These people will be in a better position to use data when making decisions, be able to win favorable contracts, and receive government subsidies or privately funded investments. This forms a positive feedback loop where the more technologically equipped become more empowered, and the less resources one has, the more they fall behind. This split eventually imperils not only the fairness of economies but also the variety and sustainability of agriculture. Technological elitism is further aided by the existence of intellectual property (IP) regimes. Most of the frontier agricultural technologies, including patent seeds, closed-source farm management software, and so on, are ensured at the expense of IP laws, which do not allow them to be copied or opened up. These may either act as an incentive to innovation or may limit the transfer of technologies to those in greatest need. There are instances where smallholder farmers are trapped into a trap of dependence on third-party suppliers of tools, inputs and services hence lacked autonomy and are more prone to market fluctuations or corporate domination. Social implication of such disparity is immeasurable. In case the advantages of innovation are seen as concentrated in the hands of a rich minority, society will lose trust in the technical process, and the opposition to change can increase. Also, the continuing the divide between the tech-has and the tech-haves in rural economies may increase flows of rural-urban migration and undermine food security systems based on small-scale and diversified farmers. As a way to combat the danger of technological elitism there should be a consistent attempt to democratize access to innovation (Marcu et al., 2020). This entails coming up with inclusive policy provisions that favor participation of the small holders, subsidizing uptake of technologies by small-income smallholders, as well as sponsoring open-source technology or community-based technology options. Farmers need an equal playing field, which is facilitated by the government spending on rural broadband, farmer cooperatives, and extension services. Notably, it is necessary to collaborate with a multiple stakeholders. Governments, private sector players, civil societies, and international organizations have to collaborate in order to promote fair distribution of the technological gains. Innovation should be combined with justice and inclusion so that the agricultural sector does not become elitist and start thinking in the future, when every farmer, small or large, can prosper.

10.4.4 Cultural erosion and loss of traditional agricultural knowledge

One issue that was given little consideration as technological changes in smart agriculture continue to transform the methods of farming around the world is the threat they pose to diluting the traditional knowledge of agriculture and all that it entails (Suma et al., 2017). Conventional farming practices are important parts of the history, ecology, and spirituality of the countryside society. They are centuries of trial, survival, and close communication with nature. Nevertheless, the growing tendency toward a high-tech approach, including automated quantities of machinery or a combination of AI in crop planning and quantum modeling, threatens the survival and awareness of such heritage. The traditional agricultural knowledge is achieved through various practices such as seed saving, mixed cropping, seasonal rotation, organic composting, rainwater harvesting plus use of local remedies to keep pests at bay. Such techniques would mostly come down within families or through practical training in society. They are closely connected with climate, soils, plant and animal life, and local cultural acts. With the substitution of intuitive, observation-based farming by using mechanization and data-based decision-making, there is the threat of the ancient conventions being lost or labeled as primitive and inefficient. Such a loss has a significant cultural meaning. To most native and rural societies, agriculture not only forms a source of economy but also a lifestyle. It has spiritual values, socialization standards, service to the community, and a profound sense of identity. Community culture is composed of seed-sharing rituals, agriculture festivals, and land stewardship traditions. When these traditions are ignored and even trampled over by modern technology, the result may involve alienation of people and generations and cultural homogenization. Besides, the substitution of cultural norms is also ecologically significant. A large number of such practices are sustainable in nature and exist in collaboration with natural processes. As an example, conventional water collection methods tend to work better in arid regions, compared to the imported irrigation designs. In the same way, local varieties of seeds (which often take centuries to develop) may prove to be more resistant to pests and weather changes as compared to the commercial hybrids. The replacement of these systems with standardized external inputs and technologies means less biodiversity and possibly more exposure to external shocks by farming systems. Ethical obligation is hence not only to preserve cultural practices as a symbol of things but also in the realization that they are useful, scientific as well as in ecology. Innovation in morality must seek to coexist, not to replace. That implies the use of conventional wisdom in the planning of intelligent farming systems, where technologies would be proper and relevant to the local context and values (Krishna et al., 2017). There are practical measures that could be adopted to preserve and enliven traditional knowledge. Among these are documentation work, local archives, and putting the indigenous methods of farming into school curricula and agricultural extension schemes. Intergenerational knowledge transfer can be of the essence through farmer cooperatives and NGOs. Traditional knowledge should be acknowledged as the IP that needs protection and reward, especially in regard to climate adaptations and food sovereignty by governments. Besides, local farmers are to be involved in participatory design designed by technology developers and researchers. This team evaluation develops mutual respect and creativity, resulting in hybrid models of combining the merits of convention and inventiveness (Fig. 10.7).

10.5 Governance, policy, and regulation

10.5.1 Ethical responsibility of stakeholders: Governments, technologists, legislators

The digital and data-driven nature of technologies in agriculture, deploying AI, quantum computing, smart instrumentation, and others, makes the ethical responsibility of the key

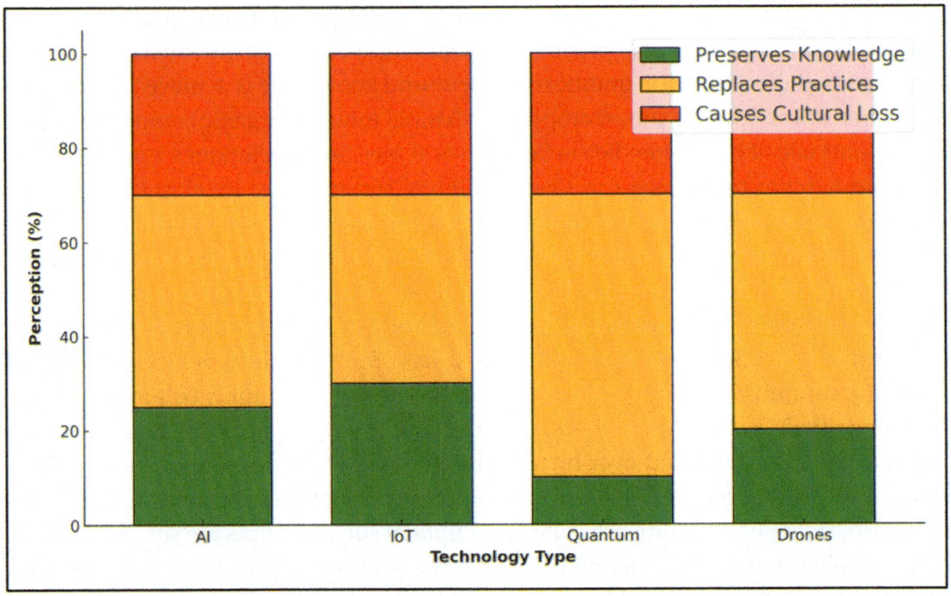

FIGURE 10.7 Perceived impact of smart technologies on traditional practices.

stakeholders, including governments, technologists, legislators, and other ecosystem players, more prominent than ever. Although innovation is a key to enhancing food security, productivity, and environmental sustainability, it should be developed ethically foresighted and with all-inclusive governance. This is because the neglect of ethical implications at every step of the way of decision-making and execution may result in becoming exclusionary, disparate, data use abuse, and loss of trust in the community. Governments have a central role in dictating the ethics and the control over agricultural innovation. It is not their only role to make policies, but they must also make sure that technologies are accessible, affordable and equitable among all farmers, the ones who are smallholder, women, and marginalized groups. Government agencies are well placed to play a pivotal role of custodian to the interest of the people. One way in which they can achieve this is by investing into digital infrastructure of underserved regions, developing incentives to promote inclusive innovation, and ensuring that there are effective data governance systems to ensure farmers rights. Governments should also sensitize the masses on the need of informing the farmers on their rights, available technologies, and usage ways without jeopardizing themselves. Moral policymaking involves foreseeing unforeseen effects, broad consultations of the relevant communities, and the technology should not substitute the local farming skills and traditions. Developers, engineers, and designers are technologists which hold a different ethical responsibility. Being the developers of the algorithms, software platforms, and hardware tools, they can directly affect the way technologies work and which people will get benefit. They should be ethically designed, and the principles of fairness, inclusiveness, transparency, and environmental responsibility have to be integrated into it. Technologists are expected to attempt to prevent biases in the algorithmic models, make them accessible to different languages and

Roles	Role and responsibilitties
Government	Establish regulations and policies
Developers	Design ethical technologies
NGOs	Promote equity and inclusiveness
Farmers	Participate in co-creation

FIGURE 10.8 Stakeholder responsibility matrix with icons.

literacy levels, and provide the tools that fit the various agro-ecological conditions. Moreover, they have to be open to farmer feedback at design and pilot trials and make their innovations to respond to the real world and not push top-down solutions. In so doing, they can assist in bridging the technological potential and social reality gap. It is the duty of policymakers and the courts to make sure that the law code and regulatory systems not only keep up with technological advances but also evolve in a similar fashion. Current laws are not usually up to dealing with data ownership issues, AI responsibility, patenting in bioengineering, or quantum-based decision-making. Lawmakers should come up with dynamic, progressive policies that protect the interests of farmers, stem off against monopoly and clarify the responsibility in case of technological malfunctions. This involves revising the data protection legislation, developing reference points of algorithm accountability and promoting global cooperation on ethical standards and technology transfers. Besides these main stakeholder categories, civil societies, academic institutions, and farmer groupings need to be directly involved in ethical governance of agriculture. These organizations also play an important role of mediators- they create awareness, represent the rights of farmers and bridge the gap by engaging in inclusive discussion between innovators and users. Finally, smart agriculture is a shared activity as ethical responsibility overall. It requires openness, responsibility, and cooperation in every sector of the farming environment. Through mutual recognition and collaboration, governments, technologists, and legislators will develop an environment of innovation that will not just be advanced technologically but also a socially just, environmentally sustainable, and ethically robust environment. The visual representation in Fig. 10.8 offers clues as to what the matrix means and can better explain how the responsibility of being ethical is broken down into the agricultural innovation ecosystem. A graphic schedule that will relate each stakeholder to particular ethical tasks.

Table 10.5 shows an outlined map of ethical obligations of different actors: governments, developers, farmers, and NGOs. The functions and the role of these players are briefly introduced

Table 10.5 Key roles of stakeholders.

Stakeholder	Ethical responsibility
Government	Regulation, subsidies, digital inclusion
Developers	Ethical design, transparency, and environmental safeguards
Farmers	Informed decision-making, local knowledge sharing
NGOs/Academia	Capacity building, participatory innovation

in the table of stakeholder matrix, which identifies ethical obligations of different principles, such as regulation to sharing of knowledge.

10.5.2 Need for inclusive and adaptive legal frameworks

Due to the blistering digitalization of agriculture (to be achieved by the development and application of sophisticated technologies, including AI, quantum computers, instrumentation based on the IoT, and blockchain), in the context of the growing complexity of ethical and economic as well as social issues, the current law and regulations are becoming more and more inadequate. Conventional agricultural policies were set up to deal with analog systems, which have fairly simple inputs and outputs, institutional arrangements of ownership, and land-focused affairs. The latest stage of data- and rule-driven farming, however, requires rethinking the laws and how they are framed, interpreted, and enforced in a fair, accountable, and inclusive way toward everyone. Open and dynamic laws are required to resolve all new issues: data ownership and sovereignty, algorithmic decision-making, privacy protection, intellectual property rights (IPR), and liability in case of defective automation or unequal distribution of benefits (Table 10.6). These obstacles also interrelate and impact diverse groups of stakeholders that include the smallholder farmers, rural populations, technology developers, agribusiness corporations, and governmental agencies. Unless substantive legal frameworks are in place to guide it, smart agriculture may turn out to be the realm of ambiguity, exploitation, and injustice. An essential element of this digital transition is the issue of data governance, which is one of the most significant legal questions. Who is in possession of the data produced in farms? Who will be allowed to utilize, have access, and benefit thereof? In numerous situations, the farmers lack legality and security concerning the data harvested by the outside sensors, drones, or software platforms. With no clear rights to data, AgriTech companies can foster a data monopoly and operate without the transparency and fair returns to the originator of the data. A legal framework that recognizes farmers as the first owners of agricultural data and develops legally binding procedures of informed consent, data anonymization, access controls, and fair sharing of the benefits of data analytics should be implemented. Of equal importance is the control of AI PM that is driven by quantum computers. In their effectiveness, such systems are usually functioning with an oblique algorithm that can lead to prejudiced or detrimental results. Laws and regulations should be changed to require that the systems in agricultural ethical technology systems should be transparent, fair, and explainable. There must be a channel through which aggrieved parties can appeal against automatic decisions, seek redress, and hold developers and service operators

Table 10.6 International models for ethical governance in AgriTech.

Country	Ethical regulation in AgriTech	Key initiatives
India	Draft Digital Agriculture Policy	AgriStack, eNAM, FPO-based data governance
Kenya	Data Protection Act (2019)	FarmDrive, DigiFarm
Netherlands	AgriTech Ethics Council	RAIA (Responsible AI in Agriculture)
Brazil	Digital Inclusion in Rural Law	Tech-access subsidies for family farming

accountable. Besides, the statutes regulating IP should not only provide a greater incentive to innovate but also serve the interests of the community. Unreasonably protective patents of seeds, algorithms, or a data environment may restrict the usage of smallholder farmers and preclude joint studies. Open-source licensing systems, collective ownership forms, and the PPPs may serve the purposes of innovation and equity: they can be promoted by the adaptive legal systems. It is also important that legal processes are inclusive. There are many occasions where rural and marginalized groups are left out of the political arena of legislative discussions and policy decisions. It leads to the development of laws that are not based on the reality of the affected people. The governments need to have participatory legal design processes where the farmers, the local organizations, and the underrepresented groups are incorporated. Legislation must be flexible and subject to amendment along with the development of technology, and should be made in such a way that it is easy to explain. Last but not least, international collaboration will be critical given that agri-tech could be crossing borders via cloud computing, AI methods, and international data transfer. Convergence of legal norms between the countries may contribute to ethical research and development, protect small farmers in the global south, and avert regulatory arbitrage by influential forces.

10.5.3 Models for equitable access and infrastructure development

Good and fair availability of technology and underlying infrastructure is the underlying demand to successful deployment of smart agriculture. Digital is seen to have so much potential with AI-based crop planning, sensor-based irrigation, and satellite-based field monitoring but must they be developed to their potential by ensuring that everyone in the farming community has the ability to employ, use and even gain access to the tools they can deliver. Unless deliberate efforts are made to employ inclusive forms of infrastructure, smart agriculture would be a source of strengthening systemic inequality by centralizing innovation solely in those groups that are already privileged. Inequality in the distribution of infrastructure, or the lack of equality in terms of electricity, internet connectivity, access relating to, and using smart devices, and proper transport systems, rank high as being the cause of the digital divide in agriculture. In a number of developing nations, the rural areas have no or very unreliable broadband wireless service or power. In the developed world as well, smallholder farmers in the isolated regions could have problems connecting to high-speed networks or sustaining complex equipment. Technological elitism should therefore be avoided by ensuring that infrastructure development is done simultaneously with digital innovation in order to eliminate the problem of exclusion. The creation of PPPs should be one of the effective models to address this access gap. Technology firms, governments, and civil societies can join forces and implement infrastructure solutions that are cost-effective and scalable. As another example, partnerships may invest in solar-powered digital kiosks in off-grid agricultural villages, or they may provide subsidized products such as smart sensors or 4G/5G mobiles in agricultural areas. Such joint ventures limit the amount of money spent on one organization while ensuring maximum reach and sustainability. Another seemingly great model is rural innovation hubs. They are village-level facilities with collective technical capabilities, including weather sensors, AI access terminals, soil testing kits and kilometers of agronomic advice. They incorporate smart tools into the routine of farmers by training, troubleshooting, and data-sharing services, given locally via local cooperatives, NGOs, or extension agencies. They are also part of the co-creation platform because farmers can collaborate with developers and researchers to adapt technologies against the local context. Another new model is called TaaS, as farmers can subscribe to or rent out the high-priced machinery or the data services as they need. One can get access to drones, GPS-guided tractors, or yield-monitoring platforms via farmer cooperatives or local agri-tech centers without fully owning them. It is a model that works particularly well in a resource-limited setting, and it fosters inclusive innovation at scale. The provision of mobile-based extension services is also instrumental in making sure that there is access and equity (Gagliardi et al., 2021). Mobile phones are more common than broadband or computers in large sections of the world. Governments and agri-startups are utilizing SMS, voice-enabled applications, and local language interfaces to provide local recommendations, market prices, weather forecasts, and training material to farmers, even in remote areas (Antonacci et al., 2018). Nevertheless, infrastructure can be solved only in a physical way. The key soft infrastructure element is the human capacity itself, crucial to the process of building the national digital literacy, institutional support, and creating localized content. There is a need to lay more emphasis on training programs, workshops, and field demonstrations to make sure that it is not only the technology that is available but also that the farmers understand how to be critical and use their technology effectively.

10.5.4 Intellectual property, data sharing, and benefit distribution

In the new environment of smart farming, in which the productivity and decision-making process are dominated by the use of data, algorithms, and novel biological technologies, the IP, data sharing, and fair distribution of the benefits pose a burning ethical and legal question. How such factors are managed decides how agricultural innovation is in control, who benefits, and whether there is exchange of improved practices within farming groups and countries. Historically, IPR were related to tangible resources, and connected with machines or plant breeds. Nonetheless, digitalization of farming has resulted in the appearance of new types of assets that are not tangible but very valuable, including AI models, crop genome databases, software platforms, and recommendations based on gathering, analyzing, and acting on data. These resources are being enclosed more than ever by private compartments and research establishments, making use of patents, copyright, as well as trade secret over resources (Prathibha et al., 2017). Though IPR regimes are needed to stimulate innovation and to provide certain protection to creators, too demanding IP regimes may disrupt knowledge flows, drive up the access price, and place smallholder farmers in dependency relations to proprietary tools and services. Arguably, proprietary rights on seed technologies and genetically modified organisms (GMOs) are one of the disputable topics. In most instances, the seeds patented cannot be salvaged or resown by the farmers, and they have to buy new seeds each year. This compromises the usual traditions of freedom of seed and sovereignty, especially in the indigenous people who have been nurturing and maintaining biodiversity for centuries. Moreover, farm data is processed by patented algorithms that make recommendations and are, in most cases, black-box and unavailable to the end users. This forms an information discrepancy in which the developer makes money out of the information created by the user but does not give much in terms of feedback to the farmer. Another tricky dimension is data sharing. This explosion of data in farms, which includes soil, weather variability, pest data, and market data, has tremendous potential value as this data is accumulated and analyzed. Nevertheless, small farmers do not understand how their data can be used, or by whom, as well as how this data can be used to develop greater innovations in agri-tech. Ethical data sharing demands the existence of clear protocols regarding consent, anonymization, access control, and transparency. The farmers should keep ownership of their data by having the right to participate or not in the data sharing contracts and by guaranteeing that their data will not be used against them. The last pillar, which connects IP and data ethics, is equitable distribution of benefits. Farmers, specifically those who made donations in the form of indigenous knowledge, distinctive genetic material, or longitudinal data deserve to be known and rewarded. The forms of benefit-sharing mechanisms may include the following: revenue-sharing contract, discounted services, capacity-building, or community investment. International systems, like Nagoya Protocol on Access and Benefit-Sharing provide templates of how the genetic and cultural resources should be utilized responsibly, but more regional and inclusive systems are needed to deal with daily data and innovation transfer. The alternatives to the top-down, IP-intensive models of agri-tech that do exist are likely to be data cooperatives owned by the community, open-source software projects, and innovator networks led by farmers. Such strategies focus on shared responsibility with the knowledge and fair compensation of innovation. These initiatives have to be promoted by the policymakers and the institutions in terms of legal waivers, subsidies, and capacity building.

10.6 Inclusive and sustainable innovation

10.6.1 Promoting digital literacy and infrastructure in rural areas

Widely propagated digital literacy and the creation of powerful infrastructure, in particular rural and underserved regions, are also one of the cornerstones of the successful implementation of smart agriculture technologies. As much as technologies like AI-based advisory systems, IoT-enabled sensors, drones, and data analytics platforms promise precision, efficiency, and sustainability in agriculture, these advantages would be of no use unless the farmers possess the knowledge, skills, and resources to access and use them. Digital literacy is a concept of the use of digital tools and platforms with sufficient confidence and criticality, including concepts of operating mobile applications, data visualizations, browsing online marketplaces, and interacting with automated systems. The digital literacy in most of the rural communities is very low because of several obstacles, such as low levels of education, language incompatibilities, poor access to technology, and gender discrimination. In cases where they have hardware like smartphones or tablets, farmers have not received the training and support needed to access and use agricultural applications, read instructions in digital form, or ascertain the validity of what they are receiving. This disparity poses a major threat of technological marginalization, and only the small group of digitally savvy or affluent farmers are taking advantage of these new technologies at the expense of the others. In order to counter this, digital literacy should be viewed as part and parcel of agricultural development and not an extra (Bachuwar et al., 2018). The governments and schools should invest in extensive digital training programs that should be local context-specific and actors in the NGO and the private sector should play a role in the same. Those programs are supposed to be conducted using regional languages, with culturally related examples and they should pay attention to both men and women farmers. Training programs may be community based, mobile, farmer's fields schools among others. This can be accomplished by use of local champions or youth who have digitally and technology skills as trainers, generating relationships of trust and bringing in more engagement. Technology demonstrations in practice are especially powerful (i.e., in real-life farming situations) as farmers can experience the benefits of digital solutions firsthand (i.e., healthier crops, reaching new markets, or adapting to changes in the weather). Similar to digital literacy, infrastructure is also important to deliver stable and unifying access to digital services. Most infrastructure that is fundamental live, such as electricity, internet, and stable communication networks, is poorly installed or completely lacking in numerous places in the rural setting. In the absence of these prerequisites, even the most progressive and favorable to the user technologies cannot be achieved. The infrastructure solutions need to comprise not only hard solutions (expansion of broadband, solar-powered charging stations for devices, support maintenance systems, and a locally based help desk over which troubles can be resolved). The mobile-first strategies are particularly useful in areas where smartphones are widely used compared to desktop computers (Mukherji et al., 2019). The barrier challenges can be addressed with tools that incorporate a voice-based interface, SMS, or a plain-icon interface, compared to text-intensive interfaces. It is also important that the policy is in support. An incentive should be given to telecom companies and tech developers by the governments to increase coverage in regions that are rural and to create devices that actually fit such low-resource environments. The government-business alliance may contribute significantly to financing the infrastructure and large-scale training (García et al., 2020).

10.6.2 Participatory technology design and co-creation with farmers

It should not be a one-solution or top-down approach when embedding the emerging technologies in agriculture, including AI, the IoT, and quantum computing. Rather, these innovations are to be created by such participatory methods that fully engage the farmers at the level of conceptualization, experimentation, and improvement. A participatory technology design (also referred to as co-creation) is what achieves this to make sure that smart agricultural tools are relevant, culturally appropriate, and practically applicable to the local realities of the people who utilize them. The point behind participatory technology design is that end-users, farmers in this case, have valuable experience and knowledge, which can be used to make innovations much more useful and effective. In the case when developers operate independently, they face the threat of developing tools that would be good in technology but which will be inapplicable or useless in the actual farming environment (Zhao et al., 2024). As an example, a very advanced weather forecasting app will be hopeless in case it does not support the local dialect, does not involve any nearby farming calendars, or needs an uninterrupted internet connection, such as that not available to rural users. Likewise, tools developed to fit in monoculture production regimes might not work with the mixed and intercrop plots practiced by global smallholders in the Global South. These risks are reduced when farmers are involved in the process of designing since they are able to see direct feedback. Interviews, focus groups, field trials, and pilot tests are an opportunity through which farmers can speak up their mind concerning the likes and dislikes, complain about the challenges they are facing, and give recommendations on the available tools being developed. This leads to technologies that are more natural, flexible, and can resolve actual issues (Shaikh et al., 2022). Developers, in their turn, will receive more information in user contexts, which will make their products more relevant and easier to use. Co-creation sets up a sense of ownership and trust in the users. Farmers will tend to take advantage of the technologies and stick to them once their inputs are recognized and recorded in the final products. Additionally, participatory design can be used to overcome the digital divide through the mutual innovation environments between developers, researchers, extension workers, and farmers. The collaborative energy is particularly crucial in making sure that marginalized populations, including women, indigenous people, or landless workers, have a say in the creation of technology upon which their activities rely (Mitra et al., 2022). Successful incidences of participatory design within the agricultural sector can be found climatically across global regions. Community-based video systems, such as Digital Green, are also used in India to enable farmers to develop local content and to disseminate good practices. Farm Drive was one of the mobile financial tools that were co-created with the farmers to enhance their useability and credibility in Kenya. Such initiatives, however, show that where farmers are talked with as co-creators and not as victims who are benefitting gratuitously, there is an increase in the level of adoption and there seems to be an increase in the social impact of technology (Aleotti et al., 2018). Institutions have to develop enabling environments that allow people to scale up participatory innovation. NGOs, research centers and agricultural universities may act as the facilitators of co-creation hubs. Policymakers and funders ought to give priorities to grants and incentives in favor of inclusive research models. The developers must also be trained on social inclusion strategies, participatory techniques and not necessarily on technical skills (Kanchi et al., 2019). This is further corroborated by Fig. 10.9 that shows that the greater the involvement of the farmer in the design process, then the greater is the outcome of the implementation process. Displays the positive correlation between participatory design and the rate of technological adoption success.

A cycle diagram depicting codesigning, testing, feedback, and scaling. Such a cyclic approach has been depicted in Fig. 10.10 that presents the cycle of inclusive innovation-co-design to localized deployment.

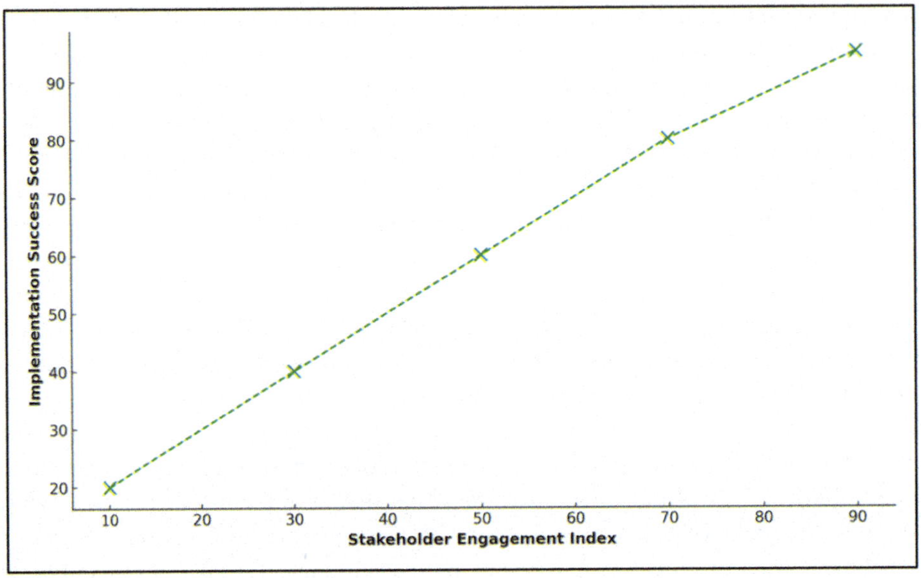

FIGURE 10.9 Stakeholder engagement vs. implementation success.

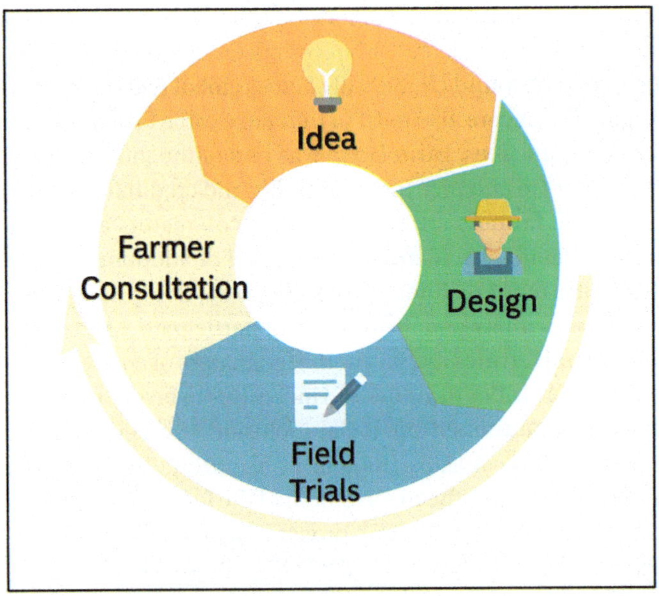

FIGURE 10.10 Inclusive innovation cycle for smart agriculture.

10.6.3 Ethical guidelines for developers and implementers

The ethical duties of developers, programmers, and planners of digital technologies are emerging as they further constitute the core of agricultural innovation. The people actually developing and implementing systems are no longer wizards in the corner, carrying out purely technical work: rather, they are designing systems that directly influence livelihoods, food security, environmental sustainability, and social equity. Therefore, there is great urgency in setting up a solid set of ethics that will help agri-tech become something that is beneficial to everyone, with special attention to those who are most likely to be excluded and exploited. Engineers who make software, data scientists, and AI modelers need to incorporate ethics in the design process at the first stage (Roy & George, 2020). The latter involves recalling the social and cultural heterogeneity of farming communities; not making assumptions about single user needs; and making sure that technologies are not fixed to specific contexts. As one of the examples, mobile apps created to implement smart irrigation must be able to work either online or offline to support regions that do not have good coverage. It needs to have interfaces that support local languages, have graphical or audible instructions for less-literate users, and have opt-in/opt-out options on data sharing and automation. Ethical codes involved in the creation of digital agricultural tools are the ideas of accepting fairness and transparency. Recommendation systems, financial analysis, or yield forecast models should be explainable and cannot be biased or discriminatory. As an example, in the case of an AI systematically downplaying the crops cultivated by women or those of minorities, it underlines social injustices. Ethical impact assessments deserve to be used by developers in order to reveal these biases and address them by using various sources of data and participatory testing (Adesipo et al., 2020). Data protection and privacy are very important as well. The farmers need to be made aware of the kind of data that is being gathered, how the data will be utilized, as well as who will have access to the data. System developers need to make sure that such systems are subject to data protection laws and follow data sovereignty. Anonymization of data is to be eliminated whenever possible, it is to be put under secure storage, and at no time sharing is to be done without consent. Also, farmers can be enabled by the use of open-source platforms and decentralized data structures that allow farmers to maintain ownership of their data. There are other ethical obligations that operators should have, complementary implementers, who are the ones who have to implement technology in action. These are making sure that technologies are implemented in a culturally ways, delivery of good training, and follow-ups. They (implementers) should not be sales agents or technicians; they must be facilitators. They should be in a position to tell the advantages as well as the shortcomings of the technology, assist in interpretations of outputs, and canvass views on how improvements can be made in the future (Bouali et al., 2021). Capacity building should be an essential part of any deployment strategy. The implementers will collaborate with local extension officials, cooperatives, and farmer associations in the implementation of digital literacy teaching and appropriate use. Additional consideration ought to be paid to reaching out to the marginalized population who would have been otherwise left unheard because of the language, gender, or socioeconomic constraints (Rajesh et al., 2020). Ethical responsibilities may be formalized with the help of institutional support mechanisms. Developers and implementers can be accountable through ethics review boards, codes of conduct, and certification programs. The ethical corporations ought to be mandatory in the approvals and evaluations of projects by governments, universities, and their funding agencies.

10.6.4 Case studies of inclusive smart agriculture initiatives

The evidence in real-life case studies is convincing that an inclusive, ethical, and transformative smart agriculture can effect and implement using the active contribution of the local communities (Liu, 2023). These are innovative projects that not only display technological ingenuity but also illustrate how the model of inclusion can accommodate the needs of socioeconomically challenged communities, honor the traditional wisdom, and give a voice to the vulnerable populations. Through such examples, policy makers, researchers, and technologists may learn important lessons in constructing equitable agricultural systems (Shadrin et al., 2019). A good example of a template of technology platforms that focuses on local knowledge and community engagement is Digital Green (India). Instead of the top-down introduction of difficult-to-use digital technologies, Digital Green works with local communities to produce short, local videos in the local language. The videos are created by farmers and about farmers, showing the best practices in issues like composting, pest control management, conserving water resources, and organic farming. The style makes the content more timely, easy to read, and believable (Rathod et al., 2020). There are local facilitators screening at village meetings and responding to questions, so the technology does not go alone without human contact and the establishment of trust. Proper research has revealed that farmers who watch Digital Green videos have a much more chance of engaging in sustainable agriculture. Another good illustration is provided by the Climate-Smart Villages (CSV) initiative by the CGIAR Research Program on Climate Change, Agriculture and Food Security (CCAFS). Adjusted in India, Nepal, Kenya, and Vietnam, the CSV model is concerned with the combination of the latest scientific tools and traditional wisdom to develop climate change resilience. Vulnerability assessments are done, villages are chosen, and farmers codesign interventions. Such climate-resistant options will entail the use of climate-resilient varieties, weather forecasts, soil analysis, irrigation practices that can save water, and retraining. The participatory nature of the model is an essential part of it—farmers provide their observations and preferences, which steer project priorities to make them locally relevant and owned. FarmDrive has started in Kenya, where it has been running a trial where alternative data, including phone usage patterns, weather information, and farming history, could be used to determine the creditworthiness of small farmer applicants who had no formal financial records. Conventional financial institutions tend to find such farmers risky and unbankable (Abdullah et al., 2020). The inclusive model of FarmDrive eliminates the divide between the two parties, namely the smallholders and financial institutions, by having the digital data turned into an accessible, real-time risk analysis. Through this, the farmers are able to access loans and insurance products, which are important in order to undertake farm investments. The application of mobile technology to reach people, even those in rural regions, and the concept of financial inclusion show how the cleverness of the technology can dismantle the long-established systemic barriers. The other successful example is SmartFarm of Pixatel in India that delivers voice-based recommendations to farmers, soil health analytics, and accessibility of inputs at affordable prices (Vadapalli et al., 2020). This effort has been led to the creation of trust via grass-roots arrangements to the community cooperatives and the village leaders. Notably, the applications of SmartFarm are user-friendly, even when it comes to people with low literacy levels, and are receiving in one of the local vernaculars, an aspect that is significant to the women and elderly farmers. These cases show that agricultural inclusive innovation can be achieved and is scalable. The main success factors are participatory design, support of local language, affordability, as well as integration with community structures. The emphasis on cocreating, access, and collective advantage through such initiatives breaks the pattern that high-tech farming has to be inequitable or culturally irrelevant (Hakhrani & Hamid, 2024).

10.7 Conclusion and Future Scope

10.7.1 Summary of ethical and social concerns

The use of high technologies in farming, especially AI, quantum computing, and precision equipment, has been shown to have acute success in transforming food production systems. But with these opportunities of revolution come severe ethical and social issues that have a persuasively thought-spring nature, along with a powerful state of governance. Technical ability should not be the only driver in the quest to achieve a smart agriculture; the search should also have a larger ethical hit big enough to ensure that the interests of all parties involved are taken care of, and none other is made vulnerable due to marginalization or isolation. Data ownership and privacy are two of the most urgent ethical issues. With the digitalization of farms and the abundance of data, more and more sensitive information is accumulated by sensors, drones, and software tools. This information usually includes information regarding agricultural production and land ownership, harvests, soil fertility, and even agricultural practices and the financial patterns of farmers. In the absence of clear and binding regulations, they may take this data into the hands of private corporations or third parties, marginalizing the farmers who create it. Often, the farmer can relinquish the right to the data in a nontransparent consent process, making it easy to see how the data can be abused or that the farmer may be left out when the benefits of the analytics are later shared. The other problem that should not be ignored is the lack of clarity and responsibility in AI and quantum-based decision systems (Rayhana et al., 2021). Such technologies are becoming popular in producing recommendations on such things as input usage, irrigation timings, selection of crops, among others, and even financial services. However, most of these systems are very much like black boxes, and we are given very little insight into how it comes up with decisions. Such obscurity does not give farmers the confidence and makes them paranoid of bias, incorrectness, and absurdity of not having a way to appeal to automated judgments, when such judgments involve the cost of money or the environment. Digital divide and technological exclusion are also social issues of concern. Smart agriculture presupposes the presence of digital infrastructure (connection to the Internet, power, smartphones, and cloud platforms). Nonetheless, such access is unavailable to millions of smallholder farmers, especially to those in the developing world. There is a danger that this gap will further widen the current inequalities in the benefits of the technological advances, since only those who are well-connected and well-capitalized farmers will take advantage of it. Economically marginalized communities, those in indigenous groups, and women in particular, are prone to being left behind. It also, however, becomes a moral and environmental issue when the traditional knowledge and practices associated with agriculture are washed away. A lot of smart technologies are designed in such a way that they meet standard uniformity, which do not fit locally adaptive, sustainable, and culturally critical modes of farming. Replacing centuries-old traditional ways of doing things with an automated system not only poses a threat to biodiversity and local resilience but also makes communities lose respect for their heritage and identity. In addition, the threat of technological elitism, which captures the benefits of innovation by a few actors, should be overcome. Agricultural enterprises and technological companies can engulf entire markets, controlling prices and marginalizing smallholders can dominate their markets because they facilitate IPR, data monopolies, and high-tech tools. Such a concentration of power jeopardizes inclusive growth, as well as the fact that it poses a threat to food system equity. Overall, the moral and social issues linked to digital agriculture are wide-reaching and integrated. They cover the concerns of justice, autonomy, participation, equity, sustainability, and cultural preservation. Policies on these issues must involve farmers, technologists, governments, civil society, and researchers, who represent a multi-stakeholder approach to their solutions. The agricultural industry can take advantage of technology and respect the rights and dignity of the food producers through inclusive governance, principled design, and participatory innovation models.

10.7.2 Vision for equitable tech-enabled agriculture

The really radical concept of food in the digital space cannot be limited only to technology development, but must guarantee equity, sustainability, and inclusion. Fairfield tech-enabled farming predicts the potential of its future with the innovations of AI, quantum computing, big data analytics, applying smart instrumentation as well as affecting all people, and not only large business operations, as they are isolated now, no matter the geographical or any other links of a farmer, his/her financial or cultural backgrounds, and gender. This vision harmonizes the technological advancement with the standards of justice, environmental integrity, and social empowerment. The core condition of this vision is open access to technology. It implies making sure that all farmers, regardless of whether they live in a remote rural village or the biggest farm on the edge of urban development, possess the means of enjoying smart agricultural systems infrastructures, training, and tools. This needs investment in rural broadband, access to electricity, mobile, and community-based shared technology centers. Equitable access also implies the development of scalable economic models that lower up-front cost of technology, i.e., pay-as-you-go systems, micro finance schemes, and sharing ownership of high-tech equipment. One of its main principles is cocreation and participatory design. Farmers should not be considered only end users- they are a part of developing knowledge and must be at the heart of innovation. Every technological solution ought to be built on their customary knowledge, experience in the field, and situational awareness. In such a vision, the development of the technology would be a shared enterprise that would be defined through continued interaction among the technologists, the researchers, and the farmers. The tools are made relevant to the local needs, languages, climates, and cropping systems, making them more adaptable and adopted, and at the same time maintaining cultural diversity. Algorithmic transparency and data sovereignty also encourage a vision of a more equitable agriculture. Quantum-enabled and AI models making a decision should be explainable, unbiased, and transparent. The data produced by the farmers must be their property to which they have the rights of access and control, and which should benefit them. Transparency systems based on blockchain protocols, data cooperatives, and open-source can guarantee ethical use of data and equitable data-owner benefit-sharing. Environmental sustainability is also a design parameter that has not been compromised in this future. The objective of smart agriculture must not only be the increased yield but also the ecological balance minimal wastage of inputs, maintaining the biodiversity, and countering the effects of the climate. Management of technologies should not just be evaluated by their profitability but also by their carbon footprint, water efficiency, and compatibility with regenerative practices. The technology-based farm system also involves gender justice and youthfulness. Women farmers should be granted the same authority over the equipment and training and leadership positions since it contributes significant number of workers in the agricultural realm. Similarly, the youth in rural areas need to be targeted as digital natives and cheerleaders of innovation, turning the cycle of farm depopulation and rural–urban movement. The area of public policy is crucial to the fulfillment of this idea. Governments have to enforce laws that advance the rights of farmers, control the influence of corporations, and reward inclusive innovation. Funding agencies and international organizations should ensure that investments are in accordance with the concepts of equity and long-term resilience.

10.7.3 2025–2035 ethical outlook for smart agriculture

- Quantum-enhanced crop modeling becomes standard in research institutions.
- AI in agriculture is regulated in over 20 countries with independent audit bodies.
- Community-owned agri-data cooperatives emerge in Africa, Latin America, and South Asia.
- Ethics-based certification becomes mandatory for agri-tech product deployment.
- Global treaty proposed for fair sharing of agri-genetic and AI-derived datasets.

10.7.4 Call to action: Aligning innovation with justice and sustainability

With the world agricultural industry revolutionizing at a rapid pace, driven by quantum computing, AI, IoT, and automation, the stakes are increasing, not only regarding food production but also toward justice, equity, and environmental sustainability. The future is being created in the present day, and the decisions that are being crafted can make this technological revolution stimulate exclusivity, so that it will be the cause of further separations. It is time to walk the talk; it is time to act collectively and decisively because even innovations should come in line with values, which are sustainable and to which human beings should rally above all, in the interests of humanity. To begin with, the policymakers need to be proactive and on point. Local, national, and international governments should develop sound ethical and regulatory frameworks that safeguard the rights of farmers, more so the smallholders, women, and indigenous people. Regulations should be revised to gravitate around data sovereignty, IP on the digital front, and algorithm responsibility. Moreover, policies ought to encourage participatory innovation by providing incentives, government financing, or cooperation between the government and companies to make it clear that the high-tech advantages are widely shared rather than centralized in small groups. Meanwhile, technologists and developers need to have an ethical responsibility as future designers. They should make sure that innovation is not only operative but also equitable- one that has been made up envisioning transparency, inclusiveness, and ecological awareness. This implies making the AI systems explainable, wherever possible employing open-source models, and including customer reviews in all phases of development. Ethical review systems must become the norm in the research and development pipeline, much as safety testing is common in pharmaceutical and aviation. Colleges and universities should also come out as custodians of moral wisdom. Every curriculum in the sciences of agriculture, computer science, and engineering should include ethical, sustainable, and participatory design training. The interdisciplinary research with the wedge of social sciences, technology, and ecology studies is demanded to investigate nontrivial effects of digital agriculture in the real world and suggest contextually sensitive responses. The agri-tech innovation is a significant element that is carried out by the private sector, and it should be accountable regarding the goals of the public interest. Companies cannot thrive on profit-making concepts but have to adopt corporate responsibility, openness, and community interventions (Gensler, 2012). This involves fair price policies, local cooperation, and investment in rural infrastructures and capacity building. Purpose and profit have to go together. The role of the international bodies and donor organizations is also very essential. They should endorse international agreements and collaborative mechanisms that regulate the trade of data, genetic resources, as well as climate-smart agriculture projects. The ethical standards must be both internationally organized and locally flexible, which considers the sovereignty and cultural diversity. Lastly, farmers need to be identified as innovators rather than innovators. Their expertise, strength, and experience cannot be dispensed with. Stakeholder farmers are to be involved as co-inventers, co-selectors, and co-testers of technologies determining their lives. It is not just ethically right to empower them; without their empowerment, transformation cannot take place in an effective and sustainable manner.

The prospect of smart agriculture that will be enabled by sophisticated instrumentation and quantum computing holds extensive possibilities and new ethical and social issues. With quantum hardware coming of age and made available via cloud computing resources, its use in the fields of crop genetics, climate modeling, and supply-chain optimization is likely to experience much more growth. The future studies should be aimed at creating explainable and transparent quantum-AI hybrid models that farmers and policymakers can trust and understand. The other important future path is the enhancement of ethical-by-design approaches, in which the privacy of data, equity, and inclusivity are considered as the guiding principles in agricultural innovations. It will need higher focus on the farmer-focused models of data governance, like data cooperatives and decentralized platforms that would warrant ownership and benefit-sharing.

The policies should be developed dynamically to deal with the cross-border data flows, IP issues, and responsibilities of autonomous systems. Socially, more focus in future efforts should be on making the digital divide smaller by accessing cost-effective infrastructure, by locating digital literacy programs, and innovative approaches that are gender-sensitive. The combination of indigenous and traditional agricultural knowledge with sophisticated analytics will be the key to preserving the cultural heritage, as well as enhancing climate change resiliency. In general, the future scope of this chapter lies in the direction of establishing an agricultural ecosystem on a global scale that is equitable, ethically regulated, and technologically empowered, and that follows the principle of productivity and social justice and environmental sustainability.

References

Abdullah, N., Durani, N. A. B., Shari, M. F. B., Siong, K. S., Hau, V. K. W., Siong, W. N., & Ahmad, I. K. A. (2020). Towards smart agriculture monitoring using fuzzy systems. *IEEE Access, 9*, 4097–4111.

Adesipo, A., Fadeyi, O., Kuca, K., Krejcar, O., Maresova, P., Selamat, A., & Adenola, M. (2020). Smart and climate-smart agricultural trends as core aspects of smart village functions. *Sensors, 20*(21), 5977.

Aleotti, J., Amoretti, M., Nicoli, A., & Caselli, S. (2018). A smart precision-agriculture platform for linear irrigation systems. In *2018 26th international conference on software, telecommunications and computer networks (SoftCOM)* (pp. 1–6). IEEE.

Antonacci, A., Arduini, F., Moscone, D., Palleschi, G., & Scognamiglio, V. (2018). Nanostructured (Bio) sensors for smart agriculture. *TrAC Trends in Analytical Chemistry, 98*, 95–103.

Bachuwar, V. D., Shligram, A. D., & Deshmukh, L. P. (2018). Monitoring the soil parameters using IoT and android based application for smart agriculture. *AIP conference proceedings*: 1989. AIP Publishing LLC.

Bouali, E. T., Abid, M. R., Boufounas, E. M., Hamed, T. A., & Benhaddou, D. (2021). Renewable energy integration into cloud & IoT-based smart agriculture. *IEEE Access, 10*, 1175–1191.

Gagliardi, G., Lupia, M., Cario, G., Cicchello Gaccio, F., D'Angelo, V., Cosma, A. I. M., & Casavola, A. (2021). An internet of things solution for smart agriculture. *Agronomy, 11*(11), 2140.

García, L., Parra, L., Jimenez, J. M., Lloret, J., & Lorenz, P. (2020). IoT-based smart irrigation systems: An overview on the recent trends on sensors and IoT systems for irrigation in precision agriculture. *Sensors, 20*(4), 1042.

Gensler, W. G. (2012). *Advanced agricultural instrumentation: Design and use*: 111. Springer Science & Business Media.

Gondchawar, N., & Kawitkar, R. S. (2016). IoT based smart agriculture. *International Journal of advanced research in Computer and Communication Engineering, 5*(6), 838–842.

Hakhrani, A. A. M., & Hamid, S. B. A. (2024). Smart agriculture: Precision farming through sensor-based crop monitoring and control system. *Mekatronika: Journal of Intelligent Manufacturing and Mechatronics, 6*(2), 52–65.

Ibba, P., Crepaldi, M., Cantarella, G., Zini, G., Barcellona, A., Rivola, M., & Lugli, P. (2021). Design and validation of a portable AD5933-based impedance analyzer for smart agriculture. *IEEE Access, 9*, 63656–63675.

Kaewmard, N., & Saiyod, S. (2014). Sensor data collection and irrigation control on vegetable crop using smart phone and wireless sensor networks for smart farm. In *2014 IEEE Conference on Wireless Sensors (ICWiSE)* (pp. 106–112). IEEE.

Kanchi, R. R., Sreeramula, V. P., & Palle, D. V. (2019). Implementation of smart agriculture using Cloud IoT and its geotagging on android platform. In *International conference on intelligent computing and communication technologies* (pp. 520–528). Singapore: Springer Singapore.

Keswani, B., Mohapatra, A. G., Mohanty, A., Khanna, A., Rodrigues, J. J., Gupta, D., & De Albuquerque, V. H. C (2019). Adapting weather conditions based IoT enabled smart irrigation technique in precision agriculture mechanisms. *Neural Computing and Applications, 31*(Suppl 1), 277–292.

Krishna, K. L., Silver, O., Malende, W. F., & Anuradha, K. (2017). Internet of Things application for implementation of smart agriculture system. In *2017 International Conference on I-SMAC (IoT in Social, Mobile, Analytics and Cloud)(I-SMAC)* (pp. 54–59). IEEE.

Lin, Y. B., Lin, Y. W., Lin, J. Y., & Hung, H. N. (2019). SensorTalk: An IoT device failure detection and calibration mechanism for smart farming. *Sensors, 19*(21), 4788.

Liu, W. (2023). Smart sensors, sensing mechanisms and platforms of sustainable smart agriculture realized through the big data analysis. *Cluster Computing, 26*(5), 2503–2517.

Madhumathi, R., Arumuganathan, T., & Shruthi, R. (2021). A survey on wireless sensor networks and instrumentation techniques for smart agriculture. In *Mobile computing and sustainable informatics: Proceedings of ICMCSI 2021* (pp. 453–467). Singapore: Springer Nature Singapore.

Marcu, I., Suciu, G., Bălăceanu, C., Vulpe, A., & Drăgulinescu, A. M. (2020). Arrowhead technology for digitalization and automation solution: Smart cities and smart agriculture. *Sensors, 20*(5), 1464.

Mitra, A., Vangipuram, S. L., Bapatla, A. K., Bathalapalli, V. K., Mohanty, S. P., Kougianos, E., & Ray, C. (2022). Everything you wanted to know about smart agriculture. *arXiv preprint arXiv:2201.04754.*

Morchid, A., Jebabra, R., Alami, R. E., Charqi, M., & Boukili, B. (2024). Smart agriculture for sustainability: The implementation of smart irrigation using real-time embedded system technology. In *2024 4th international conference on innovative research in applied science, engineering and technology (IRASET* (pp. 1–6). IEEE.

Mukherji, S. V., Sinha, R., Basak, S., & Kar, S. P. (2019). Smart agriculture using internet of things and MQTT protocol. In *2019 international conference on machine learning, big data, cloud and parallel computing (COMITCon)* (pp. 14–16). IEEE.

Mukhopadhyay, S. C., & Mason, A. (2013). *Smart sensors, measurement and instrumentation.* Massey: School of Engineering and Advanced Technology.

Pandey, P. C., Tripathi, A. K., & Sharma, J. K. (2021). An evaluation of GPS opportunity in market for precision agriculture. In G. P. Petropoulos, & P. K. Srivastava (Eds.), *GPS and GNSS technology in geosciences* (pp. 337–349). Elsevier.

Prathibha, S. R., Hongal, A., & Jyothi, M. P. (2017). IoT based monitoring system in smart agriculture. In *2017 international conference on recent advances in electronics and communication technology (ICRAECT)* (pp. 81–84). IEEE.

Rajesh, T., Thrinayana, Y., & Srinivasulu, D. (2020). IoT based smart agriculture monitoring system. *International Research Journal of Engineering and Technology (IRJET), 7*(6), 1767–1770.

Rathod, N., Panigrahi, S., & Pinjarkar, V. (2020). Smart farming: IoT based smart sensor agriculture stick for live temperature and humidity monitoring. *International Journal of Engineering and Technology, 9*, 664–669.

Rayhana, R., Xiao, G., & Liu, Z. (2021). RFID sensing technologies for smart agriculture. *IEEE Instrumentation & Measurement Magazine, 24*(3), 50–60.

Rayhana, R., Xiao, G. G., & Liu, Z. (2021). Printed sensor technologies for monitoring applications in smart farming: A review. *IEEE Transactions on Instrumentation and Measurement, 70*, 1–19.

Roy, T., & George, K. J. (2020). Precision farming: A step towards sustainable, climate-smart agriculture. In M. L. Jat, J. C. Dagar, & Y. Singh (Eds.), *Global climate change: Resilient and smart agriculture* (pp. 199–220). Singapore: Springer.

Rueda-Delgado, D., Cuellar-Torres, F., Martinez, D., & Narducci, M. S. (2025). *Instrumentation system for monitoring of soil variables in precision agriculture applications.* IEEE Access.

Saini, A. K., Banerjee, S., & Nigam, H. (2020). An IoT instrumented smart agricultural monitoring and irrigation system. In *2020 international conference on artificial intelligence and signal processing (AISP)* (pp. 1–4). IEEE.

Senoo, E. E. K., Akansah, E., Mendonça, I., & Aritsugi, M. (2023). Monitoring and control framework for IoT, implemented for smart agriculture. *Sensors, 23*(5), 2714.

Shadrin, D., Menshchikov, A., Somov, A., Bornemann, G., Hauslage, J., & Fedorov, M. (2019). Enabling precision agriculture through embedded sensing with artificial intelligence. *IEEE Transactions on Instrumentation and Measurement, 69*(7), 4103–4113.

Shaikh, F. K., Karim, S., Zeadally, S., & Nebhen, J. (2022). Recent trends in internet-of-things-enabled sensor technologies for smart agriculture. *IEEE Internet of Things Journal, 9*(23), 23583–23598.

Suma, N., Samson, S. R., Saranya, S., Shanmugapriya, G., & Subhashri, R. (2017). IOT based smart agriculture monitoring system. *International Journal on Recent and Innovation Trends in Computing and Communication, 5*(2), 177–181.

Vadapalli, A., Peravali, S., & Dadi, V. (2020). Smart agriculture system using IoT technology. *International Journal of Advance Research in Science and Engineering, 9*(9), 58–65.

Zhao, T., Zhang, M., He, L., & Li, F. (2024). MADRL-based collaborative computation offloading and resource orchestration for multi-task data sharing in smart agriculture. *IEEE Internet of Things Journal, 12*(4), 3887–3904.

PART VI

Strategic vision

11

Roadmap for quantum agriculture: Challenges and future directions

Gaganpreet Kaur[a], Ramandeep Sandhu[b], Amandeep Kaur[a] and Deepika Ghai[c]

[a]DEPARTMENT OF MANAGEMENT, ADVANCED CENTRE FOR RESEARCH AND INNOVATION, CHANDIGARH SCHOOL OF BUSINESS, CHANDIGARH GROUP OF COLLEGES, MOHALI, PUNJAB, INDIA [b]SCHOOL OF COMPUTER SCIENCE AND ENGINEERING, LOVELY PROFESSIONAL UNIVERSITY, PHAGWARA, PUNJAB, INDIA [c]SCHOOL OF ELECTRONICS AND ELECTRICAL ENGINEERING, LOVELY PROFESSIONAL UNIVERSITY, PHAGWARA, PUNJAB, INDIA

11.1 Introduction

11.1.1 Overview of agriculture in the age of smart technology

The introduction of modern technologies in the traditional agricultural system predetermines the disruption of the agricultural environment of the 21st century, which can be characterized as significant. This change is also referred to as smart agriculture, or agriculture 4.0, applying new types of digital technologies, such as artificial intelligence (AI), machine learning (ML), Internet of Things (IoT), big data analytics, geographic information systems (GIS), autonomous machinery, and blockchain, to simplify seed-to-market farming. The technologies of this sort contribute greatly to the assistance of farmers in balancing between complex information, managing the real-time monitoring of the weather and the condition of the soil, focusing on the input application, and becoming much more efficient and sustainable (Arute et al., 2019; Liakos et al., 2018; Nielsen & Chuang, 2010; Wolfert et al., 2017; Zhang et al., 2020). The mission of smart agriculture is to minimize the wastage in the agricultural projects through providing real-time conditions of essential variables such as soil moisture, nutrients, and plant health. As an example, IoT-powered sensors placed in fields can collect high-resolution data, which is further processed with AI models to assist in decision-making based on data about irrigation, fertilization, and pest control (Carolan, 2017; Gebbers & Adamchuk, 2010; Kamilaris et al., 2017; Li et al., 2014; McBratney et al., 2005; Saiz-Rubio & Rovira-Más, 2020; Wolfert et al., 2014).

Innovative agricultural practices therefore contribute toward the prevention of resource wastage, efficient cost reduction, and guarantee the quality and safety of foods. But, with

Quantum Computing and Instrumentation for Smart Agriculture. DOI: https://doi.org/10.1016/B978-0-443-44247-6.00004-X

agriculture growing heavily dependent on large and multifaceted datasets, the traditional computing systems are experiencing shortcomings in the speed of computing, storing, and analyzing the data. A lot of the issues related to agriculture are considered computationally heavy and are, in some cases, NP-hard problems, including the process of resource allocation on differing field conditions, the resolution of logistics issues in chain supply lines, and the modeling of crops that are affected by the climate. Such issues take exponents of time to get solved using classical computers, and therefore, they cannot be used in a large-scale, real-time application in agriculture. Thus, next-generation computational models with the ability to efficiently take input of complex multidimensional data are urgently needed to enter quantum computing (Feynman, 1982; Grover, 1996; Kshetri, 2018; Lobell et al., 2011; Lowenberg-DeBoer & Erickson, 2019; Montanaro, 2016; Rose & Chilvers, 2018; Trendov et al., 2019; Tzounis et al., 2017; Wolfert et al., 2015).

11.1.2 Emergence of quantum computing and its significance

Quantum computing, a revolutionary discipline that is based on the foundations of quantum mechanics, is the evolutionary process of information processing in the form of a quantum leap. In contrast to classical computing, where information is represented in binary form (0 or 1), quantum computations are represented by quantum bits (qubits), which can all simultaneously exist in a superposition of states with the co-occurring phenomena of superposition and entanglement. Through these properties, quantum computers can examine a huge number of possibilities simultaneously, which could solve complex issues at a previously unheard-of speed (Schuld et al., 2015).

The potential power of quantum computing is being developed in several areas, such as pharmaceuticals, finance, logistics, and agriculture. The agricultural sector has a distinct advantage in taking advantage of quantum innovations because of the complexity of operations, their variability, and the abundance of data used in the sector. Using quantum computing, climate modeling, and the study of soil chemistry could be implemented to predict the behavior of pests and dynamic crop yield forecasting, among others, with high levels of accuracy (Biamonte et al., 2017).

By way of example, quantum algorithms like the quantum approximate optimization algorithm (QAOA) may address supply chain issues up to millions of variables. Quantum-enhanced ML models have the potential to detect crop diseases earlier due to their ability to discern tiny and nonlinear patterns in multispectral imagery data. Also, the quantum simulation can be used to model the soil and enhance the use and composition of fertilizer by accounting for the interactions of the molecules present.

Despite being in its early days, it can be predicted that quantum computing will supplement and improve the capabilities of classical computing and not substitute it. The ever-evolving quantum, hardware, software, and algorithms with the convergence with smart agricultural technologies will open up new avenues of efficiency, resilience, and long-term sustainability (Budker & Kimball, 2013; Cerezo et al., 2021; Childs et al., 2000; Schuld & Killoran, 2019).

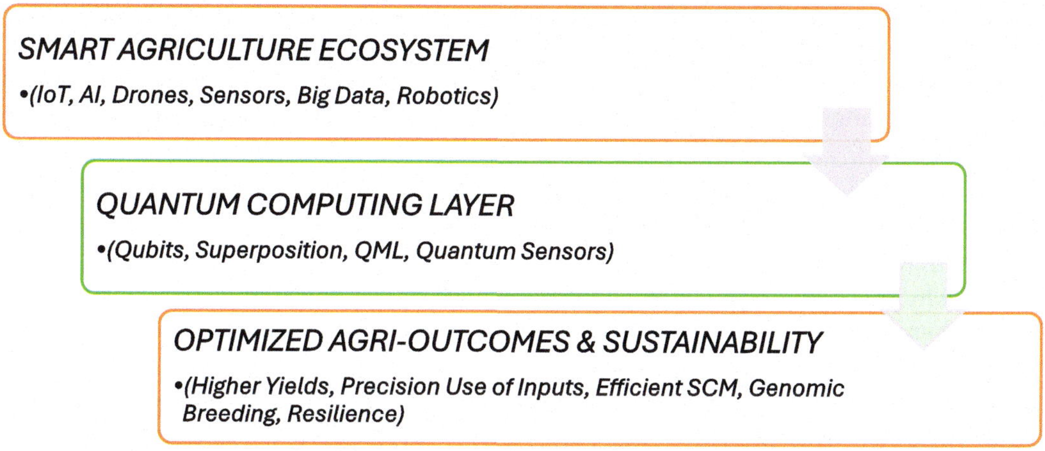

FIGURE 11.1 Conceptual framework of quantum agriculture.

11.1.3 Concept and scope of quantum agriculture

Quantum agriculture is a different concept of imposing quantum computing on agriculture; quantum agriculture is the vision of a comprehensive projection of quantum concepts in combination with digital farming ecosystems. It entails the integration of quantum computational intelligence to the agri-value chain in the quest to facilitate high-speed processing, simulation, optimization, and prediction of on-field and off-field agricultural activities.

11.1.3.1 Scope of quantum agriculture

According to research, quantum technologies will transform agriculture into data-driven and high-precision practices in the whole agri-value chain. The provided quantum-enhanced precision farming takes advantage of optimization algorithms to individualize the input application, fertilizers, water, and pesticides at the micro-plot level, maximizing the yield and reducing waste. The quantum sensors and remote sensing systems redefine the advanced soil and crop monitoring and provide previously unavailable accuracy in real-time diagnostics. In crop genomics and crop breeding, quantum machine learning (QML) can be used to quickly analyse complicated plant DNA sequences, which speeds the design of high-producing and resistant cultivars. Quantum routing algorithms can improve food distribution chains to help limit the post-harvest losses, as well as increase the efficiency of cold chains. Also, quantum simulations will be very important in illuminating the impacts of climate, soil erosion, and the carbon reclaiming capacity, which will help us shift to the regenerative approach of agriculture.

Quantum agriculture conceptual framework is shown in Fig. 11.1 that combines advanced quantum technologies and agriculture. It visualizes the interdependency of quantum computing, AI, IoT, and big data and describes how all these technologies interact to bring precision farming, resource optimization, and climate resilience through the support of ethical governance and policy infrastructure to resourceful agri-tech innovation.

This quantum computing and agricultural intersection was not just a scholar-daydream: This is a new ground-breaking sector. It needs the efforts of researchers, technologists, and policymakers in the various disciplines to facilitate the activation of its potential by addressing technological, infrastructural, and ethical issues to create scalable, sustainable, and resilient agri-systems.

Objectives of the chapter

1. To introduce basics of quantum computing and their application to the problems of agriculture;
2. To determine the technological constraints in existing smart farming procedures that can be addressed using quantum computing;
3. To investigate real-world use cases and future potential of quantum agriculture that is backed by conceptual use cases;
4. To illustrate a graphical plan which includes major problems, integration approaches, and research and practice directions to cover in the future;
5. And, to present policy and ethical advice to carry out responsive and inclusive approaches of adopting quantum technologies in agriculture.

Organization of the chapter

The rest of the chapter is organized as: Section 10.2 introduces the principles of quantum computing and their application in agricultural data and decision-making. Section 10.3 examines how the smart farming technologies have developed over the years and the shortcomings of the existing AI- and IoT-based systems in the agricultural sector. Section 10.4 defines some of the main quantum application use cases in agriculture such as yield prediction, soil monitoring, climate model, and supply chain optimization. Section 10.5 enumerates technical, infrastructural, and cybersecurity issues relating to using quantum technologies in agriculture. Section 10.6 suggests a quantum computing and classical fusion integration model to integrate quantum computing with currently existing smart farming systems. Section 10.7 presents the prospects of future research and direction of quantum-enabled sustainable agriculture in the long term. Suggestions on policy, ethical, and capacity building in responsible adoption are discussed in Section 10.8. Section 10.9 provides exemplary case studies of quantum applications in the agricultural field. And, finally, Section 10.10 concludes the chapter with future scope.

11.2 Foundations of quantum computing in agriculture

11.2.1 Basics of quantum computing for nontechnical readers

Quantum computing is a new paradigm of computation whose system of operation is founded on quantum mechanics, that is, the science that determines how particles behave at their atomic and subatomic levels. Unlike classical computers that operate on a binary basis, meaning 0 and 1 s, quantum computers operate in quantum bits (qubits), which have the opportunity of reflecting both 0 and 1 at the same time because of a phenomenon called superposition.

Intuitively speaking, quantum computing can be thought of like a light switch: a classical switch can only be either ON (1) or OFF (0). A qubit can exist in a mixed state of either ON or OFF simultaneously under observation. This leads to the ability of quantum computers to perform multiple computations in parallel and offers exponential speedups on classes of problems.

Quantum computers are notably relevant in combinatorial problems that would require classical systems years to hundreds of years to come up with the best solutions. In agriculture, much of what is done (such as maximum fertilization of different soil types or forecasting based on thousands of possible weather measurements) could benefit greatly from quantum acceleration (Shor, 1994).

11.2.2 Basic quantum rules: Qubits, entanglement, and superposition

The field of quantum computing is based on three principles, which differentiate it when compared to classical computing:

11.2.2.1 Superposition and Hilbert space expansion

A state of a register of n bits will be represented as a single point in a space of potential states of 2n. On the contrary, n qubits of a quantum system occupy a complex space of dimension 2n, which is denoted as Hilbert space. A state that comes to be the definition of a state of an individual qubit:

$$|\varphi \geq = \alpha|0 > \ + \beta\,|1\rangle$$

In the case of multiple-qubit the state of the system is n represented as-a-product of qubit states or the tensored product. A 2-qubit system is exponentially growing in its complexity (i.e., the amount of qubits grows exponentially), and as a result, the state of a qubit system can be expressed as a linear combination of all the basis states and is denoted in the equation provided below:

$$|\varphi \geq = \alpha_{00}|00 > \ + \alpha_{01}\,|01 > + \alpha_{10}\,|10 > + \alpha_{11}\,|11\rangle$$

It can be all possible combinations of the state of multiple clusters of fields including the level of nitrogen content in the soil, the moisture content, the temperature, and other factors. A massive state space, therefore, like the state of a large number of field clusters, i.e., the combination of soil level of nitrogen, moisture level, temperature, etc.

11.2.2.1.1 Superposition

A classical piece has 2 outcomes; a 0 and 1. A qubit may be found in a superposition of only states. This enables a quantum system to test a large number of possible scenarios at the same time. Consider the example of a quantum system with two qubits (the number of qubits of a system can be any integer). A qubits system can remain in four states simultaneously (00, 01, 10, 11), and a three-qubit system can exist in eight states, and so forth exponentially. When there are many qubits, the state of the system will be constructed on the basis of the product of states of every qubit. A register consists of insanely many input qubits: That is, an n-qubit register can be characterized as follows:

$$|\varphi \sum_{i=0}^{2^n-1} \alpha_i \, |i\rangle$$

11.2.2.2 Entanglement

Entanglement is a quantum phenomenon where two or more qubits become interconnected such that the state of one immediately affects the state of the other, no matter the distance between them. In agricultural models, this could allow for simultaneous optimization across geographically dispersed variables such as field clusters, climate zones, or sensor networks (Lloyd, 1996). Mathematically, an entangled state $|.\Psi\rangle$ of a bipartite system satisfies the following equation:

$$|\varphi\rangle \neq |\varphi_1\rangle \otimes |\varphi_2\rangle$$

A representative example is the maximally entangled Bell state,

$$|\emptyset^+ \geq= \frac{1}{\sqrt{2}}(|00\rangle + |11\rangle)$$

As shown in the above equation, one qubit becomes measured to instantly define the outcome of the other correlated to it, no matter the spatial distance. These are correlations due to the common quantum state and not any classical communication, and they are manifested in the form of violation of the classical correlation inequalities of Bell.

11.2.2.3 Quantum interference and measurement

The quantum amplitudes, in contrast to classical probabilities, are complex numbers, which can either constructively or destructively interfere with each other. Originally, design of quantum circuits, made up of unitary operations, amplitudes associated with the correct or optimal solution are enhanced, and those with incorrect ones are repressed.

Mathematically, the evolution of a quantum state is governed by unitary transformations. Given an initial state $|.\psi_0\rangle$, the state after applying a sequence of quantum gates represented by a unitary operator **U** is

$$|\varphi_f \boldsymbol{U}|\varphi_0 >$$

11.2.3 Differences between classical and quantum computing

To better understand the utility of quantum computing in agriculture, it is important to differentiate it from classical computing, as explained in Table 11.1.

Quantum computing is not meant to replace classical computing but to complement it, especially in solving high-dimensional optimization, pattern recognition, and simulation problems, all highly relevant to agriculture's complex and dynamic systems.

Table 11.1 Classical vs. quantum computing.

Features	Classical computing	Quantum computing
Information unit	Bit (0 or 1)	Qubit (0 and 1 simultaneously)
Data representation	Deterministic	Probabilistic and superposed
Computation model	Sequential and Boolean logic	Parallel computation via superposition
Correlation	Independent variables	Entangled states across variables
Speed for complex tasks	Linear or polynomial time	Exponential speedup (for specific tasks)
Error handling	Mature error correction	Still evolving (quantum decoherence issues)
Hardware	CMOS transistors	Superconducting circuits, ions, photons

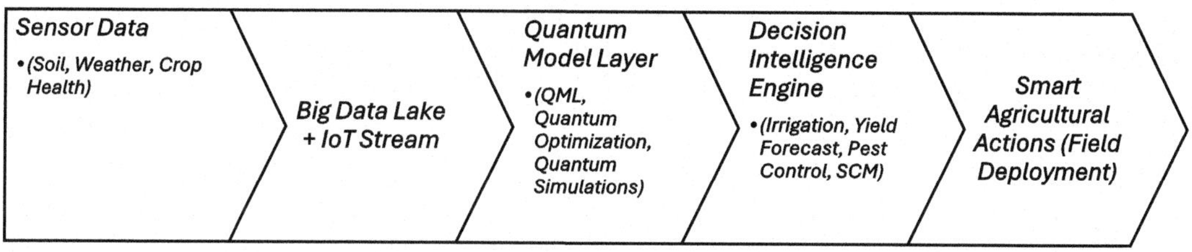

FIGURE 11.2 Flowchart of the role of quantum computing in agriculture.

11.2.4 Relevance of quantum models to agricultural data processing

Agricultural systems in the modern age collect and produce large and heterogeneous quantities of data using one or more of soil sensors, satellites, drones, weather stations, mobile devices, and genomic labs. Classical computing approaches find it difficult to process this heterogeneous and large volume of data it resulting in performance bottlenecks. Quantum computing models show potential solutions to this bigger, interconnected picture by processing large amounts of data with higher efficiencies and richness. As an example, length invariant: Quantum support vector machines (QSVM) and variational quantum classifiers (VQC) fall under the category of QML and could use multispectral imagery to classify crop diseases, predict stress conditions, and identify weeds early in development (Havlíček et al., 2019). As far as resource optimization is concerned, the quantum-inspired algorithms can dynamically distribute the available inputs (in limited quantities) such as water, fertilizers, and power in different zones of the farm that are spatially dispersed. Quantum simulations also improve climate and crop modeling by analyzing complicated weather and genetic data to produce accurate yield predictions and can model crop performance in new future climate conditions. Also, quantum algorithms will transform agricultural supply chain management and solve nontrivial routing problems, where hundreds of variables are minimized to reduce food waste, logistics costs, and to better meet delivery schedules. Fig. 11.2 highlights layered architecture that use quantum computing as the computational engine at the heart of an intelligent agricultural system with the ability to improve responsiveness, precision, or scalability.

11.3 The evolution of smart farming technologies

11.3.1 Current state of precision agriculture and smart farming

Digital agriculture (also smart farming) is a shift in the paradigm of treating the agricultural field as the unit of production to managing the farm as a site-based model with specific data. The process is facilitated by the incorporation of digital technologies such as sensing system, geospatial analytics, automation, and an intelligent decision-support platform. As opposed to the conventional age of using the same inputs in different heterogeneous fields, where individuals apply identical inputs over the various fields, precision agriculture utilizes spatial and temporal fields in fields to maximize the allocation of resources, enhance crop yields as well as minimize environmental externalities. Modern farms, at the operational scale, use the global positioning system (GPS)-controlled equipment and GIS to obtain an operational precision of centimeters in the tasks to plant, fertilize, and harvest crops.

Variable rate technology (VRT) is a dynamic technology that changes application rates of seeds, fertilizers, pesticides, and water, based on prescription maps, which are created on a basis of the soil nutrient profiles, crop yield maps, and the crop vigor indexes. This translates into high-level input-use efficiency and less agrochemical wastage. Unmanned aerial vehicles (UAVs) and satellite platforms can be used in remote sensing to offer high-resolution multispectral and hyperspectral images that have been used to monitor crops continuously. Two of the most used vegetation indices are the normalized difference vegetation index (NDVI) and the enhanced vegetation index (EVI), which are actively utilized in determining the health of plants, indicating stress conditions, and endogenizing pest or disease epidemics at their initial stages. In addition to this, in situ IoT sensor networks are used to continuously monitor soil moisture, soil temperature, soil pH levels, electrical conductivity, and microclimatic conditions in real-time and provide agronomic interventions that make decisions based on the data (Kamilaris & Prenafeta-Boldú, 2018).

Smart farming cycle is shown in Fig. 11.3, which brings in automated and autonomous technologies, such as irrigation controllers based on sensors, robot weeders, and autonomous tractors (agribots), enhancing reliance on labor and making operations more precise. The next generation of decision-support systems is based on ML and predictive analytics to implement yield projections, irrigation programs, and risk evaluation by various climate conditions. Simultaneously, traceability systems built on blockchain are being implemented to provide transparency, integrity, and confidence in the agricultural supply chain, including farm inputs and markets to consumers. Economically and at scale, the fast rate at which cloud computing, edge analytics, AI and IoT converge has increased the pace of adoption of precision agriculture on a global scale. All of these technologies enable sustainable intensification to work by maximizing productivity and reducing the impact on nature. In keeping with the development trend, the world precision agriculture industry will have a value of more than USD 20 billion by 2030 due to the increased food demand, variability in climate, and the necessity of efficient farming systems that are resource efficient (Markets & Markets, 2024).

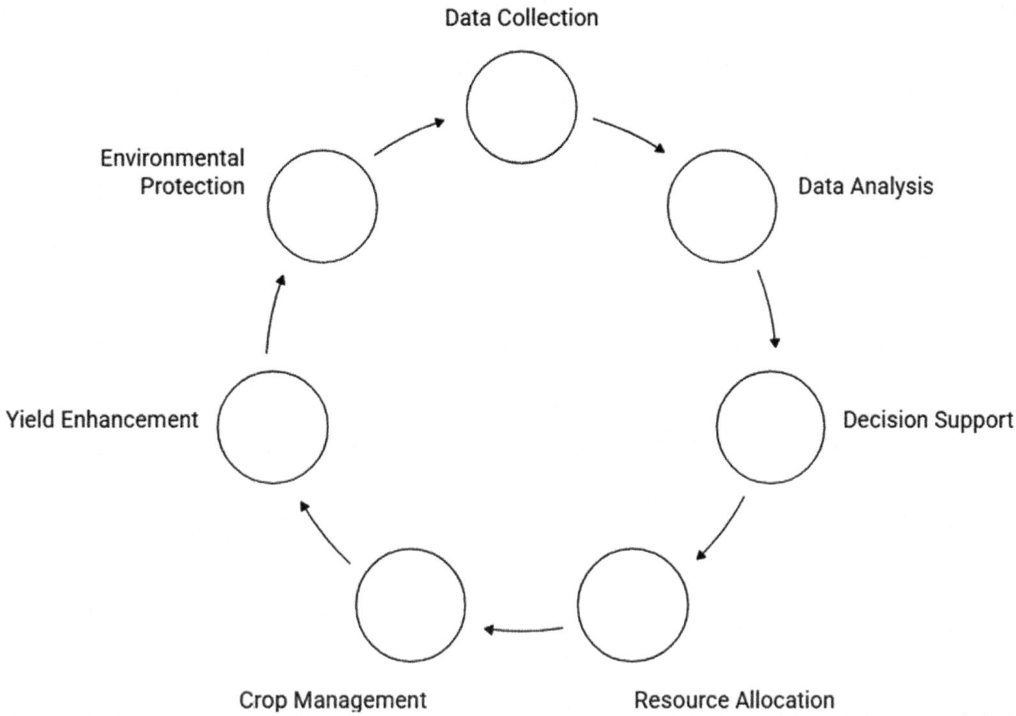

FIGURE 11.3 Smart agriculture cycle.

11.3.2 Role of AI, IoT, and big data in modern agriculture

The role of AI, IoT, and big data in modern agriculture is listed as follows:

1. **AI:** AI is at the heart of decision-making in smart agriculture. It enables predictive analytics for crop yield forecasting, computer vision for disease detection, ML for pest prediction, and natural language processing for advisory systems.
2. **IoT:** IoT systems comprise a system of sensors, actuators, and communication equipment that constantly measure parameters in the field. For example, the moisture sensors in soil may send their data to the central control systems, which will then decide to irrigate and the amount of water to irrigate to save on water wastage (Jensen, 2021).
3. **Big data analytics:** Agricultural industry produces large amounts of organized and unstructured data, such as weather prediction, market prices, remote image sensing, and genomics. The data are gathered, processed, and analysed on big data platforms to be used in precision farming. The insights obtained in such platforms are used to dictate everything as simple as the planting schedules, as well as logistics optimization. Table 11.2 describes the combination of AI, IoT, and big data in agriculture as the benefits in terms of efficiencies and sustainability of the use of the technologies. They involve controlling pests before they get

Table 11.2 Applications of AI, IoT, and big data in agriculture.

Technology	Application	Benefit
AI	Pest/disease prediction	Reduces crop losses
IoT	Soil and climate monitoring	Enables real-time decisions
Big data	Yield forecasting	Supports market preparedness
AI + Big data	Precision input use	Minimizes environmental impact
IoT + AI	Automated irrigation	Saves water and labor

to be a problem in order to aid in the facilitation of irrigation and other development due to the resource saving, reduced destruction, and a help in making informed decisions that can bring change in harvests.

11.3.3 Limitations of existing computational techniques

Even though much progress has been made in the field of digital agriculture, there is a severe limitation on the applicability of classical computing infrastructure to current agricultural data pipelines, which involve large amounts of high-dimensional real-time data. With the growth of sensor networks, the increasing number of sensor inputs inundates the traditional systems with volume and velocity of data. This advocates bottlenecks of scaling sensor networks. In addition, optimization problems tend to become computationally infeasible as multivariable resource allocation, and even genomic prediction, NP-hard. The kind of agricultural data could include satellite imaging and video, meteoconditions, soil structure, etc. is also in itself an issue (as it is not homogeneous and, thus, can be less readily incorporated into unified decision models). Besides, there are still latency issues where the classical systems are sequential hence incapable of making timely decisions in real-time situations very imperative in dynamic field situations. Finally, energy consumption and energy expense of the large-scale AI infrastructure sustainability constitute a topical sequence of inquiry within certain geographical regions with a lack of technological access. These limitations lead to the significance of more advanced computational paradigms to support greater agricultural needs on the global level, specifically in comparison to climate change, food security, and sustainable intensification issues.

11.3.4 Need for quantum-driven solutions

The quantum-driven solution has the potential to transform the agricultural sector by solving the constraints of classical computing in tackling high-dimensional, complex problems. Quantum optimization algorithms are a good way to handle constraint-intensive tasks, e.g., multizone irrigation scheduling and large-scale supply chain logistics. QML would allow quicker and more precise high-resolution agricultural datasets, including hyperspectral imagery, to be analysed and assist in the early diagnosis of crop stress and disease. Also, quantum simulation can be used to model complex soil–plant interplay at molecular scales, helping to develop more efficient fertilizers and treatments for crops. Applied together with the classical one, quantum computers

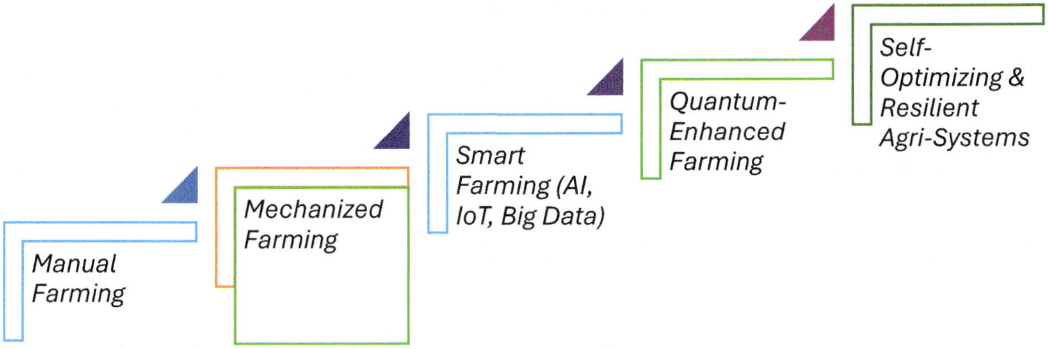

FIGURE 11.4 Technological evolution from conventional to quantum agriculture.

supplement feature selection and increase predictive accuracy and, as a result, cause more flexible and accurate decisions. With the development of quantum hardware, these quantum-classical systems are bound to be a key part of the next generation of smart farming systems. Fig. 11.4 indicates the paradigm transformation of agriculture from input-based processes to data-based and, eventually, to a quantum-smart farming structure that evolves and learns through its continuous progress.

11.4 Quantum use cases in agriculture

Quantum computing in agriculture is not just on paper. Still, it offers an opportunity to break through various complex and multi-variable problems that conventional methods of computing could not otherwise solve. This section is an outlook in the field and examines applied examples and hypothetical circumstances in which quantum models can transform agricultural systems that can improve agricultural productivity, efficiency, sustainability, and resilience.

11.4.1 Crop yield prediction and optimization using quantum models

The process of crop yield prediction is multifactorial and data rich, where factors such as type of soil, genetic factors in seeds, irrigation schedule, pressure against pests, and weather can play an important role. Classical models tend to be oversimplistic or give a partial view of these interactions or focus on nonlinear dependencies.

QML, such as QSVM or quantum neural networks (QNNs), has been promising algorithms to find hidden patterns and correlations among the high dimensionality of datasets. Millions of growth outcomes in microclimates can be estimated using these models, and planting strategies that ensure maximum output and minimum risk can be advised.

Possible use case: A farmer uploads sensor values at the field level (moisture, pH, and weather) to a quantum cloud, where QML models are used to forecast which crop variety would give the highest yield, the ideal time to plant, and the optimal set of inputs to use next season.

11.4.2 Soil health monitoring with quantum sensors

Soil health is foundational to agricultural productivity. However, traditional soil analysis methods are time-consuming, invasive, and expensive. Research on emerging quantum sensors using quantum entanglement, atomic magnetometry, or nitrogen-vacancy (NV) centers of diamonds is capable of measuring tiny variations in the composition of soil, activity of the microbes in the soil, and nutrient dynamics in real time (Preskill, 2018). Such sensors are noninvasive with ultrasensitive and highly precise measurements thus, by tracking the degradation, compaction, salinity, and trace minerals dynamics of soil. Active running case: In quantum-enabled spectroscopy technology, scans of the soil are observed in situ and micronutrient deficiencies (e.g., Zn and Fe) resolved on the nanoscale, which allows real-time and nanoscopic delivery of fertilizers.

11.4.3 Climate-resilient farming through quantum simulations

Climate change is bringing in new degrees of uncertainty in agricultural activities and shifting rain patterns, raising the extreme temperatures and changing the growing seasons. The classical climate models are frequently unable to reproduce such an overall and interacting phenomena with a high resolution and rapidity. Quantum simulations have the ability to simulate quantum dynamics of water molecules, heat flows, and atmospheric interactions with accuracy. They are especially well adapted to the simulation of plants responding to stress at a molecular level (i.e. drought and heat) and the attraction of climate-preserving qualities in plants. Application case: With quantum simulators, researchers can examine the response of root proteins in a certain crop to higher salinity level in soil when in a drought situation to help breeders create more resilient hybrids.

11.4.4 Supply chain management and logistics enhanced via quantum algorithms

Agricultural supply chains tend to be nonlinear, volatile, and spread across geographical locations meaning that they have high inefficiencies. Classical routing algorithms are considered to be less scalable when using hundreds of constraints, including perishability and storage capacity and road conditions. The quantum optimization algorithms such as QAOA and quantum annealing have the power to solve multidimensional logistics problems with increased efficiency and lower level of computational time (Tacchino et al., 2019). Example use case: Any cooperative relies on a quantum-powered logistics solution to plan ideal routing and delivery times of fresh produce of several farms to retailers, decreasing spoilage and fuel usage by 30%.

11.4.5 Case study snapshots: Conceptual illustrations of use cases

Fig. 11.5 illustrates the process in which data sent by field systems (sensors, GPS, and weather) are passed along quantum enabled engines to drive intelligent agrarian action. To demonstrate such applications in a practical setting, conceptual case studies in quantum agriculture are given in Table 11.3:

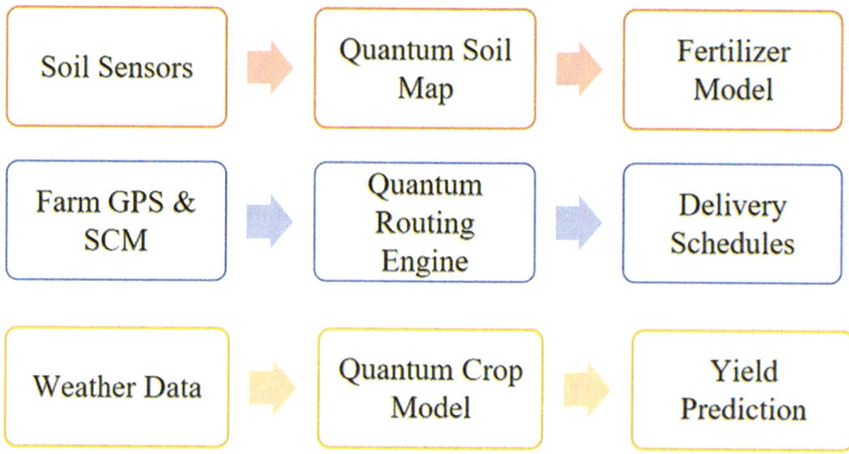

FIGURE 11.5 Quantum agriculture use case framework.

Table 11.3 Conceptual case studies in quantum agriculture.

Use case area	Technology applied	Conceptual scenario	Expected benefit
Yield prediction	Quantum ML (QSVM)	Predicting the best seed-soil-weather match	+15% yield increase
Soil monitoring	Quantum NV sensors	Real-time micronutrient scanning	Targeted fertilization
Climate modeling	Quantum simulators	Simulating drought stress response	Crop resilience insights
SCM optimization	QAOA, annealing	Optimizing cold-chain logistics	−25% spoilage reduction
Crop breeding	Quantum genomics	Quantum analysis of gene expressions	Accelerated hybrid creation

11.5 Technical and practical challenges

On the one hand, the number of potential advantages of quantum computing in agriculture is groundbreaking; in the second place, the application of quantum computing in agriculture is complicated by various technical, infrastructural, and economic factors. Mature hardware limitations go to the range of domain-specific algorithms on one end and data compatibility problems on the other. It is important to consider these issues in order to create a framework of a sustainable and scalable quantum-agricultural ecosystem.

11.5.1 Hardware limitations and costs

The state of quantum hardware is one of the biggest obstacles to the implementation of quantum computing in agriculture. The present-day quantum processors are extremely cold (<|human|>Current quantum processors are usually run at extremely low temperatures (<|human|>Modern quantum processors are hideously expensive and need special conditions such as a dilution refrigerator and a vacuum chamber to operate. The demands of quantum systems include quantum:

- Expensive to build and maintain,
- Physically delicate and nonportable,
- Highly limited in qubit count and coherence time (Harrow et al., 2009).

As of now, commercial quantum processors typically support between 50 and 100 qubits, far below the threshold needed to outperform classical computers (expected around 1,000–10,000 stable qubits). More so, quantum error correction is in its early stages, which introduces additional redundancy and limits its practical use to large-scale applications. In the case of farm applications, where it is crucial to be able to scale and afford the hardware, these hardware constraints highly limit access and usefulness.

11.5.2 Scarcity of agricultural-specific quantum algorithms

In spite of general-purpose quantum algorithms, including QAOA, Search in Grover, and quantum Fourier transform, which possess some theoretical benefits, their efficiency in agriculture is low because there are no quantum models of agriculture operationalized. Problems related to it are the absence of specialized algorithms to simulate crops, predict pests, or optimally manage resources and the lack of quantum-ready agricultural data to train on supervised or unsupervised learning. In addition, it is not that easy to code complex real world constraints like weather variability and soil heterogeneity into quantum compatible forms. This discontinuity in algorithms makes it difficult to develop practical applications of quantum computing in agriculture, and the existing work is mainly limited to theory and literature (Braunstein & van Loock, 2005).

11.5.3 Data integration barriers with classical IoT systems

Farmers are already using powerful IoT infrastructures in areas, such as field sensors, weather cameras, drones, and mobile apps, to gather real-time data on farming. But these classical sources of data have not been connected to quantum systems yet as they should be. The main challenges are data translation between classical and quantum representations (e.g. CSV, JSON, and SQL) and quantum states and circuits. The absence of a standard middleware or hybrid APIs between quantum-classical systems, and these cannot be used with real-time streams of data. The present quantum computers are mostly used in batch processing, thus not qualified in making time-sensitive farm decisions (Mosca, 2018). This, in turn, makes it impossible to implement quantum computing in smart farming processes without any major preprocessing and redesign of the system.

11.5.4 Scalability, interoperability, and deployment readiness

Quantum solutions have to be scalable in vitro to fields of smallholder to large industrial scales and dynamic to different agro-climatic environments and in heterogeneous technological ecosystems, including ERP systems, cloud platforms, and sensor networks, to have a meaningful impact in agriculture. There are, however, a number of obstacles to this scalability. The available

Table 11.4 Summary of key technical and practical challenges in quantum agriculture.

Challenge	Description	Impact on agriculture
Hardware limitations	High cost, fragile systems, low qubit counts	Limits accessibility and scalability
Lack of algorithms	A few agriculture-specific quantum models	Hinders use-case development
IoT integration issues	Data format and system incompatibilities	Prevents seamless field deployment
Scalability and interoperability	No cross-platform standards or tools	Limits ecosystem-wide adoption
Cybersecurity risks	Data breaches, quantum decryption threats	Threatens integrity and trust in systems

quantum technologies are poorly interoperable with existing digital farming devices, nonstandardized frameworks across platforms (e.g., IBM Qiskit vs. Google Cirq) and have a limited supply of skilled workforce, particularly in developing countries (Abraham et al., 2019). Moreover, the majority of quantum systems remain either experimental or constrained to a small cloud system, which reduces their usefulness in the agricultural environment.

11.5.5 Cybersecurity and data integrity in quantum agriculture

The more the technological level of quantum usage increases, the higher the risks of cybersecurity in the agricultural sector. In the future, quantum systems will be not only key assets that an adversary tries to attack but also very essential, in terms of cyberwarfare. Farm information such as harvest predictions, pricing plans, as well as proprietary seed strains is extremely delicate as well as worthwhile. With the openness of quantum platforms, they may be platforms of malicious intrusion, data breaches, or altered computation results. Additionally, quantum-based decryption poses a risk of shedding light on the existing encryption, and this may compromise secure supply chains and other monetary operations (AlSagri et al., 2025). As such, quantum-resistant cybersecurity measures are necessary to implement quantum systems in the agro-system before adopting it. Table 11.4 defines the key technical and practical challenges of quantum technologies implementation of agricultural industries. Hardware limitations and cybersecurity threats represent the most notable potential issues that demonstrate that the full potential of quantum agriculture requires both a robust infrastructure and specific algorithms and secure and scalable systems.

11.6 Bridging the gap: Toward a hybrid future

The process of quantum computing is still under development, which means that it cannot be introduced into current agricultural technologies as a drastic substitute of the classical computing. Rather, hybrid architectures are the most reasonable and useful avenue to take, with the integration of quantum components into civilized digital agriculture stables resolving particular bottlenecks. It is a hybrid approach that uses the maturity and stability of classical infrastructure but ends up gradually adding quantum acceleration on data-intensive functions such as optimization, simulation, and ML.

11.6.1 Hybrid classical–quantum architectures

Hybrid computing uses classical processing units (CPUs/GPUs), combined with quantum processing units (QPUs), to offload complex subtasks that are likely to have a quantum speedup. Scientific studies, fiscal, and logistical applications are adopting these architectures, and they are even being put forward in precision agriculture.

11.6.1.1 Key features of hybrid classical–quantum architectures

Quantum computing can be used as a co-processor in a hybrid computational model in agricultural applications. In this case, QPU will be focused on specialized functionalities, namely the solution of constraint optimization problems, and the classical system will be in charge of user interfaces, sensor data input, and storage. Workflow orchestration is essential for classical systems to pre-process data for the QPU, manage job queuing, and post-process quantum outputs into practical, field-level insights. Cloud-based hybrid environments, such as IBM Qiskit Runtime and Amazon Braket, enable seamless quantum-classical integration. For example, classical systems might collect weather and soil data, which the QPU then analyses to generate optimized, climate-resilient planting schedules, enhancing precision agriculture outcomes.

11.6.2 Role of edge computing and IoT in bridging quantum agriculture

Edge computing is used in agriculture to make real-time decentralized decisions that process data locally at a farm level, sometimes along with IoT devices. This reduces delays and central server dependencies. Although quantum computing is still limited to special labs or in the cloud, and it is not quite usable at the edge, a tiered system of computation can allow such a transition. The IoT/Sensor/Edge as part of this ecosystem will collect and do preliminary signal processing of sensor data; the classical cloud tier will perform standard analytics and control; and the quantum cloud tier will run sophisticated computations, including optimization of irrigation over multiple crop zones.

11.6.3 Proposed integration frameworks: Design and architecture

To be able to deliver quantum computing to the real-world farming system, there is a fundamental necessity to develop a viable and scalable integration system that allows smoothly integrating the existing online digital infrastructure with the quantum environments in this regard. To have a quantum-agriculture design that is optimum, it must concentrate on how quantum and classical application programming interfaces (APIs) are compatible to enable a high volume of information exchange and shared data processing between conventional farm management systems and quantum backends. That interoperability would ensure that even legacy systems, such as field systems, weather forecasting systems, or crop scheduling systems, would be able to interface with quantum algorithms, in principle with no redesign. Speaking of which, it is essential to introduce secure data transmission protocols that would send the sensitive farming data to remote quantum servers with the help of edge-based devices in the IoT. Since the

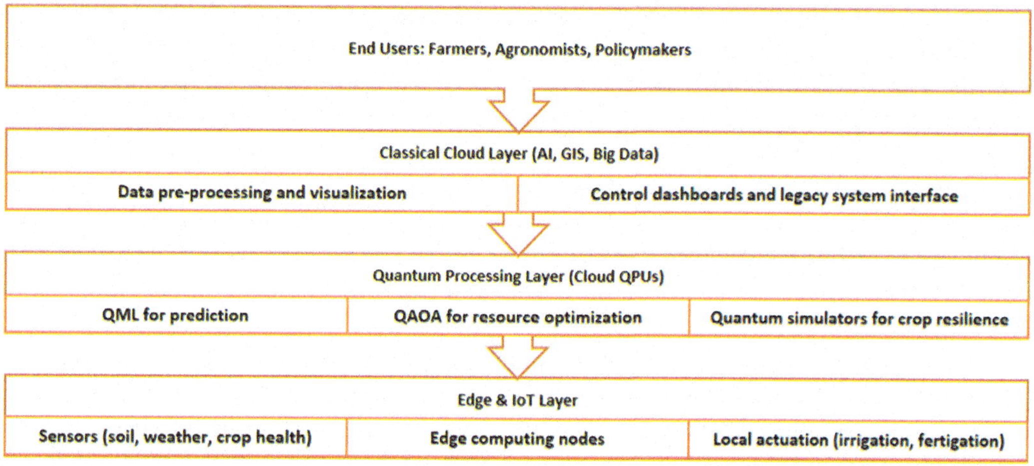

FIGURE 11.6 Layered architecture for hybrid quantum agriculture.

information on land productivity, the input rates, and proprietary seed traits are confidential and, therefore, require encryption protocols and real-time authentication, the standards should be integrated into the communication architecture. Moreover, the integration framework must contain domain-specific middleware (which may be parametrized by index or state) to encode high-level optimization problems in agriculture into quantum-readable formats, e.g., parameterized quantum circuits or Hamiltonians. This middleware is a very important interface, given the fact that the complex queries in agronomy can be encoded into quantum-friendly language. Collectively, free-market principles of these technical pillars will establish a unified, effective, and stable quantum-agricultural ecosystem. The model of operational three layers presented in Fig. 11.6 facilitates smooth functioning in which quantum computing improves decision support without substituting structures that are in place.

11.6.4 Examples of pilot research and development models

Other preliminary undertakings already underway into the interface of quantum computing and farming technologies are already centered on the proof of concept or simulation stage. The more prominent ones include Qi skit-Agri Tech demonstrations by IBM, launched in 2022 and operating on the concept of QML serving to classify drone imagery to detect pests. Such models have shown positive results; they do better than small-scale conventional models at identifying patterns and classification better in noisy farming environments (Chouhan et al., 2025). A further purely thought-based partnership has been formed by D-Wave Systems and Gro Intelligence that foresees the possibility of quantum annealing being used to optimize food supply chains in Africa. This project has concentrated on setting up a storage capacity proportional to the perishability levels within highly dynamic logistical chains, providing a theoretical basis for a quantum-enhanced supply chain resilience.

Academically, scientists have constructed a Quantum-Smart Irrigation Framework that uses traditional weather forecasting systems and quantum-tuned irrigation tactics across the field with disparate soil properties. The calculated outcomes of this prototype are up to 20% of water savings and thus display the efficacy of quantum optimization in realizing greater input efficiency and sustainability (Benedetti et al., 2019). Publications based on the use of the Xanadu PennyLane framework have also considered application to crop trait analysis using quantum convolutional neural networks (QCNNs). This paper was concerned with modeling genotype-to-phenotype maps based on complex plant breeding data, with the viewpoint of hastening the discovery of marketable/preferred characteristics and cutting the number of generations in the breeding process.

Collectively, these initiatives signal the emerging global momentum of interest in quantum agriculture, and, at least at their nascent stages, they offer the construct within which field-scale resolutions can, in the years ahead, revolutionize how the manner in which food systems are optimized, traced and managed. Even these early prototypes demonstrate that the perspectives of quantum agriculture are not only real but also that programming of quantum systems needs transdisciplinary cooperation among quantum physicists, data researchers, and farmers.

11.7 Future opportunities and research directions

The quantum agriculture roadmap is not simply the solution of the current computational bottlenecks. It opens new horizons towards a paradigm shift in the manner human being is getting food in the generation, distribution, and security. The next generation of computational agriculture: resilient, predictive, and regenerative net will come with emergence of quantum technologies that will be more and more connected to bioinformatics, earth systems science, and edge-based AI platforms.

11.7.1 Quantum-enhanced climate modeling for agriculture

One of the spheres of the world economy, where harvests and food security under threat are observed because of the increased intensity and frequency of natural disasters, is agriculture. There should be an urgent demand of more high resolution and responsive modeling tools because the classical climate model was often not capable of offering a local agricultural decision-making solution as well as a time response to changes. Quantum simulations are an innovative approach to eliminating this gap, and they promise to offer high-quality, real-time solutions to complicated climate–agriculture interactions. Simulations on a molecular level can also be given by quantum computing where the mechanism that develops during extreme heat of soil moisture retention are manipulated, and they could be a thing that classical models fail to represent because of the coarse resolution data. Additionally, quantum-enhanced fluid dynamics would also be more precise in modeling atmospheric and hydrologic events such as precipitation, evapotranspiration, and surface runoff that would directly imply irrigation and crop choice tactics. Temporal distribution of pollen and pathogen spread and Greenhouse gas emission compounds can also be traced using quantum models offering predictive solutions that are essential in the control of disease and development of greenhouse environments. Their advanced modeling capability can endorse hyper-local climatic choice encouragement framework, that is, farmers might undertake proactive steps to adjust their practices in foreseeance of experiencing those kinds of risks as alternative to entering into such risks after they have been experienced. Such predictive intelligence would play a critical role in maximizing the planting periods, water, and stress-tolerant crop management. To optimize on this potential, strong research needs have been necessitated to come up with quantum climate models of regional agro-ecological areas. These models will be required to be federated with the present-day weather data feeds and into historical records of past cropping in order to provide context, dynamic, and actionable knowledge. This would go a long way in making the food systems climate-proven and resistant to potential future forces of nature (Arora & Sandhu, 2024).

11.7.2 Quantum-powered crop genome analysis

Advances in crop improvement form a part of the key pillars of food security in the world and assist in the creation of high-yielding, climate-adapted adapted and disease-resistant crop varieties. Nevertheless, traditional genomic analysis is a bottleneck in this process, and it is likely to be affected by the immense amount of data and complexity related to these data. Massive genome-wide association studies (GWAS), epigenomic mapping, and gene trait prediction of large-scale sequences, such as the DNA of several generational organisms, necessitate huge amounts of computational power and time to process, but significantly decrease the time that new breeds of crops may take before they are born and placed into the market.

Quantum computing presents an effective solution to this problem, as it is expected to dramatically speed up genomic analysis by using custom quantum algorithms. An example could be the improvement of quantum Fourier transforms to speed up the computation of sequence alignment, as one of the most computationally demanding processes of genomics. On the same breadth, application of quantum feature extraction methods can increase the power and effectiveness of association mining among genomes and phenotypic traits, thus endowing researchers with more swift and precise identification of major genetic determinants of crop performance. Moreover, by simulating the environmental influence on the plant during its development, the specific conditions in terms of temperature, soil pH, and water stress can be integrated into the epigenomic environment with the help of quantum computing capabilities, and the adaptive trait can be better understood.

Emerging quantum algorithms, such as Harrow–Hassidim–Lloyd (HHL) to solve linear systems, and the quantum principal component analysis (Quantum PCA) to perform dimensionality reduction, offer special promise as scale-up solutions to GWAS. Such techniques have the potential of providing exponential speed-ups compared to classical methods and would allow researchers to analyse larger datasets and find target genes more efficiently. Finally, quantum computing applied in crop genomics may drastically decrease crop breeding cycles and facilitate the creation of more resilient crop variants following resource-efficient breeding, which is beneficial to global food security in an environment of resource scarcity and climate variability (Coppersmith, 1994).

Research need: Partner the might of quantum algorithm innovators with plant geneticists to build genome-to-field prediction models to assist drought/salinity/pest resistance breeding.

11.7.3 Quantum machine learning for pest and disease prediction

At the stage of crops, the detection of diseases and outbreaks of pests is one of the most important interventions in the reduction of yields and agricultural sustainability. Although recently developed AI models using drone images and multispectral sensors have achieved remarkable results, they are likely to be faced with limitations concerning data noise, image aberration, and the inability to identify new or rare pest morphology. Such limitations make the whole AI-based disease detection system weaker and less generalizable in general, especially under different and unpredictable conditions in the field.

QML offers a possible solution because it can improve the performance level in complex and high-dimensional datasets to recognize patterns in images. Compared to classical models, QML algorithms are naturally suited to detecting weak correlations and in distinguishing informative features out of sparse and incomplete data, which is typical of practice in agricultural situations. This makes QML especially useful in detecting low-frequency, high-impact biological phenomena, e.g., early occurrence of plant diseases or insect attacks that could otherwise not have been realized until it is too late. As an example, a QML model fitted to hyperspectral imagery could be used to detect the earliest signs of physiological changes that lead to potato blight in leaves long before visual symptoms have manifested, making them targets of focused treatments and prevention. To reap such benefits, there is an urgent research requirement to develop end-to-end QML pipelines to detect pests and disease. Such pipelines should include all steps of the workflow, which include data collection and preprocessing, quantum-based algorithm analysis of the images, and real-time decision-making resulting in field-level interventions. Furthermore, such systems should be tested and proven to work in different cropping systems to determine whether they are scalable, reliable, and applicable in different geographies and climatic regions. Crop protection methods might emerge that would significantly reduce reliance on chemical inputs and achieve economic and environmental aspirations in agricultural production (Lloyd et al., 2013).

11.7.4 Scalable frameworks for sustainable resource management

One of the widespread issues of the complex allocation of resources that are commonplace in the agricultural landscape is solved using the quantum optimization algorithms, in particular, QAOA and quantum annealing. As is water management, nutrient distribution, land planning, carbon budgeting, etc, tend to be the variables of a host of interdependent variables and constraints and are therefore not easy to compute with a classical system. Another option that can be rather effective in a scenario of such type is quantum optimization since this method will allow identifying their solutions quicker and more precise, which can create a new avenue of the efficiency and sustainability of agricultural procedures. The next research step would be trying to develop multiobjective quantum optimization models to balance various agricultural interests, such as producing the largest crop, making profits, and preparing no harm to the environment.

Among the most promising ones is in the creation of quantum digital twins, virtual copies of actual farm ecosystems to allow being able to simulate, predict, and optimize some processes occurring in real time in the agricultural sector. Such quantum models together with IoT and edge computing systems could be deployed to establish the backbone of dynamically adaptive control systems under precision irrigation, fertigation, and power use in response to real-time readings of the field. In order to ensure general usability and practical implementation, the open-source quantum optimization frameworks must be architecture to ensure that they can be reconfigured to larger agrarian operatives such as smallholder farms, cooperative societies, and all other large agricultural agribusiness supply chains. These frameworks should support the distributed calibration, interfaces alternative to specialists, and scaling modulus, under this technical means that inclusive and context-dependent implementation of quantum capabilities in a diversity of agricultural settings is competent.

11.7.5 Long-term vision: Quantum agriculture and global food security

Agricultural industry is also facing a situation of getting a lot of food with less resources as the growing world population approaches the mark of 10 billion by 2050 and dealing with the increasing global warming impacts. The fact that the demand will go up will not only determine the need to improve but also the innovative changes quantum agriculture is about to become one of. The quantum-enabled agriculture long-term vision involves the development of robust agro-ecosystems, which will then be self-optimizing to the changing environments not only in the economic environment but also owing to variation in environmental conditions. It would be founded on quantum enhanced simulating; these systems would model some complex biophysical feedback loops such that real time adjustment can be achieved. Further, the predictive food systems will allow the stakeholders to make precise predictions regarding the change in food supply and demand at a regional level. These systems could foresee a change in yield, truck congestion, and the market trend so they can take timely intervention at any part of the food value chain using cloud-based quantum simulations. As revolutionary as decentralized Agri-intelligence is one which establishes data spaces, which are nondivisible, among the institutions, but local control is maintained over data sovereignty. On this model, in partnership with forecasting and organized resource distribution would be practiced without infringing ethical standards of data possession and confidentiality. Lastly, quantum agriculture should be merged with the policy frameworks so that the maximum potential can be exploited. The quantum-derived insights would need to inform national policies governing Agricultural development as well as national plans of rural development, climate adaptation policies and not forgetting other farming-related policies. It would imply such regional development-related goals as long-term sustainability, food security, and equitable progress in the technical domain would be aligned. The latter is one such vision not only of an innovative method of doing things in the agricultural sector but also such would open a way to a stronger, more democratic and more information-savvy global food system. Table 11.5 visualizes the appealing emerging research frontiers in quantum agriculture and the way the main areas can be changed by quantum technologies. Occurring in criticality to climate resiliency and food machinery the world over these innovations

Table 11.5 Research frontiers in quantum agriculture.

Research domain	Quantum opportunity	Societal impact
Climate resilience	Localized quantum climate modeling	Risk reduction and resilience
Genomics and breeding	Quantum-assisted gene analysis	Accelerated crop development
Pest prediction	QML image classifiers	Reduced pesticide use
Resource optimization	Quantum annealing	Sustainable input use
Global food systems	Quantum simulations	Proactive food policy planning

Table 11.6 Policy and ethical focus areas for quantum agriculture.

Dimension	Policy priority	Action path
Regulation	Model certification and standardization	National quantum-agri standards
Data ethics	Privacy, consent, transparency	Quantum data governance boards
Collaboration	Multisector consortia	PPPs and global alliances
Education	Capacity and curriculum development	University programs, field training

can render agriculture more dependable and refined alongside policies based on them created in the finest interest of the entire society.

11.8 Policy and ethical considerations

Nevertheless, all such goodies bestowed to the agricultural sectors, caution is to be taken to incorporate quantum technologies with a responsible innovation approach. This will include transparent regulatory instructions, ethical risk remediation, formation of public-edi alliance, and the establishment of human capital in a quantum-literates agricultural labor force as shown in Table 11.6. In their absence, the application of quantum computing to agriculture can lead to digital inequity, data exploitation, and technological elitism that can be opposed to the objective of sustainability and inclusive development.

11.8.1 Regulatory and policy frameworks for emerging agri-tech

The invention of ways of mechanization, biotechnology, and electronic farming tools have, to a great extent, affected the formation of modern policy in agriculture. But with the introduction of quantum-informed decision support systems, the entirely quantum-powered genome modelling, and analytics of the supply chain, a new tier of responsive and periscopic policy frameworks must be erected. The frameworks are not only referred to in order to address the requirements of the special dynamic of quantum technologies in computation but also the consequences of the technologies in fundamental aspects of existence, including food security and biosecurity. The primary regulatory priorities are associated with certification and validation procedures of quantum models because it is essential to certify quantum models as reliable, accurate, and safe procedures before their use in the real life possibilities of the sphere of agriculture. Meanwhile, there are high-impact areas of quantum simulation of crop genetics, climate modeling, and optimization of inputs that are also supposed to elaborate the extensive standard of risk evaluation as the unfavorable outcomes might be of the lasting impact. The other regulation feature is that of adopting effective policies of data governance, which can regulate the complex interactions between quantum and classical computing systems. These protocols should ensure transparency, accountability, and safety in the process of handling data in its life cycle, both during its purchase and as a result of its procurement. Further, quantum agriculture needs to be integrated specifically in national innovation road maps and AI plans in a complementary effort between technologies. More need not be said to illustrate that on the example of the Indian National Mission on Quantum Technologies and Applications (NMQTA) just recently, they released a strategic plan, which may help incorporate some of its applications in their innovation clusters, over and above integrating the quantum developments themselves to local goals of food independence, rural empowerment, and cyberspace inclusivity. Such a proactive regulation is paramount such that quantum agriculture would expand at a safe rate, fairly, and goal orientedly (Department of Science and Technology 2023; Floridi & Cowls, 2019; Kumar et al., 2025; Rao & Sandhu, 2024; Rao et al., 2025; Somayajula et al., 2022; World Economic Forum 2018).

11.8.2 Ethical concerns in quantum data usage and privacy

Any information that is related to agriculture, such as genomic data on crops, land workability scores, proprietary seed combinations, and farm behavioral patterns, among others constitute sensitive agricultural data that can be processed and analysed using quantum computing with unparalleled possibilities. It is not only unleashing a lot of potential to enable the use of the data in reaching data-based maximization in the domain of farming, but it is also leading to deteriorous ethical anxieties. It will take just one of these powerful computing capabilities to be handled inadequately and used to invade the privacy of farmers and rob them of data control unless enhanced security measures are enforced. The corporate possession of access to quantum-derived wisdom can give rise to a new model of agricultural surveillance capitalism where major choices and benefits are concentrated in the few at the possible detriment of the farming communities that supply the data. Besides, the impossibility to understand quantum algorithms, particularly as QML evolves, may create the situation when farmers receive a recommendation and an order without having the slightest idea of how and why they work. In order to make sure that such are pacified, it is paramount to arrange an integrated strategy into ethical leadership: such that would influence explainability and transparency of the QML decision pipelines and ensure that the information that came through on a quantum scale are clarifiable and clarificable. Informed consent should be utilized in the use of the data, which should have visible restrictions of ownership and which should have objectives. Another issue that needs to be addressed is the establishment of ethical review boards, which are empowered directly, as part of ensuring that quantum research and development of the agri-food industry is monitored. On top of that, the gains of quantum technological revolution should not be monopolized, and the smallholder farmers, indigenous people, and stakeholders of the Global South should not be left on the wrong side of this technological revolution because of the absence of access to quantum technologies and infrastructure. The agricultural ethical discourses must therefore transcend the conventional discourse of digital rights to quantum rights where individuals and societies have a stake with respect to their data, decisions, and lives in the world that is increasingly quantum-enabled (Mahidar & Ghai, 2021).

11.8.3 Cross-sectoral collaboration: Academia, industry, government

With the highly multidisciplinary character of quantum agriculture, any meaningful and sustainable development critically involves the creation of effective collaborative ecosystems that incorporate both academia, industry, and government. Universities and colleges have a critical role in advancing the foundational research such as developing quantum algorithms, design of simulation in agriculture, and development of capacity building program to educate the generation of professionals skilled in quantum research. The industrial sector, in its turn, is critical in the creation of quantum hardware, real-world pilot testing, and the scale of successful models to impact society and the world at large. The government agencies are likewise essential, and they are the sources of regulation arrangements, funding, and policy orientation required to instruct ethical and inclusive application of technologies (Ghai et al., 2024). Such stakeholders should collide with the help of strategic engagement between them and the industry through public–private cooperation that promotes experimentation and innovation on the field. Applied research can be tested in agencies like field-deployable quantum pilot programs, e.g., precision farming, soil analytics, Agri-supply chain logistics, etc. In addition, quantum sandboxes and Agri-tech accelerators should be established to enable startups and innovators to interact with real data on agriculture and test their proof-of-concept models in vitro. These collaborations might also be used to support the horizontal knowledge transfer between quantum research labs and agricultural institutions to allow them to localize and contextualize quantum solutions (Ghai et al., 2022). An international "Quantum for Food Security Consortium should perhaps be initiated to catalyse greater action globally through the activities of a global entity like the Food and Agriculture Organization (FAO) or the World Economic Forum. Not only would this consortium bring innovation in line with Sustainable Development Goal 2 (Zero Hunger), but quantum agriculture would also be a global focus with the same aim of equity, sustainability, and scientific integrity (Chouhan et al., 2025; Faiz et al., 2023; Kaur et al., 2026).

11.8.4 Capacity building and training for quantum readiness in agriculture

With the highly multidisciplinary character of quantum agriculture, any meaningful and sustainable development critically involves the creation of effective collaborative ecosystems that incorporate both academia, industry, and government. Universities and colleges have a critical role in advancing the foundational research such as developing quantum algorithms, design of simulation in agriculture, and development of capacity building program to educate the generation of professionals skilled in quantum research. The industrial sector, in its turn, is critical in the creation of quantum hardware, real-world pilot testing, and the scale of successful models to impact society and the world at large. The government agencies are likewise essential, and they are the sources of regulation arrangements, funding, and policy orientation required to instruct ethical and inclusive application of technologies (Ghai et al., 2024). Such stakeholders should collide with the help of strategic engagement between them and the industry through public-private cooperation that promotes experimentation and innovation on the field. Applied research can be tested in agencies like field-deployable quantum pilot programs, e.g., precision farming, soil analytics, Agri-supply chain logistics, etc. In addition, quantum sandboxes and Agri-tech accelerators should be established to enable startups and innovators to interact with real data on agriculture and test their proof-of-concept models in vitro. These collaborations might also be used to support the horizontal knowledge transfer between quantum research labs and agricultural institutions to allow them to localize and contextualize quantum solutions (Ghai et al., 2022). An international "Quantum for Food Security" Consortium should perhaps be initiated to catalyse greater action globally through the activities of a global entity like the Food and Agriculture Organization (FAO) or the World Economic Forum. Not only would this consortium bring innovation in line with Sustainable Development Goal 2 (Zero Hunger), but quantum agriculture would also be a global focus with the same aim of equity, sustainability, and scientific integrity (Chouhan et al., 2025; Faiz et al., 2023; Kaur et al., 2026).

11.9 Case studies

11.9.1 Case study 1: Crop planning and resource allocation quantum optimization (Case Study: North America)

A significant example of a real-life case study of how quantum computing can be applied in agriculture is the Canadian pilot project on quantum optimization of agricultural operations, sponsored by Digital Technology Supercluster and industry-academic (IA) partners. This paper was a complex research that focused on the large-scale crop planning and resources allocation where farmers need to make decisions on the distribution of crop between the heterogeneous field zones and meet various constraints related to soil suitability, crop rotation policies, water, fertilizer, economic costs, and sustainability goals. These have complexity of NP hardness as such and as more and more fields and constraints are in place these problems exponentially increase in problem size and hence classical optimization is impractical or ineffective. The agricultural problem of planning in this case study was created in the form of a quadratic unconstrained binary optimization (QUBO) model in which the yields maximization and resource consumption and penalty constraints were represented by a cost matrix and binary decision variables of crop-field assignments. At QUBO was translated into an Ising model and executed by run through one of a set of hybrid quantum-classical workflow on a D-Wave quantum annealer. While experimental findings showed a better convergence behavior and quality of the global solutions were successfully achieved than the classical heuristics, especially in dense constraint situations. Notably, the analysis showed the potential of the study in showing how quantum annealing can augment instead of substitute classical decision-support systems, providing a scaling wrath for optimization of precision agriculture. The case offers an experimental paradigm to be followed by scholars who need to use quantum optimization to optimize irrigation schedules, fertilizer distribution, and land-use models and sustainability-oriented agricultural decisions (https://digitalsupercluster.ca/projects/optimizing-agricultural-operations-with-quantum-computing).

11.9.2 Case study 2: Quantum machine learning in crop stress and disease detection (Asia-Pacific)

The other technically important case study examines application of QML in crop stress and disease identification through multispectral and NDVI-based agronomic data in Asia-Pacific set up. This research concentrated on the weaknesses of the classical models of ML in the case of using agricultural imagery with high dimensions of the team, sparse labels, and environmental interference. In order to overcome these difficulties, the researchers introduced the use of variational quantum classifier (VQC), in which the information of classical agricultural parameters was coded in quantum states and was processed by using parameterized quantum circuits. The model took the advantage of quantum feature embedding to classify nonlinear correlations which classical models could not easily articulate. The structure applied was a hybrid training framework (quantum circuit evaluation) coupled with classical optimization of circuit parameters. Based on experimental tests performed on quantum simulators and quantum devices available in the NISQ regime, the quantum model was found to classify data at a level as good as or better than the classical support vector machines on both small and large datasets, with increased resistance to noise. The case study demonstrates that QML can facilitate crop stress and disease detection at an earlier stage and implement required interventions timely to minimize the use of chemicals. Regarding the research, the transferability of the methodology to other agricultural sensing challenges, including weed detection, yield forecasting, and multi-modal sensor fusion is valuable in reference to other studies of quantum-enabled smart farming in the future (Jagtap et al., 2025).

11.10 Conclusion and Future Scope

The advent of quantum computing takes the agriculture field into a new era. In the agricultural spheres, as the theory presented in this chapter has shown (along with the evidence presented in the previous chapter), the possibilities to employ quantum technologies are not only theoretically viable, but they are also strategic endeavors necessary to meet the dynamic and complex high-dimensional parameters that agriculture is confronted with on a global scale in the contemporary world. In its essence, quantum agriculture is a combination of next-generation computational science and precision agriculture, environmental sustainability, and the global food security agenda. Such a broad vision of quantum computing in agriculture has many applications, as stated by this list of use cases, which proves that this kind of computation can take the agricultural sector to a new level of innovation, whereas traditional systems have few advantages compared to it. It was a thorough analysis of the existing situation, chances, and challenges in the future development of a quantum-facilitated agricultural ecosystem. It started with an explanation of key principles of quantum computing, such as superposition, entanglement, and quantum interference, and applied them to demonstrate their application to the practical use of quantum computing to handle the complexity of agricultural information. The discourse was then shifted into the development of smart agriculture in a manner that demonstrates that AI, the IoT, and Big Data have revised farming activities. Nevertheless, these computing innovations brought new requirements of computational models that process large volumes of heterogeneous and real-time information, which reveal the shortcomings of the classical systems. The chapter discussed some of the convincing quantum applications in agriculture. They were highlighted by quantum-enhanced crop yield forecasting, real-time assessment of soil health with quantum sensors, improved climate forecasting, and logistics with optimized algorithms powered by quantum. Examples of concepts of pilots were presented, which demonstrated how these innovations may be feasible and how they may have an impact. Although applications seemed to be appealing, the roadmap recognized several technical and infrastructural challenges. These are the constraints of existing quantum hardware, the lack of any dedicated quantum algorithms involved in the agricultural domain, integration challenges with the existing digital infrastructure, vulnerabilities in cybersecurity, and scalability and interoperability problems. To overcome these shortcomings, a three-layer hybrid integration model was suggested in the chapter. It is an architecture that connects edge-IoT devices to cloud-based quantum processors and classical AI platforms to establish the technological basis for intelligent farming systems that are future-ready. Last, the chapter stressed the necessity of closed research and policy building. It urged more funding in QML predictive studies of the biological and environmental, quantum genomics, and climate modeling, as well as regulatory approaches to focus on ethical data use, visibility of algorithms, and fair access to emerging tech.

Future Scope

Quantum agriculture has an ambitious and then achievable long-range vision, namely, a world with a global agricultural ecosystem, in which quantum computers can become reality in predictive, regenerative, and equitable food systems. Farms (large or small) will be intelligent in this future, self-adaptive ecologies, and will use real time, quantum-based advisor platforms to make decisions during the planting to harvest cycle. At the systemic level, quantum-enhanced supply chain analytics will organize the national and international food networks, which will lead to a significant reduction in waste, the use of the logistics process to the utmost efficiency, and a reduction in CO2 emissions. By using innovative molecular simulations, climate-resilient agriculture will be able to understand deeply the plant–soil–water interactions and make data-driven optimization decisions regarding the irrigation, land use, and seed design. The model of quantum computing will support the creation of rich, multidimensional datasets, which will influence the agricultural policies and structure the governance as anticipatory, evidence-based and equitable in addressing the issues of food security and environmental stewardship. It is not only an evolvation in the future of technology but also the paradigm change over an agrarian model that is smarter, more participatory, and much more sustainable. It leaves the stakeholders with a challenge of transgressing incremental innovation and to pursue radical changes based upon daily everyday ethical foresee-strategy interaction. Last but not least, quantum agriculture is not the dream of the future but a call of action that must be provided in due time. It summons the scientists, technologists, educators, policy makers, as well as farmers in their joint development of a robust and equitable forthcoming world food systems where quantum intelligence is the core essence of all things.

References

Abraham, H., AduOffei, A., Akhalwaya, I. Y., Alexander, T., Barkoutsos, P., Bello, L., & Gambetta, J. M. (2019). Qiskit: An open-source framework for quantum computing. *Zenodo*. https://doi.org/10.5281/zenodo.2562110.

AlSagri, H. S., Kumar, A., Jilani Saudagar, Kumar, A., & Raja, L. (2025). Optimizing resource allocation in precision farming using quantum enhanced algorithms and quantum sensor networks. *Journal of Cloud Computing, 14*(1), 47–76. https://doi.org/10.1186/s13677-025-00769-2.

Arora, C., & Sandhu, R. (2024). Adoption of quantum NLP for improved healthcare system. In *Proceedings of the international conference on computation of artificial intelligence & machine learning* (pp. 51–59). Springer Nature.

Arute, F., Arya, K., Babbush, R., Bacon, D., Bardin, J. C., Barends, R., Boixo, S., Brandao, F. G. S. L., Buell, D. A., Burkett, B., Chen, Y., Chen, Z., Chiaro, B., Collins, R., Courtney, W., Dunsworth, A., Farhi, E., Fowler, A., & Martinis, J. M. (2019). Quantum supremacy using a programmable superconducting processor. *Nature, 574*, 505–510. https://doi.org/10.1038/s41586-019-1666-5.

Benedetti, M., Realpe-Gómez, J., Biswas, R., & Perdomo-Ortiz, A. (2019). Quantum-assisted Helmholtz machines: A quantum–classical deep learning framework for generative modeling. *Quantum Science and Technology, 4*(4), 043001. https://doi.org/10.1088/2058-9565/ab4eb5.

Biamonte, J., Wittek, P., Pancotti, N., Rebentrost, P., Wiebe, N., & Lloyd, S. (2017). Quantum machine learning. *Nature, 549*, 195–202. https://doi.org/10.1038/nature23474.

Braunstein, S. L., & van Loock, P. (2005). Quantum information with continuous variables. *Reviews of Modern Physics, 77*, 513–577. https://doi.org/10.1103/RevModPhys.77.513.

Budker, D., & Kimball, D. F. J. (2013). *Optical magnetometry*. Cambridge University Press.

Carolan, M. (2017). Publicising food: Big data, precision agriculture, and co-experimental techniques of addition. *Sociologia Ruralis, 57*(2), 135–154. https://doi.org/10.1111/soru.12120.

Cerezo, M., Arrasmith, A., Babbush, R., Benjamin, S. C., Endo, S., Fujii, K., … Coles, P. J. (2021). Variational quantum algorithms. *Nature Reviews Physics, 3*, 625–644. https://doi.org/10.1038/s42254-021-00348-9.

Childs, A. M., Preskill, J., & Renes, J. (2000). Quantum information and precision measurement. *Journal of Modern Optics, 47*(2–3), 155–176. https://doi.org/10.1080/09500340008244031.

Chouhan, S., Sandhu, R., Kumar, S., & Naz, S. M. I. (2025). Optimizing industrial IoT predictive maintenance using quantum algorithms on google quantum AI. In *2025 international conference on emerging smart computing and informatics (ESCI)* (pp. 1–5). IEEE.

Chouhan, S., Sandhu, R., Yadav, O. P., & Islam, S. M. (2025). Quantum-powered autonomous deep-sea explorer using AI, microbiofuels, and advanced sensors for oceanic discovery. In *2025 international conference on computer, electrical & communication engineering (ICCECE)* (pp. 1–5). IEEE.

Coppersmith, D. (1994). *An approximate Fourier transform useful in quantum factoring (IBM Research Report No. RC19642)*. IBM Research Division.

Department of Science and Technology, Government of India. (2023). *National mission on quantum technologies and applications (NM-QTA)* https://dst.gov.in.

Faiz, M., Sandhu, R., Akbar, M., Shaikh, A. A., Bhasin, C., & Fatima, N. (2023). Machine learning techniques in wireless sensor networks: Algorithms, strategies, and applications. *International Journal of Intelligent Systems and Applications in Engineering, 11*(9s), 685–694. https://www.ijisae.org/index.php/IJISAE/article/view/3217.

Feynman, R. P. (1982). Simulating physics with computers. *International Journal of Theoretical Physics, 21*, 467–488. https://doi.org/10.1007/BF02650179.

Floridi, L., & Cowls, L. (2019). A unified framework of five principles for AI in society. *Harvard Data Science Review, 1*(1), 1–15. https://doi.org/10.1162/99608f92.8cd550d1.

Gebbers, R., & Adamchuk, V. I. (2010). Precision agriculture and food security. *Science, 327*(5967), 828–831. https://doi.org/10.1126/science.1183899.

Ghai, D., Rawal, K., Dhir, K., & Tripathi, S. L (2024). *Artificial intelligence techniques for sustainable development.* CRC Press.

Ghai, D., Tripathi, S. L., Saxena, S., Chanda, M., & Alazab, M. (2022). *Machine learning algorithms for signal and image processing.* John Wiley & Sons.

Grover, L. K. (1996). A fast quantum mechanical algorithm for database search. In *Proceedings of the 28th Annual ACM symposium on theory of computing* (pp. 212–219). ACM. https://doi.org/10.1145/237814.237866.

https://digitalsupercluster.ca/projects/optimizing-agricultural-operations-with-quantum-computing/, published on January 21, 2025.

Harrow, A. W., Hassidim, A., & Lloyd, S. (2009). Quantum algorithm for solving linear systems of equations. *Physical Review Letters, 103*(15), 150502. https://doi.org/10.1103/PhysRevLett.103.150502.

Havlíček, V., Córcoles, A. D., Temme, K., Harrow, A. W., Kandala, A., Chow, J. M., & Gambetta, J. M. (2019). Supervised learning with quantum-enhanced feature spaces. *Nature, 567*, 209–212. https://doi.org/10.1038/s41586-019-0980-2.

Jagtap, S. S., Sudhagar, G., & Jadhav, R. N. (2025). Reinventing smart farming using adaptive quantum machine learning model. *J Inf Syst Eng Manag, 10*(5), 2379. https://doi.org/10.52783/jisem.v10i5s.665.

Jensen, R. R. (2021). Applications of the internet of things (IoT) in smart farming: A review. *Journal of Agriculture and Food Research, 2*, 100–122. https://doi.org/10.1016/j.jafr.2020.100122.

Kamilaris, A., Kartakoullis, A., & Prenafeta-Boldú, F. X. (2017). A review on the practice of big data analysis in agriculture. *Computers and Electronics in Agriculture, 143*, 23–37. https://doi.org/10.1016/j.compag.2017.09.037.

Kamilaris, A., & Prenafeta-Boldú, F. X. (2018). Deep learning in agriculture: A survey. *Computers and Electronics in Agriculture, 147*, 70–90. https://doi.org/10.1016/j.compag.2018.02.016.

Kaur, P., Jasrai, L., & Sandhu, R. (2026). Quantum-enhanced plant breeding and genetic optimization: Quantum computing for crops. In M. F. Horng, O. I. Khalaf, & C.-S. Shieh (Eds.), *Revolutionizing sustainable food production with quantum computing* (pp. 151–176). IGI Global Scientific Publishing. https://doi.org/10.4018/979-8-3373-3957-3.ch005.

Kshetri, N. (2018). Blockchain's roles in meeting key supply chain management objectives. *International Journal of Information Management, 39*, 80–89. https://doi.org/10.1016/j.ijinfomgt.2017.12.005.

Kumar, L., Raj, A., Singh, N., Sandhu, R., Gangwar, G., & Ghai, D. (2025). Future directions in computational intelligence: Anticipating the next wave. In Mahajan, S., & Braga de Vasconcelos, J. (Eds.), Next-generation computational intelligence: Trends and technologies (pp. 51–74). Switzerland: Springer Nature. https://doi.org/10.1007/978-3-031-96871-6_3.

Li, L., Zhang, Q., & Huang, D. (2014). A review of imaging techniques for plant phenotyping. *Sensors, 14*(11), 20078–20111. https://doi.org/10.3390/s141120078.

Liakos, L. I., Busato, P., Moshou, D., Pearson, S., & Bochtis, D. (2018). Machine learning in agriculture: A review. *Sensors, 18*(8), 2674. https://doi.org/10.3390/s18082674.

Lloyd, S. (1996). Universal quantum simulators. *Science, 273*(5278), 1073–1078. https://doi.org/10.1126/science.273.5278.1073.

Lloyd, S., Mohseni, M., & Rebentrost, P. (2013). Quantum algorithms for supervised and unsupervised machine learning. *arXiv preprint arXiv:1307.0411*. https://doi.org/10.48550/arXiv.1307.0411.

Lobell, D. B., Schlenker, W., & Costa-Roberts, J. (2011). Climate trends and global crop production since 1980. *Science, 333*(6042), 616–620. https://doi.org/10.1126/science.1204531.

Lowenberg-DeBoer, J., & Erickson, B. (2019). Setting the record straight on precision agriculture adoption. *Agronomy Journal, 111*(4), 1552–1569. https://doi.org/10.2134/agronj2018.12.0779.

Mahidar, P. R., & Ghai, D. (2021). Bio-inspired optimization algorithms for machine learning in agriculture applications. In G. S. Patel, A. Rai, N. N. Das, & R. P. Singh (Eds.), *Smart Agriculture: Emerging Pedagogies of Deep Learning, Machine Learning and Internet of Things* (pp. 53–60). CRC Press. https://doi.org/10.1201/9781003138884-4.

Markets and Markets. (2024). *Precision farming market by technology, application, and geography: Global forecast to 2030* https://www.marketsandmarkets.com/Market-Reports/precision-farming-market-1243.html.

McBratney, A., Whelan, B., Ancev, T., & Bouma, J. (2005). Future directions of precision agriculture. *Precision Agriculture, 6*(1), 7–23. https://doi.org/10.1007/s11119-005-0681-8.

Montanaro, A. (2016). Quantum algorithms: An overview. *npj Quantum Information, 2*(1), 15023. https://doi.org/10.1038/npjqi.2015.23.

Mosca, M. (2018). Cybersecurity in an era with quantum computers: Will we be ready? *IEEE Security & Privacy, 16*(5), 38–41. https://doi.org/10.1109/MSEC.2018.2872841.

Nielsen, M. A., & Chuang, I. L. (2010). *Quantum computation and quantum information* (10th ed.). Cambridge University Press.

Preskill, J. (2018). Quantum computing in the NISQ era and beyond. *Quantum, 2*, 79. https://doi.org/10.22331/q-2018-08-06-79.

Rao, P. R., Sandhu, R., & Islam, S. M. (2025). ReIoT-Powered Smart Agriculture in Controlled Environments. In *2025 international conference on pervasive computational technologies (ICPCT)* (pp. 6–11). IEEE.

Rao, P. R., & Sandhu, R. (2024). Farmer well-being with metaverse innovations in agriculture. In *Proceedings of the 11th International Conference on Reliability, Infocom Technologies and Optimization (Trends and Future Directions) (ICRITO)* (pp. 1–6). IEEE.

Rose, D. C., & Chilvers, J. (2018). Agriculture 4.0: Broadening responsible innovation. *Frontiers in Sustainable Food Systems, 2*, 87. https://doi.org/10.3389/fsufs.2018.00087.

Saiz-Rubio, V., & Rovira-Más, F. (2020). From smart farming towards agriculture 5.0: A review on crop data management. *Agronomy, 10*(2), 207. https://doi.org/10.3390/agronomy10020207.

Schuld, M., & Killoran, N. (2019). Quantum machine learning in feature hilbert spaces. *Physical Review Letters, 122*(4), 040504. https://doi.org/10.1103/PhysRevLett.122.040504.

Schuld, M., Sinayskiy, I., & Petruccione, F. (2015). An introduction to quantum machine learning. *Contemporary Physics, 56*(2), 172–185. https://doi.org/10.1080/00107514.2014.964942.

Shor, P. W. (1994). Algorithms for quantum computation: Discrete logarithms and factoring. In *Proceedings of the 35th annual symposium on foundations of computer science* (pp. 124–134). IEEE. https://doi.org/10.1109/SFCS.1994.365700.

Somayajula, V. K. A., Ghai, D., Kumar, S., Tripathi, S. L., Verma, C., Safirescu, C. O., & Mihaltan, T. C. (2022). Classification and validation of spatio-temporal changes in land use/land cover and land surface temperature of multitemporal images. *Sustainability, 14*(23), 15677. https://doi.org/10.3390/su142315677.

Tacchino, F., Macchiavello, C., Gerace, D., & Bajoni, D. (2019). A quantum machine learning classifier implemented in a quantum processor. *npj Quantum Information, 5*(1), 1–8. https://doi.org/10.1038/s41534-019-0140-4.

Trendov, N. M., Varas, S., & Zeng, M. (2019). *Digital technologies in agriculture and rural areas*. FAO https://www.fao.org/3/ca4887en.pdf.

Tzounis, A., Katsoulas, N., Bartzanas, T., & Kittas, C. (2017). Internet of Things in agriculture. *Biosystems Engineering, 164*, 31–48. https://doi.org/10.1016/j.biosystemseng.2017.09.007.

Wolfert, J., Goense, D., & Sørensen, C. A. G. (2015). A future internet platform for precision agriculture. *Computers and Electronics in Agriculture, 114*, 66–80. https://doi.org/10.1016/j.compag.2015.03.010.

Wolfert, J., Sørensen, C. G., & Goense, D. (2014). A future internet collaboration platform for safe and healthy food from farm to fork. *Computers and Electronics in Agriculture, 104*, 112–125. https://doi.org/10.1016/j.compag.2014.04.004.

Wolfert, S., Ge, L., Verdouw, C., & Bogaardt, M. J. (2017). Big data in smart farming–A review. *Agricultural Systems, 153*, 69–80. https://doi.org/10.1016/j.agsy.2017.01.023.

World Economic Forum. (2018). *Harnessing artificial intelligence for the earth* https://www3.weforum.org/docs/Harnessing_Artificial_Intelligence_for_the_Earth_report_2018.pdf.

Zhang, Y., Wang, G., Wang, J., & Xu, X. (2020). Review of precision agriculture based on artificial intelligence. *Computers and Electronics in Agriculture, 178*, 105. https://doi.org/10.1016/j.compag.2020.105.

Index

Page numbers followed by "*f*" and "*t*" indicate, figures and tables respectively.

R

S

Printed and bound by CPI Group (UK) Ltd, Croydon, CR0 4YY
24/09/2026
02222290-0007